AKADEMIE DER WISSENSCHAFTEN DER DDR
FORSCHUNGSZENTRUM FÜR MOLEKULARBIOLOGIE UND MEDIZIN
ZENTRALINSTITUT FÜR MIKROBIOLOGIE
UND EXPERIMENTELLE THERAPIE JENA
ABTEILUNG LIMNOLOGIE

LIMNOLOGICA

SCHRIFTLEITER: S. JOST CASPER

HERAUSGEBER: GOTTFRIED BREITIG, HANS KNÖLL,
ALFRED RIETH, WILFRIED SCHÖNBORN,
DIETRICH UHLMANN UND GÜNTHER WALTER

BAND 11

1. Heft

AKADEMIE-VERLAG · BERLIN

1976

LIMNOLOGICA

INHALTSVERZEICHNIS
Band 11 · Heft 1
Abgeschlossen Oktober 1975

Band 11 der LIMNOLOGICA umfaßt zwei Hefte.

Das Heft 2 wird folgende Beiträge enthalten:

JESCHKE, L., & K. MÜTHER: Die Pflanzengesellschaften der Rheinsberger Seen. – REICHHOFF, L.: Die Wasser- und Röhrichtpflanzengesellschaften des Mittelelbe-Gebietes zwischen Wittenberg und Aken. – LINDNER, A.: Soziologisch-ökologische Untersuchungen an der submersen Vegetation in der Boddenkette südlich des Darß und des Zingst (südliche Ostsee). – SUCCOW, M., & A. REINHOLD: Das Vegetationsgefüge eines jungpleistozänen Klarwassersees und seine Belastbarkeit – eine Studie mit Hilfe von Farbluftbildern. – NASAR, S. A. K., & J. D. MUNSHI: Biomass production of *Potamogeton crispus* L. in a freshwater pond of Bhagalpur, India. – DOLL, R.: Drei bemerkenswerte Seen im südlichen Mecklenburg und ihre Vegetation. – JUNGNICKEL, F.: Phosphatbedarf und Mangelsymptome bei einigen axenisch kultivierten Lemnaceen. – Rezensionen.

Limnologica (Berlin)	11	1	1976	1–8

MASSOUD A. H. SAAD

Studies on the nature and composition of the sediments of two Egyptian lakes found under different local conditions

With 3 figures and 1 table

Introduction

The fish production in Egypt depends largely on the Egyptian inland water bodies. In the past time, only very few scattered data were collected from these aquatic habitats. In the last few years, however, these water bodies were subjected to further limnological investigations. It is hoped that these studies and also the future ones will help to improve the fisheries of these water bodies.

The present investigation was carried out to study the nature and composition of sediment samples collected from two Egyptian brackisch-water lakes, namely Lake Brollus and Lake Qarun, and to compare the results obtained from each lake with those of the other. Such comparison seems to be of a considerable importance, since the two lakes are subjected to various local conditions. Lake Brollus, a coastal Delta lake, is connected to the Mediterranean Sea with a narrow slit. Lake Qarun, on the other hand, is a closed basin situated in Upper Egypt with no apparent outlets.

Description of Lake Brollus and Lake Qarun

Lake Brollus, lying at the northern part of the Nile Delta, has an area of about 146,000 feddans (one feddan is equivalent to 4200 m²). It is separated from the Mediterranean Sea at the North by a strip of land covered with sand dunes, and is connected to the sea with Boughaz El-Brollus. The eastern and southern borders of this lake are irregular and surrounded by cultivated lands (Fig. 1).

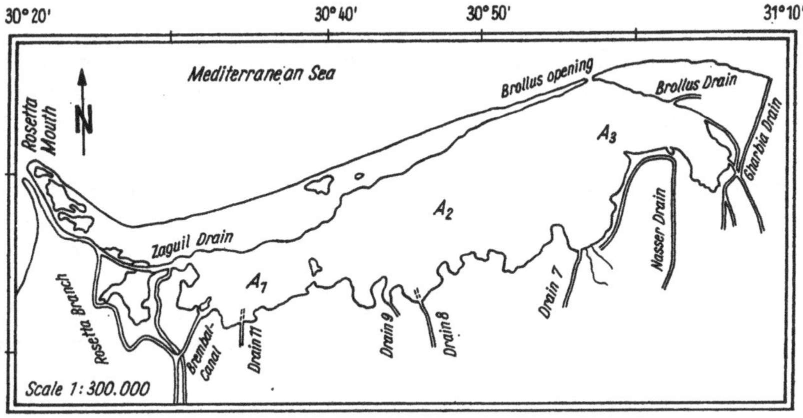

Fig. 1. Lake Brollus

Great amounts of drainage waters pour into the lake at these shores by means of several drains. These drainage waters cause the rise in the lake level, and thus creating a lake—sea current. The rise in the sea level, on the other hand, is due mainly to the wind action. This induces a sea—lake current, and the sea water enters into the lake to affect mainly the area of the Boughaz, and may also reach a great distance inside the lake.

The bottom of Lake Brollus shows various structures. It is muddy at the southern part of the lake, and sandy mixed with silty materials at the area of the lake—sea connection.

The water depth is subjected to rapid variations, and it ranges from 70 to 240 cm. It increases gradually from south to north and from east to west.

The chlorosity of Lake Brollus varies according to locality and season. It is lower in the regions near the drains, and higher in the area of the lake—sea connection. The movement of water masses, due to the effect of the prevailing wind, plays an important role in the distribution of chlorosity in the lake.

Lake Qarun, on the other hand, has no connection with the sea, and is found in the western desert at about 100 kilometers to the southwest of Cairo (Fig. 2). The lake has an area of about 50,000 feddans, and can be considered as having two basins. The eastern basin is shallower with an average depth of 3 meters, whereas the western basin is deeper ranging in depth from 2 to 9 meters.

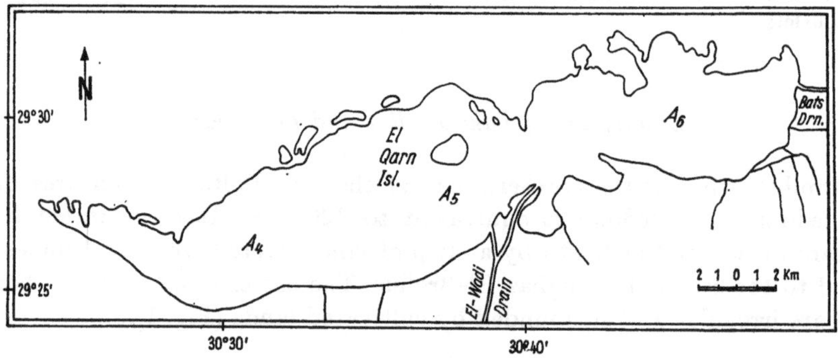

Fig. 2. Lake Qarun

Lake Qarun is bordered by cultivated lands at its south and southeastern sides, and by an uninhabited desert at its northern side. It receives drainage waters mainly from two large drains, namely, the Bats and El-Wadi drains joining the lake at the east and the middle south sides, respectively.

The bottom of Lake Qarun is characterized by the presence of great accumulations of calcareous shells and shell fragments of molluscs. The sediments near the drains contain a higher percentage of silt and clay, whereas they are blanked with a wind blown layer of sand at the northern side of the lake.

The great amounts of drainage waters discharged in Lake Qarun, and the huge quantities of water lost annually by evaporation cause a continuous accumulation of salts in the lake. GORGI (1959) gave this lake a salinity value ranging from 25 to 29 ‰. NAGUIB (1961) gave Lake Qarun a mean salinity value of 22 ‰ for the year 1954.

Materials and methods

A modified EKMAN-bottom sampler was used to collect nine sediment samples from each lake. Three stations were sampled at the west, three at the middle and the remaining three at the east of each lake. The western, the middle and the eastern samples of Lake Brollus and Lake Qarun were mixed to form the composite A_1, A_2, A_3, A_4, A_5, and A_6 respectively (Fig. 1 and 2).

The density of wet mud was determined, on the same day of collection, using a pycnometer. The density of dry mud was calculated from the density of wet mud and the water content (SAAD 1970). The amount of water was determined by drying the wet deposits in an oven at 105°C.

The organic matter was determined as ignition loss by igniting about 500 mg dry sediment in a Muffel furnace at 525°C from 4 to 5 hours (UNGEMACH 1960).

The HCl-soluble and insoluble parts of the sediments were determined by adding 12.5% HCl to the remaining inorganic fractions of the deposits in conical flasks, which were heated for one hour on an electric plate. Filtration was carried out using ashless filter paper. The dissolved parts of the sediments were considered as calcareous substances. The undissolved parts, on the other hand, were considered as allochthonous materials plus diatom frustules.

The method described by MULLIN & RILEY (1955) and modified by TESSENOW (1964) was used to determine the carbonate-soluble (diatom)-silica of the sediments.

Results and discussion

The data given in Table 1 represent the values of the different components of the composite A_1, A_2, A_3, A_4, A_5 and A_6. These data are represented as percentages per dry mud in order to give an idea about the percentage distribution of the different constituents of the sediments in the three studied regions of each lake. The data were also calculated in kg per m^2 wet mud in order to give a good picture for comparing the amounts of the different components of the sediments obtained from the different regions of each lake, and also to compare the results of one lake with those of the other. The silica content was calculated in g per m^2, due to its low value.

The values of the determined compents differently vary in the compared lakes, the exact values are given in Table 1.

The external and internal events are responsible to determine the quality and quantity of the materials precipitated down on the lake bottom. In case of Lake Brollus and Lake Qarun, the external events have a considerable effect on the nature and composition of the lake sediments. These lakes are bordered by cultivated lands, and receive continuously great amounts of drainage waters from several drains. The drainage waters, which are enriched with silt and clay particles, continuously transport to the lakes great amounts of the allochthonous materials. In addition, the two lakes are bordered at their northern sides by sand dunes. The sand particles are carried away by the effect of the prevailing wind to settle down on the bottom of the northern sides of the two lakes. Moreover, Lake Brollus is connected with the sea by a narrow slit through which the sea water may enter into the lake and reach a great distance inside it. The sea water may transport to the lake considerable amounts of sand.

The allochthonous minerogenic materials entered Lake Brollus and Lake Quarun were distributed on their bottom and covered the autochthonous organic sediments or mixed with them. Accordingly, the exchange of materials between the sediments and the free water must be greatly reduced under this condition (OHLE 1960, 1962, 1964, UNGEMACH 1960, and SAAD 1970).

Every substance mentioned in the present study was found in all sediment samples, ranging between highest and lowest values. The maximum density of wet mud record-

Table 1. Density of the wet and dry mud, as well as the percentages and amounts of some constituents deposited down on the bottom of Lake Brollus and Lake Qarun

Lake	Locality	Water depth (cm)	Density (g/cm^3)		Water		Dry matter		Org. matter		Calc. substances		Alloch. materials		Si		SiO_2	
			wet mud	dry mud	%	kg/m^2	%	kg/m^2	%	kg/m^2	%	kg/m^2	%	kg/m^2	%	g/m^2	%	g/m^2
Brollus	West (A_1)	140	1.50	3.08	50.35	7.6	49.65	7.4	13.37	1.0	51.43	3.8	34.55	2.55	0.64	50	1.37	100
	Middle (A_2)	130	1.53	2.43	41.12	6.3	58.88	9.0	18.75	1.7	45.54	4.1	34.97	3.13	0.74	70	1.58	140
	East (A_3)	100	1.25	3.50	71.63	9.0	28.37	3.5	17.37	0.6	37.44	1.3	44.35	1.57	0.84	30	1.80	60
Qarun	West (A_4)	400	1.31	3.38	66.43	8.7	33.57	4.4	9.25	0.4	42.79	1.9	46.96	2.06	1.0	40	2.14	90
	Middle (A_5)	200	1.66	2.78	37.99	6.3	62.01	10.3	4.46	0.5	39.43	4.0	55.54	5.74	0.57	60	1.22	130
	East (A_6)	400	1.33	3.36	64.31	8.6	35.69	4.7	8.66	0.4	54.15	2.5	36.23	1.75	0.96	50	2.05	100

ed from the middle of Lake Brollus and Lake Qarun was associated with the minimum density of dry mud. The maximum wet density is mainly due to the increase in the amounts of dry matter and the decrease in the amounts of water content. The great increase in the weight of dry matter in the middle regions of the two lakes is attributed principally to the maximum values of different components of the sediments; namely, the organic matter, the calcareous substances, the allochthonous materials and the diatom-silica.

The minimum density of wet mud found in the sediments of the eastern side of Lake Brollus and the western side of Lake Qarun was associated with the maximum density of dry mud. This minimum wet density is due mainly to the great decrease in the amounts of dry matter and the increase in the amounts of water content. These minimum values of dry matter are attributed to the great decrease in the amounts of different constituents of the sediments.

The quantitative distribution of the different components of the sediments at various localities of the lake bottom depends upon several factors. The amounts of organic matter found in the lake sediments were affected by the amount of autochthonous organic matter produced inside the lake and also the quantity of allochthonous organic matter entered the lake via drainage waters. The intensity of mineralization of the organic matter and the particle composition of the sediments must be also considered.

The bottom of Lake Brollus and Lake Qarun is characterized by the great accumulations of calcareous shells and shell fragments of calcareous organisms. The abundance of these shells increased markedly the amount of calcareous substances in the sediments (El-Wakeel 1964 and Saad 1974). The relatively lower amounts of calcareous substances found in the sediments of the eastern side of Lake Brollus and the western side of Lake Qarun coincide with the relative decrease in the amounts of calcareous shells at these localities. Such decrease of the calcareous shells may be due to the environmental conditions, which might be unfavourable for the growth of calcareous organisms, and the increase in the rate of supply of noncalcareous substances via drains. The solution of calcium carbonate, which might be occurred after death of the organisms, must be also considered.

The maximum amounts of diatom-silica recorded from the sediments of the middle region of Lake Brollus and Lake Qarun reflect the richness of these sediments with diatom shells (Saad 1971). It may be concluded that the ecological conditions necessary for the growth of diatoms were favourable at the time of deposition. Also, the diatom frustules in these sediments were subjected to a high degree of preservation. In addition, the conditions might be unfavourable to accelerate the release of silica from these sediments into the free water.

Comparison of Lake Brollus and Lake Qarun sediments

A comparison of Lake Brollus and Lake Qarun sediments seems to be of special interest from the limnological point of view, since the two lakes are found under different local conditions. Lake Brollus is connected with the sea, whereas Lake Qarun is a closed basin.

The maximum and minimum values of wet density of Lake Brollus sediments were slightly lower than those of Lake Qarun sediments. The maximum amount of water content in Lake Brollus deposits was slightly higher than that maximum in Lake Quarun deposits. However, a similar minimum value of water content was recorded from the sediments of both lakes. The average amounts of water content in the sediments of Lake Brollus and Lake Qarun reached approximately similar values of 7.6 and 7.9 kg/m², respectively (Fig. 3). The highest and lowest values of dry matter in Lake Brollus deposits were slightly lower than those in Lake Qarun deposits. The average value of dry matter calculated from the sediments of Lake Brollus was nearly similar to that of Lake Qarun sediments (Fig. 3).

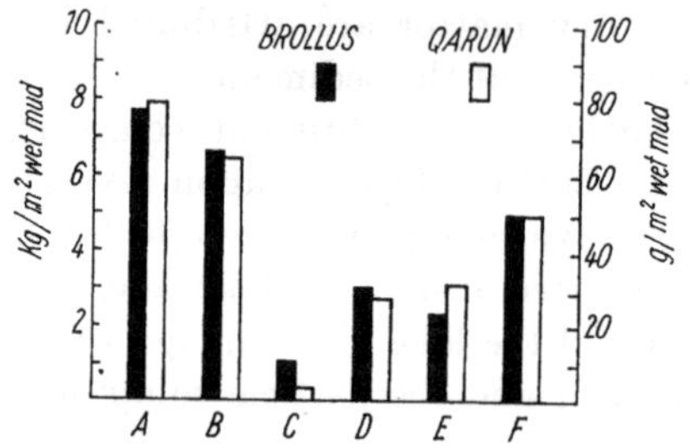

Fig. 3. Average values of the different components of the sediments of Lake Brollus and Lake Qarun

A – water; B – dry matter; C – organic matter; D – calcareous substances; E – allochthonous materials (kg/m² wet mud); F – Si (g/m² wet mud)

The maximum values of calcareous substances in the deposits of both lakes were approximately similar. On the other hand, the minimum value of Lake Brollus deposits was lower than that minimum of Lake Qarun deposits. The average value of calcareous substances in the sediments of Lake Brollus was slightly higher than that of Lake Qarun sediments (Fig. 3)

The highest and lowest amounts of allochthonous materials in the deposits of Lake Brollus were lower than those in the deposits of Lake Qarun. The calculated average value of these materials from the sediments of Lake Brollus was lower than that of Lake Qarun sediments (Fig. 3).

The maximum amount of diatom-silica in the deposits of Lake Brollus was slightly higher than that maximum of Lake Qarun deposits. On the other hand, the minimum amount of silica in the deposits of Lake Brollus was lower than that minimum of Lake Qarun deposits. A similar average value of diatom-silica was calculated from the sediments of both lakes (Fig. 3).

It is also of particular importance to compare the amounts of organic matter precipitated down on the bottom of both lakes. The highest amount of organic matter in the sediments of Lake Brollus was about three times higher than that of Lake Qarun sediments. Moreover, the minimum value of Lake Brollus sediments was slightly higher than the maximum value of Lake Qarun sediments. The calculated average value of organic matter from the deposits of Lake Brollus was about three times higher than that average of Lake Qarun deposits (Fig. 3). The great decrease in the organic content of Lake Qarun sediments may be due mainly to the high intensity of decomposition of the organic matter favoured by the special environmental conditions in the region (SAAD 1972). The intensity of mineralization of the organic matter in the sediments of Lake Qarun must be higher than that in Lake Brollus sediments.

Summary

Eighteen sediment samples were collected from Lake Brollus and Lake Qarun in order to study the nature and composition of their deposits. These samples were subjected to some physical and chemical investigations.

The maximum density of wet mud found in the sediments of the middle region of Lake Brollus and Lake Qarun was associated with the minimum density of dry mud. On the other hand, the minimum density of wet mud recorded from the sediments of the eastern side of Lake Brollus and the western side of Lake Qarun was associated with the maximum density of dry mud.

The organic matter, the calcareous substances, the allochthonous materials, and the diatom-silica were deposited down in variable amounts on the bottom of both lakes. The quantitative distribution of these components at various localities of the lake bottom was found to depend principally upon certain factors which were discussed.

Comparison of the results obtained from Lake Brollus sediments with those from Lake Qarun sediments gave considerable variations, since the two lakes are found under different local conditions. Lake Brollus, lying at the northern part of the Nile Delta, is connected to the sea with a narrow slit. Lake Qarun, on the other hand, has no connection with the sea, and is found in the western desert.

Comparison of the organic content of the sediments of both lakes showed a remarkable variation. The calculated average value of organic matter from the deposits of Lake Brollus was about three times higher than that average of Lake Qarun deposits. The great decrease in the organic content of Lake Qarun sediments may be attributed mainly to the high intensity of decomposition of the organic matter favoured by the special environmental conditions in the region.

References

EL-WAKEEL, A. K., 1964: Recent bottom sediments from the neighbourhood of Alexandria, Egypt. — Marine Geol. **2**, 137—146.

GORGI, S., 1959: The use of energy equation in calculation of the rate of evaporation from Lake Qarun. — Notes & Mem. Alex. Inst. Hydrobiol. Fish. **42**, 1—26.

MULLIN, I. B., & RILEY, I. P. 1955: The colorimetric determination of silicate with special reference to sea and natural waters. — Analytica chim. Acta **12**, 162—176.

NAGUIB, M., 1961: Studies on the ecology of Lake Qarun (Faiyum-Egypt). Part II. — Kieler Meeresforsch. **17** (1), 94—131.

OHLE, W., 1960: Fernsehen, Photographie und Schallortung der Sedimentoberfläche in Seen. — Arch. Hydrobiol. **57**, 135—160.

— 1962: Der Stoffhaushalt der Seen als Grundlage einer allgemeinen Stoffwechseldynamik der Gewässer. — Kieler Meeresforsch. **18**, 107—120.

— 1964: Interstitiallösungen der Sedimente, Nährstoffgehalt des Wassers und Primärproduktion des Phytoplanktons in Seen. — Helgol. wiss. Meeresunters. **10**, 411—429.

SAAD, M. A. H., 1970: Entwicklungsgeschichte des Schöhsees aufgrund mikroskopischer und chemischer Untersuchungen. — Arch. Hydrobiol. **67** (1), 32—77.

— 1971: Diatom-silica in the sediments of Lake Mariut and the Nozha Hydrodrome, Egypt. — Bull. Fac. Sci. Univ. Alex. **11** (1), 215—229.

— 1972: Investigations of the surface sediments of the Nozha Hydrodrome, near Alexandria, Egypt. — Rapp. Comm. int. Mer Médit. **20** (4), 515—517.

— 1974: Calcareous deposits of the brackish-water lakes in Egypt. — Hydrobiologia **44** (4), 381—387.

TESSENOW, U., 1964: Untersuchungen über den Kieselsäurehaushalt der Binnengewässer. — Arch. Hydrobiol., Suppl. 32, 1—136.

UNGEMACH, H., 1960: Sedimentchemismus und seine Beziehungen zum Stoffhaushalt in 40 europäischen Seen. — Diss., Univ. Kiel, 1—420.

Received December 5, 1974.

Address of the author:

Ass. Prof. Dr. MASSOUD A. H. SAAD
Department of Oceanography, Faculty of Science, University of Alexandria
Moharem Bey, Alexandria, Egypt

Limnologica (Berlin)	11	1	1976	9—16

BINOY KRISHNA MANDAL

The fluctuation and distribution of plankton in a freshwater fish pond at Burdwan, West Bengal, India

With 2 Tables

Abstract

Regular weekly plankton hauls were made in a freshwater fish pond at Burdwan, West Bengal, India, during the period of January to December, 1971. Qualitative and quantitative observations on the seasonal distribution and fluctuations of the plankton components have been made. Total plankton volume varied bimodally with the peaks occurring in spring and monsoon while the throughs appeared in summer and autumn. The highest percentage of zooplankton and phytoplankton was recorded in March and Deoember respectively. The maximum density of the rotifer population was noted in January while that of the crustaceans appeared in March. The algal population was highest in December and the peak population of the diatoms was recorded in July. An inverse correlation between the zooplankton and phytoplankton volumes has been recorded 13 species of crustaceans, 7 species of rotifers, 5 species of algae and 7 species of diatoms have been recorded in the present studies.

Introduction

Studies on the seasonal fluctuations of plankton in freshwater bodies is rather scanty in India. The observations of PRASAD (1916) on the seasonal conditions governing the pond life in Punjab is the first limnological study in India. A more detailed account of similar ecological studies of some ponds and reservoirs was reported from time to time (PRUTHI 1933, GANAPATI 1940, 1941, 1943, CHACKO & KRISHNAMURTHY 1954, DAS & SRIVASTAVA 1955, 1956, 1959, GONZALVES & JOSHI 1946, ALIKUNHI et al. 1955, GEORGE 1966; SREENIVASAN 1964, 1965, 1968, 1970, MOITRA & BHATTACHARYA 1965; MOITRA & MUKHERJEE 1972). However, there is no adequate information on the seasonal variations of plankton in freshwater bodies of West Bengal.

In view of such paucity of information, the present investigations were undertaken to study the seasonal fluctuation and distribution of the planktonic milieu in a freshwater fish pond at Burdwan, West Bengal, India.

Description of the pond

The freshwater fish pond studies is perennial and is located at Burdwan University Campus, Burdwan. The pond is locally known as 'Laldighi'. It is rectangular in shape and the average depth of the water varies from 3.5 m during the rainy season to 2.0 m during the summer. There is a general scarcity of aquatic vegetations. However, a dense growth of *Eichhornia crassipes* SOLMS appears in winter months.

Materials and methods

Regular weekly plankton hauls were made in the fish pond from January to December, 1971 with on-meter NENSON's plankton net. The total plankton volume was determined by the "Standard Displacement" method (EDDY 1934) after fixing the organism in five percent formalin. The organisms were identified and the percentage composition was determined by the "Points Method" (HYNES 1950).

Observations

The seasonal cycles of the planktons, the total plankton volume and the percentage composition of the plankton have been shown in Tables 1 and 2.

Total plankton volume: The total plankton volume varied markedly ranging form 7.0 cc to 15.0 cc during the period of studies. The variations followed a bimodal pattern of distribution. The maxima were recorded in spring and monsoon while the minima appeared in summer and autumn (Table 1).

Table 1. Variations of plankton volume and the percentage composition of the plankton components during the period of January to December, 1971

Months	Total plankton volume (cc)	Zoo-plankton (%)	Phyto-plankton (%)	Rotifers (%)	Crusta-ceans (%)	Algae (%)	Diatoms (%)
January	8.4	72.0	28.0	47.5	24.5	16.0	12.0
February	14.5	75.0	25.0	36.72	38.28	11.5	13.5
March	9.5	80.0	20.0	23.6	56.4	8.6	11.4
April	9.0	61.5	38.5	21.0	40.5	5.2	33.3
May	10.8	55.6	44.4	13.6	42.0	7.0	37.4
June	12.5	48.0	52.0	15.2	32.8	10.8	41.2
July	13.8	40.8	59.2	8.2	32.6	18.0	41.2
August	15.0	33.0	67.0	15.0	18.0	35.4	31.6
September	9.5	36.0	64.0	15.0	21.0	48.0	16.0
October	8.6	42.0	58.0	31.4	10.6	46.0	12.0
November	7.0	29.5	70.5	21.5	8.0	65.0	5.5
December	7.4	25.0	75.0	20.0	5.0	72.5	2.5

Zooplankton: The zooplankton population varied from 25.0 percent to 80.0 percent during the course of studies. In January, the zooplankton formed 72.0 percent of the total plankton which increased in February and March comprising 75.0 and 80.0 percent respectively of the total plankton. The population declined from April onwards and persisted till August sharing 33.0 percent of the total plankton. The percentage of zooplankton increased again in September (36.0 percent) and October (42.0 percent). But the population declined in November and December comprising 29.6 and 25.0 percent respectively (Table 1).

Rotifers: Rotifers were abundant throughout the period of studies and the population varied from 8.2 percent to 47.5 percent of the entire zooplankton population.

The maximum density of the rotifer population was recorded in January which declined sharply in February (36.72 percent) and continued till May comprising 13.6 percent of the entire population. The population increased slightly in June (15.2 percent) but declined abruptly in July (8.2 percent). In August, the rotifer populations increased rapidly and persisted till October sharing 31.4 percent of the plankton. From November the populations began to decline (Table 1).

Among the rotifer plankters *Brachionus diversicornis* and *Brachionus falcatus* were present throughout the year. *Brachionus forficula* was absent from May to August. *Asplanchna priodonta* was abundant in February and March but was absent from August to October. *Filinia brachiata* was abundantly present in October, November and December but was absent from April to August. *Trichocerca cylindrica* appeared as stray sample (Table 2).

Table 2. Occurrence of plankters in the fish pond during the period of January to December 1971

Name	J	F	M	A	M	J	J	A	S	O	N	D
Crustaceans:												
Cyclops viridis JURINE	p	p	p	pp	pp	pp	pp	pp	pp	p	p	p
Diaphanosoma sp. FISCHER	—	—	—	—	—	p	p	p	p	—	—	—
Ceriodaphnia rigaudi RICHARD	p	p	—	—	—	—	—	—	p	p	p	p
Alonella diaphana KING	—	—	—	p	p	p	—	—	p	p	p	—
Alonella nana BAIRD	—	p	p	—	—	—	pp	pp	pp	—	—	—
Bosmina longirostris MÜLLER	p	p	p	p	p	—	—	—	p	p	p	p
Chydorus sphaericus MÜLLER	—	—	—	p	p	p	p	—	—	—	—	—
Diaptomus cadacus LIGHT	p	p	p	—	—	p	p	—	—	p	p	p
Leptodora kindti FOCKE	p	p	—	—	—	—	—	—	—	—	p	—
Daphnia carinata KING	—	—	—	p	p	pp	pp	—	—	—	—	—
Cypricercus affinis JURINE	pp	p	—	—	p	p	—	—	—	—	pp	pp
Herpetocypris maridana FURTOS	—	—	—	—	p	p	p	p	pp	pp	—	—
Stenocypris fontinalis VAVRA	p	p	—	—	—	—	p	p	—	—	—	p
Rotifers												
Brachionus diversicornis DADAY	pp	pp	pp	pp	p	p	p	p	p	p	p	pp
Brachionus falcatus ZACHARIAS	pp	pp	pp	p	p	p	p	p	p	p	p	pp
Brachionus forficula WIERZEJSKI	p	p	p	p	—	—	—	—	p	p	p	p
Asplanchna priodonta GOOSE	p	pp	pp	p	p	p	p	—	—	—	p	p
Filinia brachiata ROUSSELET	pp	p	p	—	—	—	—	—	p	pp	pp	pp
Trichocerca cylindrica LAMARCK	p	p	p	—	—	—	—	—	—	—	p	p
Keratella tropica APSTEIN	—	—	—	—	—	—	p	—	p	p	p	p
Algae:												
Anacystis cyania KÜTZ.	pp	p	p	p	—	—	—	—	p	p	pp	pp
Microcystis aeruginosa KÜTZ.	pp	pp	p	p	—	—	—	p	—	pp	pp	pp
Volvox aureus EHRENBERG	p	p	—	—	—	—	p	—	—	p	—	p
Ulothrix zonata KÜTZ.	—	—	—	p	p	p	p	—	—	—	p	p
Lyngbya sp.	p	p	p	—	—	—	—	p	p	p	p	p
Diatoms:												
Diatoma vulgare BARY	p	p	p	p	p	—	—	—	—	p	p	p
Synedra ulna EHRENBERG	pp	p	p	—	—	p	p	p	p	p	p	p
Melosira granulata RALF	—	—	—	p	p	p	—	—	—	p	—	—
Navicula rosida KÜTZ.	—	p	p	—	—	—	—	—	—	—	—	p
Cymbella cistula KIRCHH.	p	p	p	p	—	—	—	—	—	—	p	p
Frustulia rhomboides DET.	—	—	p	—	—	—	—	—	p	p	p	p
Bacillaria paradoxa GMEL.	p	—	p	—	—	—	p	p	—	—	—	—

Abbreviations: p — present; pp — abundant; — — absent

Crustaceans: The crustaceans fluctuated from 5.0 to 56.4 percent of the entire plankton population. In January, the crustaceans formed 24.5 percent of the plankton which increased rapidly and recorded the highest population density in March (56.4 percent). The population decreased from April (40.5 percent) and culminated in August comprising 18.0 percent of the plankton. The crustaceans increased again in September (21.0 percent) but began to decline from October and produced the minimum density in December (Table 1).

Among the crustacean plankters, *Cyclops viridis* was abundantly present throughout the year. *Alonella diaphana* was present only during the summer and autumn months while *Alonella nana* was abundantly present during the monsoon months. *Bosmina longirostris* was absent only in June, July and August. *Chydorus sphaericus* was present as stray sample only from April to July. *Leptodora kindti* was recorded in January, February, November and December. *Herpetocypris maridana* was abundant in September and October (Table 2).

Phytoplankton: The phytoplankton varied from 20.0 to 75.0 percent of the total plankton during the period of studies. The highest population density was recorded in December while the lowest phytoplankton population was noted in March. The percentage of phytoplankton was high from June to December (Table 1).

Bluegreen andgreen algae: These algae and diatoms were the main components of the phytoplankton. The algal population fluctuated from 5.2 percent in April to 72.5 percent in December of the entire plankton. In January the algae comprised only 16.0 percent which declined gradually till April. The percentage of the algae began to increase from May and culminated in December comprising the highest percentage (Table 1).

The algal plankters were represented by five species only. *Anacystis cyania* and *Microcystis aeruginosa* were abundantly present during winter months. *Volvox aureus*, *Lyngbya* sp. and *Ulothrix zonata* were present as stray samples (Table 2).

Diatoms: The diatoms varied from 2.5 percent to 41.2 percent of total plankton. In January, the population formed 12.0 percent only and improved slightly in February (13.5 percent). It declined again in March consisting 11.4 percent of the entire plankton. The diatom population increased rapidly in April (33.3 percent) and culminated in the peak population in July (41.2 percent). There after, the diatom population began to decline steadily (Table 1).

The diatom population comprised seven species only. *Synedra ulna* was present more or less throughout the year and was abundantly present during the winter months. *Melosira granulata* was recorded only in April, May, June, and October. The other diatom plankters recorded were seasonal in appearance (Table 2).

Discussion

The total plankton volume varies from season to season in all aquatic situations although the degree of waxing and waning may differ. In tropical waters, the general pattern of plankton productivity is of the bimodal type (WELCH 1952). The occurrence of the maxima and minima however, may not appear in regular sequence during the successive years. In the present studies, the variations of plankton volume followed a bimodal pattern of distribution. The maxima were recorded in spring and monsoon while the minima appeared in summer and autumn. It was further observed that spring maxima was predominated by the zooplankton and the monsoon maxima by the phytoplankton. DAS & SRIVASTAVA (1955, 1956, 1959) observed a bimodal pattern of plankton produktion in freshwater lakes of Uttar Pradesh, India. The peaks were recorded in spring and monsoon while the troughs in summer and autumn. BIRGE & JUDAY (1922) also observed a bimodal pattern of distribution of the total plankton volume in lake Mendota. These phenomena correspond with the bluegreen algal bloom during the fall and together contribute to the formation of the highest maximum. On the other hand, rotifers reproduce profusely in winter due to the commencement of the 'winter egg' production period and probably contribute to the second plankton volume peak.

Tropical freshwater fauna greatly resembles the European forms specially as regards Rotifera, Copepoda and Cladocera (THIENEMANN 1954). Among the different organisms recorded during the present investigations, *Cyclops*, *Diaphanosoma*, *Ceriodaphnia*, *Keratella*, *Filinia*, *Brachionus* and *Asplanchna* are common in North American freshwater also (PENNAK 1957).

During the course of present studies, 7 species of rotifers, and 13 species of crustaceans along with their nauplii have been recorded.

SEWELL (1934) reported 10 species of rotifers and 26 species of crustaceans from a tank in Bengal. ALIKUNHI et al. (1955) recorded 11 rotifers and 5 crustaceans with occasionally occurring ostracods as important components of the plankton from some nursery ponds in Orissa. DAS & SRIVASTAVA (1959) recorded 12 species of crustaceans and 6 species of rotifers in the plankton samples from Lake Kathauta, Lucknow, India. MOITRA & MUKHERJEE (1972) reported 10 species of crustaceans and 7 species of rotifers from a freshwater pond at Kalyani, West Bengal.

Many more species, however, have been recorded by limnologists abroad (BAYER & RAJKOV 1929, KRAATZ 1941, HARDA 1933, DAMANN 1945, 1960, 1966, PENNAK 1949). MOORE (1950) observed the scarcity of zooplankton in lake Providence in which rotifers were the predominant group. JOLLY (1952) found zooplankton to be the richest in an Australian lake being represented mainly by the entomostracan crustaceans. PRÓSZYŃSKA (1962) found 44 species of crustacean plankton of which 23 species were cladocerans and 21 species of copepods from small water bodies in Poland. She distinguished two groups according to the period of occurrence viz. those occurring throughout the year and those during the warm season.

The commonest zooplankters recorded from the fish pond at Burdwan during the course of present investigations have been shown in Table 2. The bulk of the zooplankton consisted of copepods and cladocerans and rotifers, an observations similar to that in South Indian freshwater ponds (GANAPATI & CHECKO 1951, CHECKO & KRISHNAMURTHY 1954). Changes in the composition of species and their appearance and disappearance are most characteristics in spring and autumn, on the other hand summer and winter are periods of stabilization. Identical categories in the crustacean fauna, however, could not be distinguished in the present investigations but two groups appeared to be characteristic, one occurring throughout the year and the other seasonally.

Marked fluctuations in the phytoplankton components have been recorded during the present investigations and the phytoplankton predominated the plankton sample from monsoon to winter. GANAPATI & CHECKO (1951) recorded only three species of bluegreen algae in the phytoplankton of freshwater ponds in South India. ALIKUNHI et al. (1955) reported 34 species of phytoplankton in nursery ponds in Orissa. DAS & SRIVASTAVA (1959) have reported 17 phytoplankton components from some freshwater bodies at Lucknow. In the present investigations, only 5 species of algae and 7 species of diatoms have been recorded thus indicating a low diversity of species in the fish pond studied. The seasonal variations of the algae have been shown in Table 1. The algal populations declined during the summer and monsoon but increased rapidly in autumn and winter. The green algae were represented by 2 species only, viz. *Volvox aureus*, *Ulothrix zonata* and their seasonal occurrence have been shown in Table 2. The bluegreen algae were represented by *Anacystis cyania*, *Microcystis aeruginosa* and *Lyngbya* sp. only. They were abundant in winter months and formed the blooms.

The commonest diatoms recorded during the course of the present studies have been shown in Table 2. The population of diatoms increased during the monsoon but declined in winter. CHACKO & KRISHNAMURTHY (1954) observed *Melosira* and *Synedra* to predominate during November to March in South Indian ponds. ALIKUNHI et al. (1955) recorded *Navicula* and *Melosira* as the most common diatom components in the nursery ponds at Cuttack. Observations made by WELCH (1952) that the bluegreen algae and the green algae virtually diasappear during winter does not hold true for Indian freshwaters as observed by DAS & SRIVASTAVA (1959), MOITRA & BHOWMICK (1968) and further confirmed in the present investigations. An inverse correlation between the zooplankton and phytoplankton volume was recorded during the present studies. Zooplankton attained a peak in March when the phytoplankton registered a minimum and again when the phytoplankton registered a peak in December, the zooplankton produced the minimum (Table 1). A similar correlation between zooplankton and phytoplankton has also been observed by DAS & SRIVASTAVA (1956) in North Indian freshwater bodies and MOITRA & MUKHERJEE (1972) in the freshwaters of West Bengal.

Acknowledgements

The author is grateful to the Council of Scientific and Industrial Research, Government of India, for awarding a Junior Research Fellowship to carry out the present investigations. Grateful acknowledgements are also due to Dr. S. K. MOITRA, Reader in Zoology, Burdwan University, Burdwan, for his constant guidance troughout the period of studies.

Summary

Regular weekly plankton hauls were made in a freshwater fish pond at Burdwan, West Bengal, India, during the period of January to December, 1971. Qualitative and quantitative observations on the seasonal distribution and fluctuations of the plankton have been made. Total plankton volume varied from 7.0 cc to 15.0 cc with a bimodal pattern of distribution. The peaks were recorded during spring an monsoon while the troughs appeared in summer and autumn. The population density of zooplankton varied from 25.0 percent to 80.0 percent while the phytoplankton population fluctuated from 20.0 percent to 75.0 percent of the total plankton. An inverse correlation between the zooplankton and phytoplankton volume was recorded during the period of present investigations. Rotifers and crustaceans were the main components of the zooplankton. Rotifer population increased during winter months while the crustaceans increased during the summer months. The peak of the algal population was recorded in winter while that of diatoms appeared in summer. 13 species of crustaceans, 7 species of rotifers, 5 species of algae and 7 species of diatoms were recorded and the plankters showed a seasonal appearance during the course of the present studies.

References

ALIKUNHI, K. H., S. V. GANAPATI & V. RAMACHANDRAN, 1955: On the mortality of carp fly in nursery ponds and the role of plankton and their survival and growth. — Ind. J. Fish. **2**, 257—313.

BAYER, E. M., & A. L. RAJKOV, 1929: Hydrological investigations of lakes near Lendice. I. Quantitative studies of the plankton. — Bull. Ecol. Silp. Agron. Brno, Rcs. FAO, Silvicult. **D 14**, 1—165.

BIRGE, E. A., & C. JUDAY, 1922: The inland lakes of Wisconsin. The plankton I. Its quantity and chemical composition. — Wis. Geol. Nat. Hist. Surv., Bull. **64**, Sci. Ser. **13**, 1—222.

CHACKO, P. I., & J. B. KRISHNAMURTHY, 1954: On the plankton of three freshwater fish-ponds in Madras City. — Indian Symp. Mar. Freshwater Plank., Indo-Pacific Fish. Counc. (UNESCO), 103—107.

DAMANN, K. E., 1945: Plankton studies of Lake Michigan I. Seventeen years of plankton data collected at Chicago. — Amer. Midl. Nat. **34** (3), 769—796.

— 1960: Plankton studies of Lake Michigan II. Thirtythree years of continuous plankton and coliform bacteria data collected from lake Michigan at Chicago. — Trans. Amer. Micr. Soc. **72**, 397—404.

— 1966: Plankton studies of Lake Michigan III. Seasonal periodicity of total plankton. — Pub. No. **15**, Great Lake Res. Div., Univ. Michigan, 9—17.

DAS, S. M., & V. K. SRIVASTAVA, 1955: Quantitative studies of freshwater plankton I. Plankton of a fish tank at Lucknow, India. — Nat. Acad. Sci. **26**, 82—89.

— & — 1956: Some observations on plankton from freshwater ponds and tanks of Lucknow, India. — Sci. & Cult. **21**, 466—467.

— & — 1959: Studies on the freshwater plankton III. Qualitative composition and seasonal fluctuations in plankton components. — Ibid. **29** (4), 174—179.

EDDY, S., 1934: A study of freshwater plankton community. — Illinois Biol. Monogr. **21**, 7—92.

GANAPATI, S. V., 1940: The ecology of the temple tank containing a permanent bloom of *Microcystis aeruginosa* (KÜTZ.). — J. Bombay Nat. Hist. Soc. **42** (1), 65—77.

— 1941: Studies on the chemistry and biology of ponds in the Madras city: Seasonal changes in the physical and chemical conditions of a garden pond containing aquatic vegetations. — J. Madras Univ. **13** (1), 55—59.

— 1943: An ecological study of a garden pond containing abundant zooplankton. — Proc. Ind. Acad. Sci. **B 17** (2), 41—58.

GANAPATI, S. V. & P. I. CHACKO, 1951: Investigations on plankton production in a pond of Chetput fish-farm. — Proc. Internat. Assoc. Theor. & Appl. Limnol. **2**, 111—112.

GEORGE, M. G., 1966: Comparative plankton ecology of five fish-tanks in Delhi, India — Hydrobiologia **27** (1—2), 81—108.

GONZALVES, E. A., & B. D. JOSHI, 1946: Seasonal variations in the composition of phytoplankton of Bombay Harbour. — J. Bom. Nat. Hist. Soc. **46**, 154—176.

HARDA, I., 1933: Zur Biologie des Zooplanktons in Zitugetutan. — Mem. Fac. Sci. & Agri. Taihoko Imp. Univ. **7** (2), 71—139.

HYNES, J. B. N., 1950: The food of freshwater sticklebacks (*Gasterosteus aculeatus* and *Pygosteus pungitius*) with a review of the methods used in the studies of food of fishes. — J. anim. Ecol. **19** (1), 36—58.

JOLLY, V. H., 1952: A preliminary study of limnology of lake Hayers. — Austr. J. Mar. Freshwater Res. **3**, 74—91.

KRAATZ, W. C., 1941: Quantitative plankton studies in Turkeyfoot lake near Akron, Ohio. — Ohio J. Sci. **41**, 1—22.

MOITRA, S. K., & B. K. BHATTACHARYA, 1965: Some hydrological factors affecting plankton production in a fish pond at Kalyani, West Bengal, India. — Ichthyologica **4** (1—2), 8—12.

MOITRA, S. K., & M. L. BHOWMICK, 1968: Environmental factors affecting the fish food in the fish-farm at Kalyani, West Bengal. — Ph. D. thesis, Kalyani University.

MOITRA, S. K., & S. K. MUKHERJEE, 1972: Studies on the freshwater plankton of a fish pond at Kalyani, West Bengal, India. — Vest. Čs. Spol. Zool. **36** (1), 23—28.

MOORE, W. G., 1950: Limnological studies in Louisiana Lakes. I. Lake Providence. — Ecol. **31**, 86—99.

PENNAK, R. W., 1949: Annual limnological cycles in some Colorado reservoirs lakes. — Ecol. Monogr. **19**, 233—267.

— 1957: Species composition of limnetic zooplankton communities. — Limnol. & Oceanogr. **2**, 222—232.

PRASAD, B., 1916: The seasonal conditions governing the pond life in Punjab. — J. Asiat. Soc. Beng. **2**, 1—75.

PRÓSZYŃSKA, M., 1962: The annual cycle in occurrence of Cladocera and Copepoda in small water Bodies. — Polsk. Arch. Hydrobiol. **10** (23), 379—422.

PRUTHI, H. S., 1933: Studies on the bionomics of freshwaters in India I. Seasonal changes in the physical and chemical condition of the waters of the tanks in the Indian museum compound. — Internat. Rev. ges. Hydrobiol. Hydrogr. **28**, 46—67.

SEWELL, R. B. S., 1934: Studies on the bionomics of freshwaters in India II. On the fauna of the tank in the Indian museum compound and seasonal changes. — Ibid. **31**, 203—238.

SREENIVASAN, A., 1964: Limnological studies of tropical impoundments II. Hydrobiological features and plankton of Bhavanisagar reservoir for 1961—1962. — Proc. Ind. Acad. Sci. **59 B**, 53—71.

— 1965: Limnological studies of tropical impoundments III. Limnology and productivity of Amravathy reservoir. — Hydrobiologia **24** (4), 514—539.

— 1968: Limnology of tropical impoundments V. Studies of two upland impoundments in Nilgiris, Madras State (India). — Phykos 8 (2), 86—92.

— 1970: Limnology of tropical impoundments: A comparative study of the major reservoirs in Madras State (India). — Hydrobiologia **36** (3—4), 443—469.

THIENEMANN, A., 1954: Tropical freshwater plankton. — Indo-Pac. Fish Com. (UNESCO) Bangkok, 85—89.

WELCH, P. S., 1952: Limnology. — McGraw Hills, Publ. Inc., New York, 2nd Ed.

Received November 7, 1974.

Address of the author:

Dr. BINOY KRISHNA MANDAL
P. G. Department of Zoology
Bhagalpur University
Bhagalpur-7, Bihar, India

Limnologica (Berlin)	11	2	1976	17–46

BRIGITTE KOCH-ALTHAUS

Die Rotatorienfauna der Saidenbachtalsperre und der oberen Neunzehnhainer Talsperre

Mit 14 Abbildungen und 6 Tabellen

Inhalt

1. Einleitung

In der vorliegenden Arbeit wird über Untersuchungen berichtet, die an Rotatorien zweier Trinkwassertalsperren in den Jahren 1969 und 1970 durchgeführt wurden. Es sollte versucht werden, an Hand des Artenspektrums und auf Grund der Häufigkeit für jede Talsperre die charakteristische Rotatoriengemeinschaft zu ermitteln. Erhofft wurde ein Zusammenhang zwischen Trophiegrad und der Art des Rotatorienvorkommens.

In den Arbeiten von BEUSCHOLD (1961), KLAPPER (1957/58, 1961), RUDOLF (1966) ist über die Lage und das Einzugsgebiet der Saidenbachtalsperre ausführlich berichtet worden. Es wird hier auf diese Arbeiten verwiesen. Das gilt auch für die obere Neunzehnhainer Talsperre[1]), deren Lage von HEDLICH (1959), HOHMANN (1963) und HÖHNE

[1]) Im weiteren Verlauf der Arbeit wird die obere Neunzehnhainer Talsperre nur als Talsperre Neunzehnhain bezeichnet.

(1965) besprochen wird. Zum allgemeinen Verständnis sei nur folgendes erwähnt: Es handelt sich um zwei verschiedene Talsperrentypen. Saidenbach ist mäßig eutroph, während Neunzehnhain oligotroph zu nennen ist. Für den Trophiegrad ist das Einzugsgebiet verantwortlich. Bei Saidenbach ist es besiedelt und landwirtschaftlich genutzt. Die Zuflüsse führen also nährstoffreiches Wasser. Die Umgebung der Neunzehnhainer Talsperre ist bewaldet und unbesiedelt. die dem Staubecken zufließenden Wasser sind nährstoffarm. Ein typisches Litoral ist bei beiden Talsperren nicht ausgebildet.

In Tab. 1 sind die wichtigsten Daten beider Talsperren zusammengestellt.

Tab. 1. Physiographie der untersuchten Talsperren

	Saidenbach	Neunzehnhain
Bauzeit	1929–1933	1911–1914
Einzugsgebiet (km^2)	60,6	12,8
Stauinhalt (Mio m^3)	22,4	3,0
Wasserfläche (ha)	146	28,9
Maximaltiefe (m)	48	33
mittlere Tiefe (m)	15,3	10,4

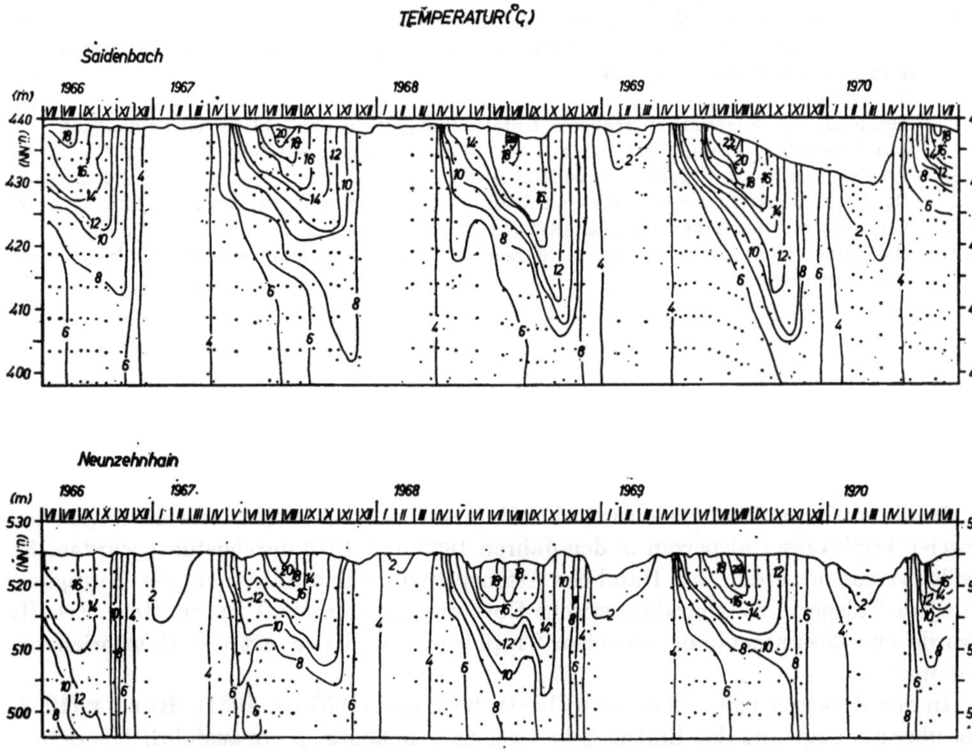

Abb. 1. Temperaturkurven der Talsperren Saidenbach und Neunzehnhain für die Jahre 1966–1970 (nach HÖHNE, unveröff.)

Herrn Professor Dr. D. UHLMANN und Herrn Dr, E. HÖHNE danke ich für die allseitige Förderung der vorliegenden Arbeit. Beim Schreiben dieser Arbeit gedachte ich mit Dankbarkeit meines verehrten, im Jahre 1970 verstorbenen Lehrers, Herrn KURT WULFERT.

2. *Systematischer Teil*

2.1. Methodik

Die Proben wurden mit einem Planktonnetz (Müllergaze 25) entnommen. Ein 14tägiger Untersuchungsrhythmus wurde im Normalfall eingehalten. Die Untersuchungsdauer erstreckt sich über die Zeit von November 1968 bis Dezember 1970. Für die Horizontalproben wurde das Netz an der Oberfläche geschleppt. Die Vertikalproben wurden an der jeweils tiefsten Stelle der Talsperre entnommen. Die Häufigkeitsangaben der Arten sind geschätzt. Für die quantitative Arbeit wurde die Wassersäule berechnet, die das Planktonnetz umfaßt. Die Zählergebnisse beziehen sich auf 1 l Wasser (s. Tab. 4). Die Planktonkammer nach Kolkwitz, 1 ml Fassungsvermögen, diente zum Zählen. Von jeder Probe wurden 3 ml quantitativ ausgewertet und ihre Werte gemittelt. Formol war das Fixierungsmittel für das zum Auszählen bestimmte Plankton.

In großen zeitlichen Abständen wurden mit dem Herbst-Schöpfer in 0, 5, 10, 15, 25 und 35 m Tiefe (Saidenbach) bzw. in 0, 5, 10, 15 und 25 m Tiefe (Neunzehnhain) Proben entnommen. Je 5 l wurden durch ein Planktonnetz (Nr. 25) filtriert und auf 30 ml eingeengt. Von jeder Tiefe wurden zwei Proben ausgezählt. Die Ergebnisse sind in Tab. 5 zusammengefaßt.

2.2. Jahreszeitliche Verteilung und relative Häufigkeit der Rotatorien im Plankton der Talsperren Saidenbach und Neunzehnhain

Die relative Häufigkeit der verschiedenen Arten wurde durch folgende Symbole gekennzeichnet:

1 — selten = ~ 3 Exemplare je Probe; 2 — häufiger = ~6 Exemplare je Probe; 3 — häufig = ~20 Exemplare je Probe; 4 — sehr häufig = ~50 Exemplare je Probe; 5 — massenhaft = unzählbar viele Exemplare

Ferner bedeuten:

E — Rotator trägt Embryonen oder Eier; + nur abgestorbenes Tier; V — Vertikalplankton;
H — Horizontalplankton; — — keine Probe entnommen.

In den Tabellen 2 und 3 sind die jahreszeitlichen Verteilungen der eu- und tychoplanktischen Rotatorienarten der beiden Talsperren sowie die relative Häufigkeit zu den einzelnen Terminen für den Zeitraum vom November 1968 bis Dezember 1970 wiedergegeben.

Für die Talsperre Saidenbach können als Hauptvertreter der Planktonrotatorien gelten:

Keratella cochlearis }
Keratella quadrata } perennierend
Kellicottia longispina }
Polyarthra dolichoptera
Polyarthra vulgaris
Synchaeta pectinata
Synchaeta oblonga
Trichocerca porcellus

Diese Arten werden in den beiden Untersuchungsjahren ziemlich regelmäßig registriert. Allerdings zeigt sich beim Betrachten der Tab. 2, daß die Artenzahl im Jahre 1970 abnimmt: *Asplanchna priodonta, Collotheca mutabilis, Conochiloides natans, Conochilus unicornis* erscheinen wohl im Jahre 1970, aber mit viel geringerer Beständigkeit als im Vorjahr. *Collotheca mutabilis* ist nur noch spärlich vorhanden, im Gegensatz zu häufigem Vorkommen 1969. Die beiden *Gastropus*-Arten werden 1970 überhaupt nicht mehr gesichtet, und *Ploesoma hudsoni* ist nur noch einmal im August im Plankton. Die graphische Darstellung der quantitativen Ergebnisse läßt die Abnahme der Artenzahl nicht so deutlich erkennen, da die meisten Arten, die offenbar im Jahre 1970 verschwunden sind, 1969 den Grenzwert 0,1 Tier pro ml nicht erreichen und dadurch nicht in der Tabelle erscheinen.

Die Einzelfunde sollen in diesem Zusammenhang unberücksichtigt gelassen werden, obgleich auch hier das Jahr 1969 mehr Arten aufzuweisen hat.

Sehr viel deutlicher wird die Abnahme der Artenzahl in der Talsperre Neunzehnhain. Beim Betrachten von Tab. 3 sehen wir den großen Unterschied. 1969 sind 11 Arten häufiger bis häufig im Plankton vertreten, diese sind:

Ascomorpha saltans
Asplanchna priodonta
Collotheca mutabilis
Conochilus unicornis
Gastropus stylifer
Keratella quadrata
Keratella cochlearis
Kellicottia longispina
Ploesoma hudsoni
Polyarthra dolichoptera
Polyarthra vulgaris

Dazu kommen 14 Arten, die weniger häufig, aber doch mit einer gewissen Beständigkeit immer wieder gefunden werden. Im Jahre 1970 hat sich das Bild gewandelt, nur *Keratella cochlearis* wird von den oben genannten Arten registriert, aber in weit geringerer Konzentration. *Microcodon clavus* wird im Plankton neu gefunden. *Lecane flexilis*, *Lecane lunaris*, *Trichotria tetractis* und *Rotaria*-Vertreter, Tychoplankter, fallen in dem verarmten Biotop durch ihr relativ beständiges Auftreten auf.

2.3. Besprechung der einzelnen Arten

2.3.1. Allgemeines

Im folgenden werden die Rotatorien in alphabetischer Reihenfolge aufgeführt. Soweit es die Seltenheit oder die Besonderheit einer Art wünschenswert erscheinen ließ, ist die Beschreibung ausführlicher gehalten und durch Abbildungen erläutert. Die Angaben zur Verbreitung der einzelnen Art sind der einschlägigen Literatur entnommen. Für alle Arten werden die Maße, die zur Bestimmung wichtig sind, angegeben. In diesem Teil der Arbeit werden auch die Einzelfunde berücksichtigt, deren Bestimmung teilweise mit Schwierigkeiten verbunden war. Außerdem ist es durchaus möglich, daß sich unter diesen Einzelfunden ein häufigerer Bewohner der Talsperren verbirgt.

Ist eine Beschreibung einer Art erforderlich, so wird der Gattungsname im Text mit seinem ersten Buchstaben abgekürzt. Die Arten sind alphabetisch angeordnet.

2.3.2. Die Arten der Talsperre Saidenbach

1. *Ascomorpha ecaudis* PERTY, 1850

Maße: Lg. 138 µm; größte Br. 80 µm
Literatur, die zur Bestimmung benutzt wurde: VOIGT (1957), WULFERT (1960b)
Verbreitung: Im See- und Teichplankton; Kosmopolit.

2. *Ascomorpha minima* von HOFSTEN, 1909 (Abb. 2)

Maße: Lg. 85 µm; Br. 54 µm; Kauerlg. 42 µm; Fulcr. 34 µm; Eigr. 92×68 µm
Literatur: VOIGT (1957)

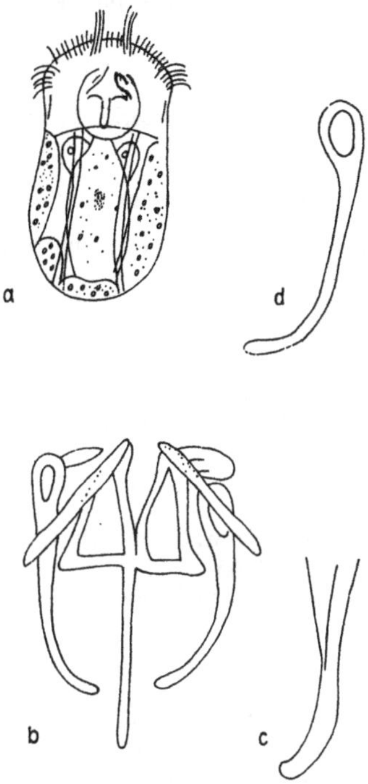

Abb. 2. *Ascomorpha minima*

a – Aufsicht des Tieres; b – Kauer; c – Fulcrum seitlich; d – Manubrium seitlich

Bemerkungen: In der Literatur werden Lg. 43 µm–51 µm; Br. 30 µm angegeben. CARLIN (1943) fand im Motalaström eine *Ascomorpha*, die er als *A.* cf. *minima* bezeichnet, diese ist 80 µm groß (Anmerkung bei VOIGT). Bei den vorliegenden Funden wurden deutlich die für *A. minima* charakteristischen zwei Apikaltaster gesehen. Auffällig war das flinke Schwimmen des Tieres. Trotz der Abweichung in den Maßen, wird der Fund der Art *A. minima* zugerechnet. Das Fulcrum zeigt am freien Ende keine Verbreiterung, war aber dorsalwärts gebogen.

Verbreitung: In Moorgewässern. Nicht sehr verbreitet.

3. *Asplanchna priodonta* GOSSE, 1850

Maße: Lg. total 660 µm
Literatur: WULFERT (1956), VOIGT (1957)
Verbreitung: Im See- und Teichplankton; Kosmopolit.

4. *Collotheca mutabilis* (HUDSON, 1885)

Maße: Lg. total 396 µm; Körperlg. 165 µm; Fußlg. 231 µm; Eigr. 66 × 37 µm;
Literatur: VOIGT (1957)

Bemerkungen: Gallerthülle und vier Dotterstockkerne wurden deutlich gesehen.

Verbreitung: Im See- und Teichplankton; Kosmopolit.

5. *Collotheca pelagica* (ROUSSELET, 1893)

Maße: Lg. total 485 µm; Eigr. 75 × 51 µm
Literatur: VOIGT (1957)
Verbreitung: Im See- und Teichplankton.

6. *Conochiloides natans* (SELIGO, 1900)

Maße: Lg. total 170 µm; Tasterlg. 24 µm
Literatur: VOIGT (1957)

Bemerkungen: Der Kauer ist dauernd in Bewegung, auffallend.

Verbreitung: Im See- und Teichplankton; Kosmopolit.

7. *Conochilus unicornis* ROUSSELET, 1892

Maße: Lg. 200 µm; Br. 61 µm; Taster 22 µm
Literatur: VOIGT (1957)
Verbreitung: Im See- und Teichplankton.

8. *Filinia maior* (COLDITZ,1924)

Maße: Körperlg. 132 µm; Terminalborste 430 µm; Seitenborsten 505—535 µm
Literatur: VOIGT (1957)
Verbreitung: Im SSee- und Teichplankton; Kosmopolit.

9. *Gastropus hyptopus* (EHRENBERG, 1838)

Maße: Lg. 130 µm; Höhe 85 µm; Zehenlg. 14 µm
Literatur: VOIGT (1957), WULFERT (1956, 1960a, 1960b, 1961)
Verbreitung: Im See- und Teichplankton.

10. *Gastropus stylifer* IMHOF, 1891

Maße: Lg. 132 µm; Höhe 90 µm; Zehenlg. 28 µm
Literatur: VOIGT (1957), WULFERT (1956, 1960b, 1961)
Verbreitung: Im See- und Teichplankton.

11. *Kellicottia longispina* (KELLICOTT, 1879)

Maße: Lg. total 626—872 µm; Vorderdornlg. 323—417 µm; Hinterdornlg. 203—336 µm
Literatur: VOIGT (1957)
Verbreitung: Im See-, seltener im Teichplankton.

12. *Keratella cochlearis* (GOSSE, 1851)

Maße: Lg. total 148—272 µm; Hinterdornlg. 40—118 µm
Literatur: VOIGT (1957), WULFERT (1960b, 1961)
Verbreitung: Im See- und Teichplankton.

13. *Keratella quadrata* (O. F. MÜLLER, 1786)

Maße: Lg. total 130—145 µm; Hinterdornlg. 100—110 µm
Literatur: VOIGT (1957), WULFERT (1960b, 1961)
Verbreitung: Im See- und Teichplankton; Kosmopolit.

14. *Lecane closterocerca* (SCHMARDA, 1895)

Maße: Panzerlg. dorsal 54 µm; Br. 61 µm; Panzerlg. ventral 61 µm; Br. 51 µm; obere Br. 41 µm; Zehenlg. 29 µm; Fußöffnungsbr. 10 µm; Höhe 7 µm
Literatur: HARRING & MYERS (1926), VOIGT (1957), WULFERT (1956, 1960a, 1960b)
Verbreitung: Vereinzelt in größeren und kleineren Gewässern.

15. *Lecane flexilis* (GOSSE, 1889)

Maße: Panzerlg. dorsal 99 µm; Br. 83 µm; Panzerlg. ventral 86 µm; Br. 66 µm; obere Br. 60 µm; Zehenlg. 28—29 µm; obere seitliche Dornen 3—4 µm
Literatur: HARRING & MYERS (1926), VOIGT (1957), WULFERT (1956, 1960a)
Verbreitung: In großen und kleinen Gewässern; nicht selten.

16. *Lecane luna* (O. F. MÜLLER, 1776)

Maße: Panzerlg. dorsal 118 µm; Br. 148 µm; Panzerlg. ventral 140 µm; obere Br. 78 µm; Zehenlg. 55 µm
Literatur: VOIGT (1957), WULFERT (1956, 1960a, 1960b, 1961)
Verbreitung: In größeren und kleineren Gewässern; verbreitet, auch im Plankton.

17. *Lecane lunaris* (EHRENBERG, 1832)

Maße: Panzerlg. dorsal 105 µm; Br. 85 µm; Zehenlg. 65 µm
Literatur: VOIGT (1957), WULFERT (1956, 1960a, 1960b, 1961)
Verbreitung: Zwischen Wasserpflanzen, zuweilen im Plankton; verbreitet.

18. *Lecane scutata* (HARRING & MYERS, 1926) (Abb. 3)

Maße: Panzerlg. dorsal 64 µm; Br. 61 µm; Panzerlg. ventral 68 µm; Br. 48 µm; obere Br. 48 µm; Zehenlg. mit Kralle 32 µm; Kralle 5 µm; Fußöffnungsbr. 12 µm
Literatur: HARRING & MYERS (1926), VOIGT (1957)
Verbreitung: In Moorgewässern; vereinzelt.

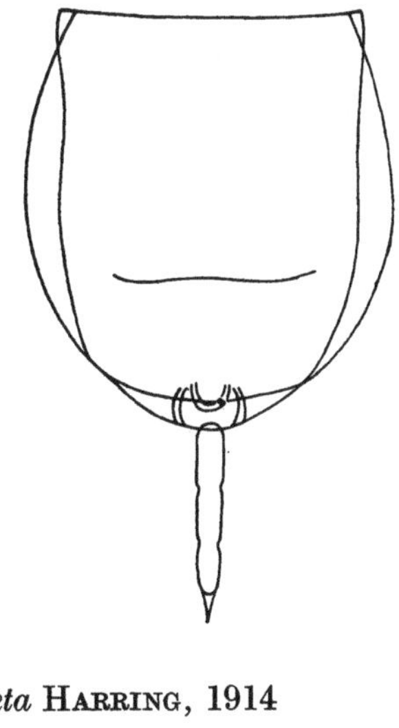

Abb. 3. *Lecane scutata*

19. *Lecane tenuiseta* HARRING, 1914

Maße: Panzerlg. dorsal 97 µm; Br. 53 µm; Panzerlg. ventral 94 µm; Zehenlg. 34—39 µm, davon Kralle 8 µm
Literatur: HARRING & MYERS (1926), VOIGT (1957)

Bemerkungen: Die Krallen sind bei den gefundenen Arten kürzer, als in der Bestimmungsliteratur angegeben ist, es bestehen 5—8 µm Unterschied in der Länge.

Verbreitung: Zwischen Wasserpflanzen; verbreitet.

20. *Lepadella acuminata* (EHRENBERG, 1834)

Maße: Lg. 100 µm mit Dorn; Br. 58 µm; obere Br. 28 µm; Zehenlg. 21 µm
Literatur: VOIGT (1957), WULFERT (1956, 1960a, 1960b)
Verbreitung: Zwischen Pflanzen größerer und kleinerer Gewässer; Kosmopolit.

21. *Lepadella ovalis* (O. F. MÜLLER, 1786)

Maße: Panzerlg. 90 µm; Br. 80 µm; Zehenlg. 23 µm
Literatur: VOIGT (1957), WULFERT (1956, 1960b, 1961)
Verbreitung: Zwischen Pflanzen größerer und kleinerer Gewässer; Kosmopolit.

22. *Lepadella patella* (O. F. MÜLLER, 1786)

Maße: Panzerlg. 105 µm; Br. 61 µm; Zehenlg. 27 µm
Literatur: HAUER (1952), VOIGT (1957)
Verbreitung: Zwischen Pflanzen größerer und kleinerer Gewässer; Kosmopolit.

23. *Lindia torulosa* DUJARDIN, 1841 (Abb. 4)

Maße: Lg. total 200 µm; Zehenlg. 11 µm; Kauerlg. 40 µm; Fulcrum 15 µm
Literatur: VOIGT (1957), WULFERT (1956, 1961)

Bemerkungen: Die Größenangaben für diese Art schwanken. Der Fund aus Saidenbach ist klein, verglichen mit vorhandenen Literaturangaben. In Abb. 4 sind alle wesentlichen, zur Bestimmung wichtigen Eigenheiten der Art eingezeichnet.

Verbreitung: Zwischen Wasserpflanzen; Kosmopolit.

24. *Mytilina ventralis* (EHRENBERG, 1832)

Maße: Lg. 203 µm; Höhe 97 µm
Literatur: VOIGT (1957), WULFERT (1956, 1960b)
Verbreitung: Zwischen Pflanzen größerer und kleinerer Gewässer; Kosmopolit.

25. *Notholca acuminata* var. *extensa* OLOFSSON, 1918 (Abb. 5)

Maße: Lg. total 325 µm; größte Br. 105 µm; obere Br. 72 µm; vordere Dornen 36 µm; Hinterende Lg. 90 µm
Literatur: VOIGT (1957)

Bemerkungen: Die Varietät *N. acuminata extensa* wurde bisher nur auf Spitzbergen und im Waterneversdorfer Binnensee (Brackwasser) beobachtet. Sie zeichnet sich durch ihr langausgezogenes Hinterende aus. Die Maße der erstmals im Mittelgebirgsraum gefundenen Exemplare entsprechen genau den Literaturangaben.

Verbreitung: s. Bemerkungen.

26. *Notholca squamula* (O. F. MÜLLER, 1786)

Maße: Lg. 145 µm; Br. 107 µm; obere Br. 61 µm
Literatur: VOIGT (1957), WULFERT (1961)
Verbreitung: Im Plankton von Seen und Teichen.

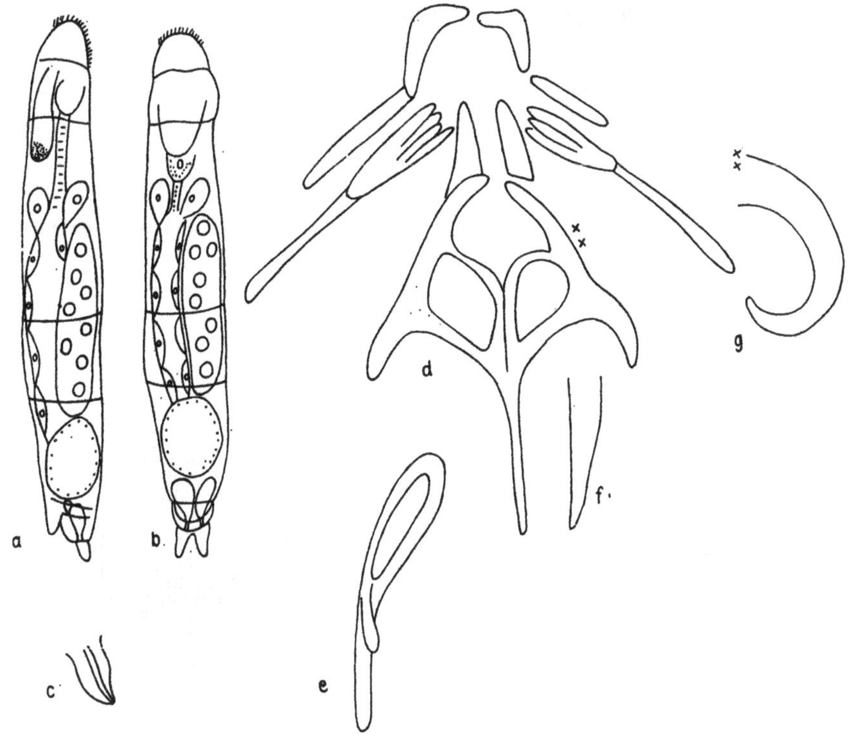

Abb. 4. *Lindia torulosa*

a – Seitenansicht; b – Aufsicht; c – Zehe mit Klebdrüsengang; d – Kauer; e – Manubrium; f – Fulcrum von der Seite; g – Haken am Ramus sitzend

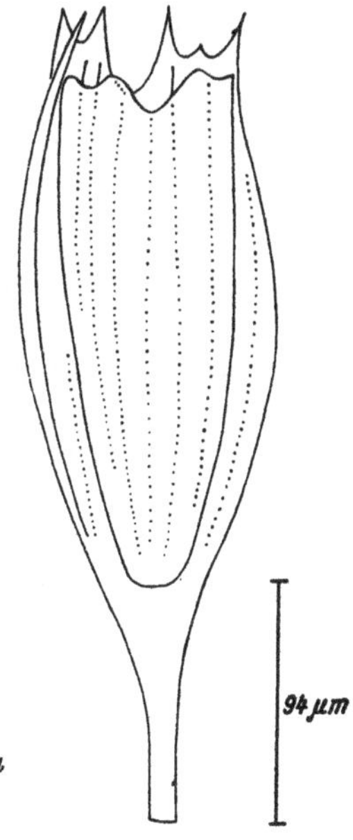

Abb. 5. *Notholca acuminata* var. *extensa*

27. *Ploesoma hudsoni* (IMHOF, 1891)

Maße: Panzerlg. 413 µm; Klauenlg. 67 µm
Literatur: VOIGT (1957), WULFERT (1956, 1961)
Verbreitung: Im Plankton von Seen, seltener in kleineren Gewässern.

28. *Polyarthra dolichoptera* IDELSON 1925 (Abb. 6, 7)

Maße: Lg. total 167 µm; Höhe 77 µm; Flossenlg. 140 µm; Flossenbr. 13—14 µm; Ventralstachellg. 41 µm; Ramus 33 µm; Fulcrum 34 µm
Literatur: DONNER (1954), VOIGT (1957), RUTTNER (1959)

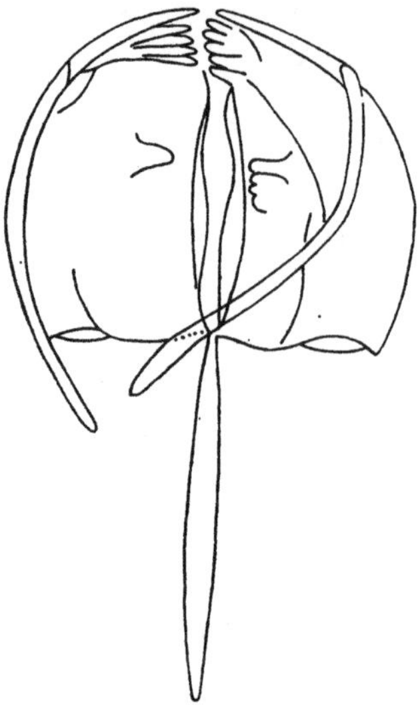

Abb. 6. Kauer von *Polyarthra dolichoptera*

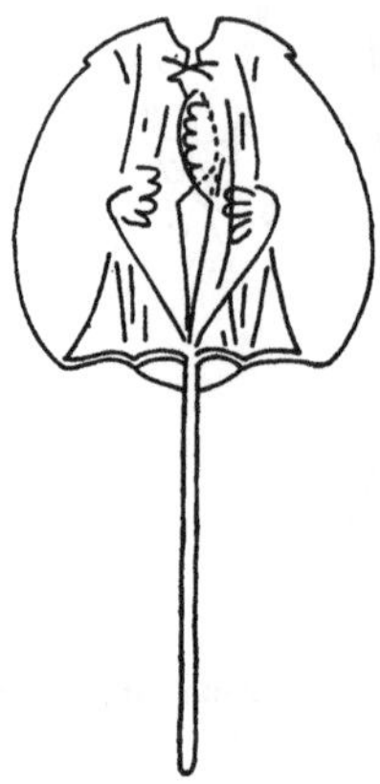

Abb. 7. Kauer von *Polyarthra dolichoptera* nach DONNER

Bemerkungen: Polyarthren zu bestimmen, ist mit großen Schwierigkeiten verbunden. Die Flossenlänge, die oft als spezifisches Artmerkmal herangezogen wird, schwankt so stark, daß sie heute nicht mehr als Merkmal gelten kann. Die Anzahl der Dotterstockkerne, bisher als gutes Artmerkmal bekannt, wird von SUDZUKI (1955) in Frage gestellt. So bleiben artspezifisch die ventralen Anhänge in Höhe des Mastax', die für *P. dolichoptera* borstenförmig, für *P. vulgaris* blattförmig sind, und die Lage der Lateraltaster. Beide eben genannten Merkmale sind nicht immer leicht zu finden. Die Beschaffenheit der Flossen: Breite, gesägt oder blattförmig, ist von Wichtigkeit. Kauerabbildungen gibt es nur wenig. Es erhebt sich auch die Frage, ob die Kauer der einzelnen Arten genügend Unterschiede aufweisen, um daran die jeweilige Art erkennen zu können. Vielleicht sind auch hier die Merkmale schwankend. Dennoch sollte ein Versuch, die Kauer gründlich zu untersuchen und abzubilden unternommen werden, es könnten sich im Laufe der Untersuchungen artspezifische Unterschiede ergeben, die Sicherheit in die Taxonomie der Gattung *Polyarthra* brächten. In der Saidenbachtalsperre kamen die beiden Arten *P. dolichoptera* und *P. vulgaris* vor. Von beiden Arten wurde der Kauer untersucht. DONNER (1954) bildet den Kauer von *P.*

dolichoptera ab, er ist in Abb. 7 dargestellt. Die Abb. 6 zeigt den Kauer einer *P. dolichoptera* aus Saidenbach. DONNER zeichnete keine Unci und Manubria, er weist darauf hin, daß sie ihm verloren gegangen sein könnten. Bei dem Saidenbachexemplar sind 5- bzw. 6zähnige Unci zu erkennen. Die Zähne am Ramus sind nicht immer deutlich zu sehen, statt des einen gezeichneten Zahnes links können auch zwei vorhanden gewesen sein (schwankende Zahnzahl ?). Die Manubria liegen den Rami seitlich auf.

Verbreitung: Im Plankton größerer und kleinerer Gewässer.

Polyarthra aptera wurde nicht besonders aufgeführt, da es sich um eine Temporalvariation der *P. dolichoptera* handelt.

29. *Polyarthra vulgaris* CARLIN, 1943 (Abb. 8)

Maße: Lg. 136 μm; Br. 97 μm; Flossenlg. 156 μm; Flossenbr. 20 μm; Fulcrum 33 μm; Rami 26 μm; Unci 11 μm; Manubria 31 μm
Literatur: VOIGT (1957), RUTTNER (1959)

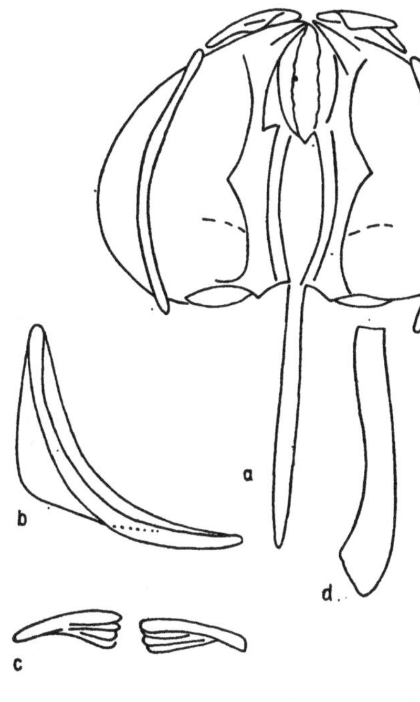

Abb. 8. Kauer von *Polyarthra vulgaris*

a – Aufsicht; b – Manubrium; c – Unci; d – Fulcrum seitlich

Bemerkungen: An den Rami kann an der Innenseite eine Raspelleiste vorhanden sein, es konnte nicht mit Sicherheit erkannt werden, wenn ja, wäre es ein Unterschied zu dem Kauer der *P. dolichoptera*. Die Unci haben wenigstens 4 Zähne. Die Manubria sind stark gekrümmt, ihre Lamellen sind sehr zart, ebenso die Lamelle, die die Rami rechts und links begrenzt. Bemerkenswerte Unterschiede konnten bisher zwischen den Kauern von *P. dolichoptera* und *P. vulgaris* nicht beobachtet werden. Die Zähne am Ramus der *P. dolichoptera* könnten eine Eigenart sein, wenn sie sich als konstant erweisen.

Verbreitung: ImPlankton größerer Gewässer.

30. *Proales fallaciosa* Wulfert, 1937

Maße: Körperlg. total 275 µm; Zehenlg. 14 µm
Literatur: Voigt (1957), Wulfert (1939, 1956, 1960a, 1960b)

Bemerkungen: Diese Art ist leicht an dem dorsal gelegenen Zapfen zwischen den Zehen zu erkennen.

Verbreitung: In größeren und kleineren Gewässern; vereinzelt.

31. *Rotaria haptica* (Gosse, 1886)

Maße: Lg. 500 µm; Dorsaltaster wenigstens 40 µm
Literatur: Voigt (1957)
Verbreitung: Zwischen Wasserpflanzen.

32. *Rotaria rotatoria* (Pallas, 1766)

Maße: 400 µm
Literatur: Voigt (1957)
Verbreitung: Zwischen Wasserpflanzen, auch frei schwimmend; verbreitet.

33. *Synchaeta kitina* subsp. *monodactyla* n. subsp. (Abb. 9)

Maße: Lg. total 195 µm, Fußlg. 15 µm; Zehenlg. 2,4—3 µm; Fulcrum 33 µm; Manubria 35 µm
Literatur: Rousselet (1902), Voigt (1957)

Beschreibung: Der vorliegende Fund aus der Saidenbachtalsperre hat Ähnlichkeit mit *S. kitina* Rousselet, 1902. Die Länge beträgt für *S. kitina* 136 µm, die Synchaeta aus Saidenbach ist größer. *S. kitina* gehört zum Verwandtschaftskreis der *S. tremula*, beide Tiere haben die Lateraltaster am Rumpfende, ihr Vorderende ist breit mit geringen Erhebungen, die Zehen werden dicht zusammen getragen. *S. kitina* ist schwach gelblich, bei der neuen Unterart ist der Magendarm bräunlich gefärbt. Die äußere Form der neuen Unterart entspricht etwa den Angaben für *S. kitina*, ihr fehlt nur der eingezogene untere Rumpfteil, sie ist in der Gestalt ohne Einbuchtungen. Die Lateraltaster sitzen am Ende des Rumpfes. Die Zehen scheinen verwachsen zu sein, das ist ein bemerkenswerter Unterschied zu *S. kitina*. Sie werden als eine Zehe bewegt und nie gespreizt. Bei starker Vergrößerung kann man zwei winzige Zehenspitzen und mitunter eine Linie zwischen den außerordentlich zarten, verwachsenen Zehen erkennen. Das Auge ist, wie bei *S. kitina*, zweiteilig, rot, dies ist schwer zu beobachten, denn die Augenflecke liegen dicht zusammen, von dunklen Granula umgeben. 8 Dotterstockkerne wurden gezählt. Der Fuß wird beim Schwimmen gern etwas eingezogen. Die Cuticula ist gestreift. Die Fußdrüsen durchziehen den Fuß. Der Fuß ist vom Rumpf deutlich abgesetzt. Vom Kauer der *S. kitina* wird im Voigt nur der Uncus erwähnt, der 5 oder 6 Zähne aufweisen soll. Der Kauer der neuen Unterart hat Ähnlichkeit mit dem der *S. tremula*. Die Rami sind wie ein Kamm gezähnt. An den Unci wurden auch wenigstens 6 Zähne gesehen, bei bestimmter Stellung erschienen die Manubria gekrückt.

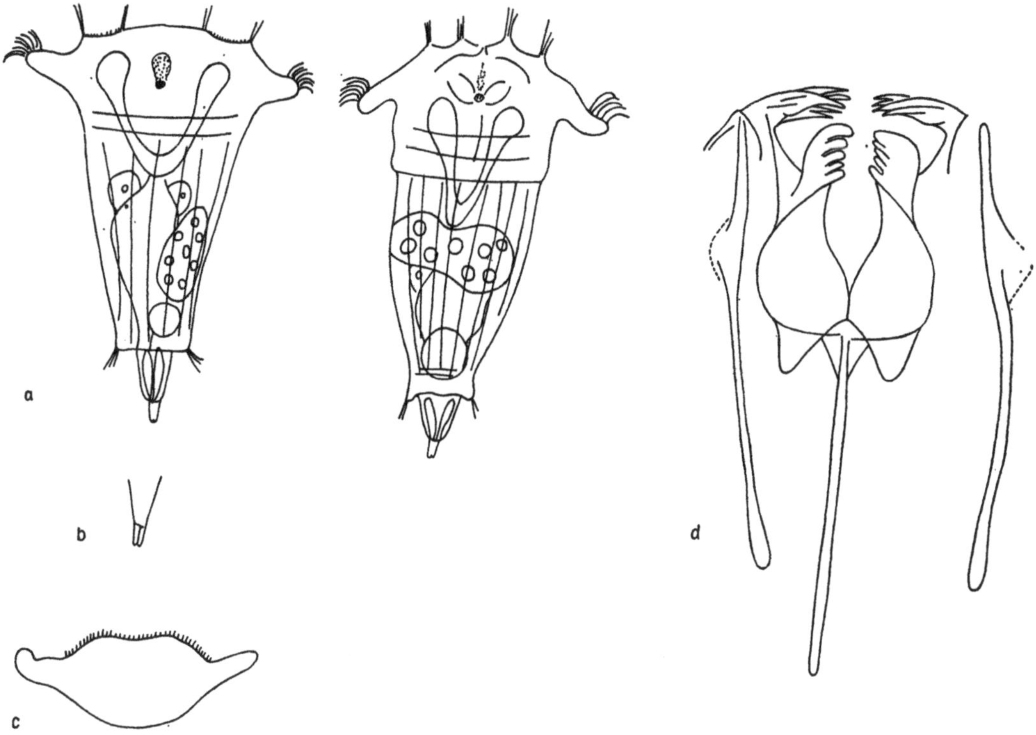

Abb. 9. *Synchaeta kitina monodactyla* nov. subsp.

a — zwei Ansichten des Tieres; b Zehe stark vergrößert; c — Tier von oben gesehen; d — Kauer

Wahrscheinlich sitzen an ihnen außen Lamellen. SLÁDEČEK (1950) gibt für *S. kitina* eine Zehenlänge von 13 µm. Es handelt sich um eine kurze Notiz über das Vokommen dieser Art in der ČSSR. Außer einer Zeichnung der *S. kitina* enthält die Arbeit keine genaue Beschreibung. Liegt hier eine Verwechslung mit *S. tremula* vor? Die Zehenlänge läßt es vermuten. Auf Grund der zweispitzigen, verwachsenen Zehe wird die neue Unterart *monodactyla* aufgestellt.

Verbreitung: Talsperre Saidenbach. — *S. kitina*: In Seen und Teichen; selten.

34. *Synchaeta oblonga* EHRENBERG, 1832 (Abb. 10)

Maße: Lg. total 170 µm; Zehenlg. 5—8 µm; Fulcrum 30—33 µm; Rami 47 µm; Manubria 50 µm
Literatur: ROUSSELET (1902), VOIGT (1957), KOCH-ALTHAUS (1963)

Bemerkungen: Die *Synchaeta* sind eine schwer zu bestimmende Gattung. In der Literatur sind die Abbildungen oft unzureichend. Kauerzeichnungen fehlen meistens, deshalb wurden eine Kauerabbildung beigefügt.

Verbreitung: Im Plankton von Seen und Teichen.

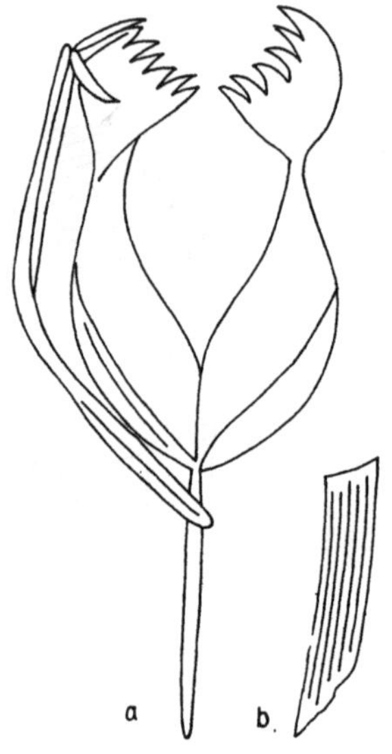

Abb. 10. Kauer von *Synchaeta oblonga*
a – Aufsicht, linke Seite vollständig;
b – Fulcrum seitlich

35. *Synchaeta pectinata* Ehrenberg, 1832

Maße: Lg. total 314 µm; Zehenlg. 8 µm; Kauerlg. 144 µm; Fulcrum 86 µm; Manubria 135 µm
Literatur: Rousselet (1902), Voigt (1957), Wulfert (1956, 1960a, 1960b, 1961)
Verbreitung: Im Plankton größerer und kleinerer Gewässer; verbreitet.

36. *Synchaeta tremula* (O. F. Müller, 1786)

Maße: Lg. total 290 µm; Zehen 10 µm;
Literatur: Rousselet (1902), Voigt (1957), Wulfert (1960a, b)
Verbreitung: Im Plankton größerer und kleinerer Gewässer; Kosmopolit.

37. *Trichocerca brachyura* (Gosse, 1851)

Maße: Lg. 90 µm; Höhe 40 µm; Zehenlg. I 28 µm; II 23 µm
Literatur: Voigt (1957), Wulfert (1960b)
Verbreitung: Vereinzelt zwischen Pflanzen stehender Gewässer; Kosmopolit.

38. *Trichocerca porcellus* (Gosse, 1886)

Maße: Lg. total 150 µm; Zehenlg. I 52 µm; II 39 µm; großer Dorn 13 µm
Literatur: Voigt (1957), Wulfert (1960b)
Verbreitung: Zwischen Pflanzen größerer und kleinerer Gewässer, im Sommer und Herbst auch im Plankton; verbreitet.

39. *Trichocerca pusilla* (Jennings, 1903)

Maße: Lg. total 65 µm; Höhe 34 µm; Zehenlg. I 32 µm; II 16 µm
Literatur: Voigt (1957)
Verbreitung: Im Plankton von Seen und Teichen; verbreitet.

40. *Trichocerca tenuior* (GOSSE, 1886)

Maße: Körperlg. mit Fuß 156 µm; Höhe 42 µm; Zehenlg. I 62 µm; II 30 µm; Dornlg. 9,5 µm
Literatur: VOIGT (1957), WULFERT (1961)
Verbreitung: Zwischen Pflanzen größerer und kleinerer Gewässer; verbreitet.

41. *Trichotria pocillum* (O. F. MÜLLER, 1776)

Maße: Lg. total 300 µm
Literatur: VOIGT (1957), WULFERT (1956, 1960b, 1961)
Verbreitung: Zwischen Pflanzen größerer und kleinerer Gewässer; verbreitet.

42. *Trichotria tetractis* (EHRENBERG, 1830)

Maße: Lg. total (ohne Zehen) 121 µm; Br. 94 µm; obere Br. 88 µm; Zehenlg. 80 µm
Literatur: VOIGT (1957), WULFERT (1956, 1961)
Verbreitung: Zwischen Pflanzen größerer und kleinerer Gewässer.

2.3.3. Die Arten der oberen Neunzehnhainer Talsperre

Die Rotatorienarten, die bereits für die Saidenbachtalsperre beschrieben, und auch in der oberen Neunzehnhainer Talsperre gefunden wurden, werden nur mit Namen angeführt.

1. *Ascomorpha ecaudis* PERTY, 1850

2. *Ascomorpha minima* VON HOFSTEN, 1909

3. *Ascomorpha saltans* BARTSCH, 1870

Maße: Gesamtlg. 160 µm
Literatur: VOIGT (1957), WULFERT (1960b)
Verbreitung: Im See- und Teichplankton.

4. *Asplanchna priodonta* GOSSE, 1850

5. *Brachionus angularis* GOSSE, 1851

Maße: Panzerlg. 152 µm; größte Br. 119 µm; obere Br. 104 µm
Literatur: VOIGT (1957)
Verbreitung: Im Plankton von Seen und Teichen.

6. *Brachionus quadridentatus* HERMANN, 1783

Maße: Panzerlg. 228 µm; größte Br. 200 µm; obere Br. 156 µm
Literatur: VOIGT (1957)
Verbreitung: Zwischen Pflanzen in Teichen und Seen, auch im Plankton; verbreitet.

7. *Cephalodella apocolea* HARRING & MYERS, 1924 (Abb. 11)

Maße: Lg. 260 µm; Zehenlg. 65 µm; Kauerlg. 60 µm; Fulcrum 34 µm
Literatur: HARRING & MYERS (1924), VOIGT (1957), WULFERT (1940)

Bemerkungen: Die Maße für *C. apocolea* betragen in der Literatur 125—160 µm für die Gesamtlänge und 30—40 µm für die Zehen. Das Tier aus der Talsperre Neunzehnhain ist größer. Es entspricht sonst gut der Beschreibung, die für *C. apocolea* vorliegt. Es ist augenlos, glasklar, die Zehen enden in einer Kralle, die Zehen stehen weit auseinander. Die Fußdrüsen sind groß, ein Schwanzanhang ist deutlich. Diatomeenfresser.

Verbreitung: Algenbewuchs in Seen, in krautigen Teichen

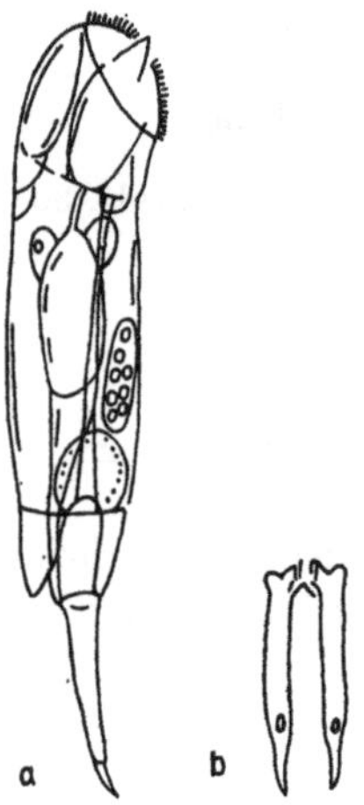

Abb. 11. *Cephalodella apocolea.*
a — Seitenansicht; b — Zehen, Aufsicht

8. *Cephalodella auriculata* (O. F. MÜLLER, 1773)

Maße: Lg. 80 µm; Zehenlg. 27 µm
Literatur: VOIGT (1957), WULFERT (1937)
Verbreitung: In Tümpeln und Teichen, in Seen zwischen Wasserpflanzen; verbreitet.

9. *Cephalodella gibba* (EHRENBERG, 1832)

Maße: Gesamtlg. 254 µm; Zehenlg. 72 µm; Kauerlg. 52 µm; Fulcrum 35 µm
Literatur: VOIGT (1957), WULFERT (1937, 1956, 1960b, 1961)
Verbreitung: In größeren und kleineren Gewässern, zuweilen im Plankton; verbreitet.

10. *Cephalodella tenuior* (GOSSE, 1886)

Maße: Gesamtlg. 119 µm; Zehenlg. 20 µm; Höhe 34 µm
Literatur: WULFERT (1937), VOIGT (1957)

Bemerkungen: Schmales, augenloses Tier. Magendarm war grün, die Magendrüsen bräunlich gefärbt, letzteres ist charakteristisch für *C. tenuior*. Die freien Enden der Manubria gekrückt.

Verbreitung: Zwischen Wasserpflanzen, besonders in fließenden Gewässern; verbreitet.

11. *Cephalodella ventripes* (DIXON-NUTTAL, 1901)

Maße: Gesamtlg. 160 µm; Zehenlg. 33 µm; Kauerlg. 36 µm; Fulcrum 21 µm; Manubria 21 µm; Ramus links 15 µm; rechts 18 µm
Literatur: VOIGT (1957), WULFERT (1937, 1956, 1960a, 1960b)

Tab. 2. Jahreszeitliche Verteilung und relative Häu

		1969																								
03. 12.		11. 02.		12. 03.		22. 04.		06. 05.		17. 05.		03. 06.		12. 08.		18. 08.		02. 09.		15. 09.		30. 09.		13. 10.		03.
H	V	—	V	—	V	H	V	H	V	H	V	H	V	H	V	—	V	H	V	H	V	H	V	H	V	H
.	1	.	.	.	.	.	.	.	.	.	.	.	.	.	2	.	3	.	3	.	2	.	1	.	.	.
2E	2	.	1	.	1	1	2	1	.	.	1+	3	4	.	.	.	.	.	.	1+	1+	.	1+	2	1	1
1E	1E	.	.	.	.	.	.	.	.	.	.	.	.	3	3	.	3	2	2	1	1+	.	1E	.	.	.
.	.	.	.	.	.	.	.	.	.	.	.	.	.	3	3	.	1	.	1	.	.	.	.	.	.	.
1	1	.	2	.	2E	3E	3	.	1E	2	2	.	1	.	.	.	.	.	.	.	.	.	.	.	.	.
.	.	.	.	.	.	.	.	.	.	.	.	.	.	.	.	.	.	.	2	.	.	.	1	.	.	.
.	.	.	.	.	1+	.	.	.	.	.	.	.	.	.	1+	.	1+	.	.	.	.	.	.	.	.	.
.	.	.	.	.	.	.	.	.	.	.	.	.	.	.	.	.	.	2	2	.	1	.	.	.	.	.
.	.	.	.	.	.	.	.	.	.	.	.	.	.	.	.	.	.	.	1	.	.	.	.	1	1	.
3	3	.	1	.	1	3	3	1	1	2	3	3E	4E	3	3	.	3	4	3	4	3	3	3	3	3	2
1	1	.	1+	.	1	1	1	.	1	3	2	3	3E	3	2	.	3	1	.	.	2	2E	3	1	1	2
1	1	.	1+	.	1	1	1	.	1	3	2	3	3E	3	2	.	3	1	.	.	2	2E	3	1	1	2
.	.	.	.	.	.	.		.	.	.	.	.	.	.	.	.	.	.	.	.	.	.	1	.	.	.
.	1	.	.	.	.	.	.	.	.	.	.	.	.	.	.	.	.	.	.	.	.	.	.	.	.	.
.	.	.	.	.	.	.	.	.	.	.	.	.	.	.	.	.	.	.	.	.	.	.	.	.	.	.
.	.	.	.	.	.	.	.	.	.	.	2	2	1	.	.	.	.	.	.	.	.	.	.	.	.	.
.	.	.	.	.	.	.	1	.	1	.	2	.	.	.	.	.	.	.	.	.	.	.	.	.	.	.
.	.	.	.	.	.	.	.	.	.	.	.	.	.	.	1	.	.	.	.	.	.	1	1	.	.	.
3	1	.	.	.	.	.	.	.	.	.	.	.	.	.	.	.	.	.	.	2	2	2	2	2	2	.
1	1	.	.	.	.	.	.	.	4	4E	3E	3	4	3	3	.	4	3	3E	3	3	2	2	1	1	.
.	.	.	.	.	.	.	.	.	.	.	.	.	.	.	.	.	.	.	.	.	.	.	.	.	.	.
2	1	.	.	.	.	.	.	.	4	3	3	.	.	.	.	.	.	.	.	.	.	.	1	.	.	.
.	.	.	.	.	.	1	.	.	.	.	.	.	.	.	.	.	.	.	.	.	.	.	.	.	.	.
.	.	.	1	.	1	2	2	.	2	2	2	.	.	.	.	.	.	.	.	.	.	.	.	.	.	.
.	.	.	.	.	.	.	.	.	.	.	.	.	.	.	.	.	.	.	.	.	.	.	.	.	.	.
4	4	.	.	.	.	.	.	.	.	.	.	.	.	3	2	.	2	3	4	1	3	3	3	1	1	1
.	.	.	.	.	.	.	.	.	.	.	.	.	.	.	.	.	.	.	.	.	.	.	.	.	.	.
.	.	.	.	.	.	.	.	.	.	.	.	.	.	.	.	.	.	.	.	.	.	.	.	.	.	.

Rotatorien im Plankton der Talsperre Saidenbach

			1970																							
11.	02. 12.		31. 01.		10. 02.		24. 03		09. 05.		02. 06.		16. 06.		22. 06.		30. 06.		06. 07.		22. 07.		13. 08.		08. 09.	
V	H	V	H	V	–	V	–	V	H	V	H	V	–	V	H	V	H	V	–	V	H	V	–	V	H	V
.	.	.	.	.	.	.	.	.	.	.	.	.	.	.	.	.	.	.	.	.	1	.	.	1	.	.
2	.	1	.	2	.	2	.	.	.	.	.	.	.	1	1	1	.	1	.	1	1	3	.	1	.	.
.	.	1	.	.	.	.	.	.	.	.	.	.	.	.	.	.	.	.	.	.	1	.	.	.	.	.
.	.	.	.	.	.	.	.	.	.	.	.	.	.	2	.	2	.	1	.	.	.	.	.	.	.	.
.	.	.	.	3	.	2	.	2	1	2E	1	1	.	.	.	2	.	2	.	4	.	2	.	.	.	.
1+	.	.	.	1	.	.	.	.	.	1	1	.	.	3	.	.	.	.	.	.	.	.	.	.	.	.
.	.	.	.	1+	.	.	.	.	.	.	.	.	.	1	.	1	.	1	.	1	.	1	.	2	.	.
.	.	.	.	.	.	.	.	.	.	.	.	.	.	.	.	.	.	.	.	.	.	.	.	.	.	.
1	1	.	.	.	.	.	.	.	.	.	.	.	.	.	.	.	.	.	.	.	.	.	.	.	.	.
2	2	2	.	2	.	2	.	2	2	1	1	1	.	2	2	2	3	2	.	2	3	3	.	2E	1	2
2	1	1	.	2	.	1	.	1	1	.	.	.	.	1	.	1	1	1	.	1	2	2	.	1	.	1
2	1	1	.	2	.	1	.	1	1	.	.	.	.	1	.	1	1	1	.	1	2	2	.	1	.	1
.	.	.	.	.	.		.	.	.	.	.	.	.	.	.	.	.	1	.	1	.	.	.	.	.	.
.	.	.	.	.	.	.	.	.	.	1	.	.	.	.	.	.	.	.	.	.	.	.	.	.	.	.
.	.	.	.	.	.	.	.	.	.	1	.	.	.	.	.	.	.	.	.	.	.	1	.	.	.	.
.	.	.	.	.	.	.	.	.	.	.	.	.	.	1	.	1	1	.	.	.	.	.	.	.	.	.
.	.	.	.	.	.	1	.	.	.	.	.	.	.	.	.	.	.	.	.	.	.	.	.	.	.	.
.	.	.	.	.	.	.	.	.	.	.	.	.	.	.	.	.	.	.	.	.	.	.	.	1	.	.
2	2	1	1	2	.	2	.	3	.	.	.	.	.	.	.	.	.	.	.	.	.	.	.	.	.	.
.	.	.	.	.	.	.	.	1	2	1	1	2	.	5	1	2	2	2	.	3	.	2	1	2	.	2
.	1	1	.	.	.	.	.	.	.	.	.	.	.	.	.	.	.	.	.	.	.	.	.	.	.	.
1	.	2	.	2	.	1+	.	.	.	.	1	2	.	.	.	.	.	.	.	.	.	.	.	.	.	.
.	.	.	.	.	.	.	.	.	.	.	2	2	.	.	.	.	.	.	.	.	.	.	.	.	.	.
.	.	2E	.	.	.	.	.	.	1	.	3	2	.	.	.	.	.	.	.	.	.	.	.	.	.	.
.	2	2	.	.	.	.	.	.	.	.	2	2	.	.	.	.	.	.	.	.	.	.	.	.	.	.
1	.	1	.	.	.	.	.	.	.	.	.	.	.	.	.	.	.	.	.	.	.	1	.	2	.	.
.	.	.	.	.	.	.	.	.	.	.	.	.	.	.	.	.	.	.	.	.	1	.	.	.	.	.
.	.	.	.	.	.	.	.	.	.	1	.	.	.	.	.	.	.	1	.	.	.	.	.	.	.	.

Tab. 3. Jahreszeitliche Verteilung und relative Häuf

1968				1969																						
19. 11.		03. 12.		11. 02.		12. 03.		22. 04.		06. 05.		18. 05.		03. 06.		11. 08.		25. 08.		02. 09.		15. 09.		30. 09.		13.
H	V	H	V	–	V	–	V	H	V	H	V	H	V	H	V	H	V	–	V	H	V	H	V	H	V	H
	.	.	.	.	.	.	.	.	.	.	.	.	.	.	.	.	1	.	.	.	1	.	.	2	1	1
	.	.	.	.	.	.	.	.	.	.	.	.	.	2	2	.	.	.	.	1	.	1	.	1	.	1
	.	.	.	.	.	.	.	.	.	.	.	.	.	.	.	.	1	.	1	.	3	1+	1	.	1	2
	.	.	.	.	.	.	.	.	.	.	.	.	.	.	.	.	2	.	1+	.	2	.	.	1	.	2
E	1E	1E	1E	.	.	.	.	.	.	.	.	.	.	1	.	2	.	.	3	3	2	3	3	2	2	2
	.	.	.	.	.	.	3	.	.	.	.	.	.	.	.	.	1+	.	.	.	.	.	.	.	.	.
	3	3	3	.	.	.	.	.	.	.	.	.	.	1	.	.	.	.	4	5	5	.	.	.	.	.
	1	.	.	.	.	.	.	.	.	.	.	3	.	.	.	.	.	.	.	.	.	.	.	.	.	1
	.	.	.	.	.	.	.	.	.	.	.	.	.	2	2	.	.	.	.	2	2	.	1	.	.	.
+	.	1	.	.	.	.	.	.	.	.	.	.	.	.	.	3	3	.	1	3	3	1	3	2	1	1
	4	4	4	.	.	.	4E	2	2	.	.	2	2E	1	1	3	3	.	3	4	4	4	4	4	3	3
E	1	1	1	.	.	.	1E	1+	1	1	3	2	3	3	4	1+	1+	.	2	.	2	.	1	1+	2	.
	3	1	1	.	.	.	1	.	.	.	.	.	.	1	.	3	3E	.	3E	4	4	3	3	3E	3E	4
	1	.	1	.	.	.	.	.	.	.	.	.	.	.	1	.	.	.	.	.	.	.	.	1	.	.
	.	.	.	.	.	.	.	.	.	.	.	.	.	1	.	1	1	.	.	1	.	.	.	1	.	.
+	.	.	.	.	.	.	.	.	.	.	.	.	.	.	.	.	.	.	1	.	.	.	1	.	.	.
	.	.	.	.	.	.	.	.	.	.	.	.	1	.	1	.	.	.	1	.	.	.	.	.	.	.
	.	.	.	.	.	.	.	.	.	.	.	.	.	.	.	.	.	.	.	.	.	.	.	.	.	.
	.	.	.	.	.	.	.	.	.	.	.	.	.	.	.	1	.	.	.	.	.	.	.	.	.	.
	.	.	.	.	.	.	.	.	.	.	.	.	2	.	3	3	3	.	2	1	3E	.	.	2	1	.
	3	3	1	.	.	.	.	.	.	.	1	.	.	.	.	.	.	.	.	.	.	.	.	.	.	.
	.	.	.	.	.	.	.	.	.	.	1	2	2	3E	3	1	2	.	1	.	.	.	1	.	1	1
	.	.	.	.	.	.	.	.	.	.	.	.	1	1	2	.	.	.	.	.	.	.	.	.	.	.
+	2	1	1	.	.	.	.	.	.	.	.	.	.	.	.	.	.	.	.	.	.	.	3	.	.	.
	.	.	.	.	.	.	.	.	.	.	1	.	.	1	.	.	.	.	.	.	.	.	.	.	.	.

Rotatorien im Plankton der Talsperre Neunzehnhain

			1970																									
11.	13. 11.		31. 01.		24. 03.		09. 05.		02. 06.		16. 06.		22. 06.		30. 06.		06. 07.		22. 07.		06. 08.		08. 09.		21. 09.	
V	H	V	–	V	–	V	H	V	H	V	–	V	H	V	H	V	–	V	H	V	–	V	H	V	H	V
1	.	1	.	.	.	.	.	.	.	.	.	.	.	.	.	.	.	.	.	.	.	.	.	.	.	.
1	.	.	.	.	.	.	.	.	.	.	.	.	.	.	.	.	.	.	.	.	.	.	.	.	.	.
2	2	1	.	.	.	.	.	.	.	.	.	.	.	.	.	.	.	.	.	.	.	.	.	.	.	.
1	2	1	.	2	.	1+	.	.	.	.	.	.	.	.	.	.	.	.	.	.	.	.	.	.	.	.
3	2E	2E	.	.	.	.	.	.	.	.	.	.	.	.	.	.	.	.	.	.	.	.	.	.	.	.
.	.	.	.	.	.	.	.	.	.	.	.	.	.	.	.	.	.	.	.	.	.	.	.	.	.	.
.	.	1+	.	1	.	1	.	.	.	.	.	.	.	.	.	.	.	.	.	.	.	.	.	.	.	.
.	.	.	.	.	.	.	.	.	.	.	.	.	.	.	.	.	.	.	.	.	.	.	.	.	.	.
.	.	.	.	.	.	.	.	.	.	.	.	.	.	.	.	.	.	.	.	.	.	.	.	.	.	.
1	1	.	.	.	.	.	.	.	.	.	.	.	.	.	.	.	.	.	.	.	.	.	.	.	.	.
3	3	3	.	2	.	3	1+	1	1	1	.	2	1	1	.	.	.	1	.	.	.	1	1	1	2	1
1	2	1	.	2	.	2E	.	.	.	1	.	.	.	.	.	.	.	.	.	.	.	.	.	1	.	.
4	3	3	.	.	.	1+	.	1	.	.	.	.	.	.	.	.	.	.	.	.	.	.	.	.	.	.
.	.	1	.	.	.	.	.	.	.	.	.	.	.	.	1	1	.	1	1	2	.	.	.	1	2	1
.	.	1	.	.	.	.	.	.	.	.	.	.	.	.	1	1	.	1	1	2	.	.	.	1	2	1
.	.	.	.	.	.	.	.	.	.	.	.	.	.	.	.	.	.	1	.	.	.	.	.	.	.	.
.	.	.	.	.	.	.	.	.	.	.	.	.	.	.	.	.	.	.	.	.	.	.	.	.	.	1
.	.	.	.	.	.	.	.	.	.	.	.	.	.	.	.	.	.	.	.	.	.	.	.	.	1	.
.	.	.	.	.	.	.	.	.	.	.	.	.	.	.	.	.	.	.	.	1	.	.	.	1	.	2
1	1	1	.	.	.	.	.	.	.	1	.	.	.	.	.	.	.	.	.	.	.	.	.	.	.	.
1	1	1	.	.	.	1	.	.	.	.	.	.	.	.	.	.	.	.	.	.	.	.	.	.	.	.
1	.	.	.	.	.	.	.	1	1	1	.	1	1	1	.	1	.	1	.	.	.	1	.	1	.	.
.	.	.	.	.	.	.	.	.	.	.	.	.	.	.	.	.	.	.	.	.	.	.	.	.	1	1
.	.	.	.	.	.	.	.	.	.	.	.	.	.	.	.	.	.	.	.	.	.	.	.	.	.	2
.	.	.	.	.	.	.	.	.	.	.	.	.	.	.	.	.	.	.	.	.	.	.	.	1	.	1

Tab. 4. Graphische Darstellung des quantitativen Vorkomn

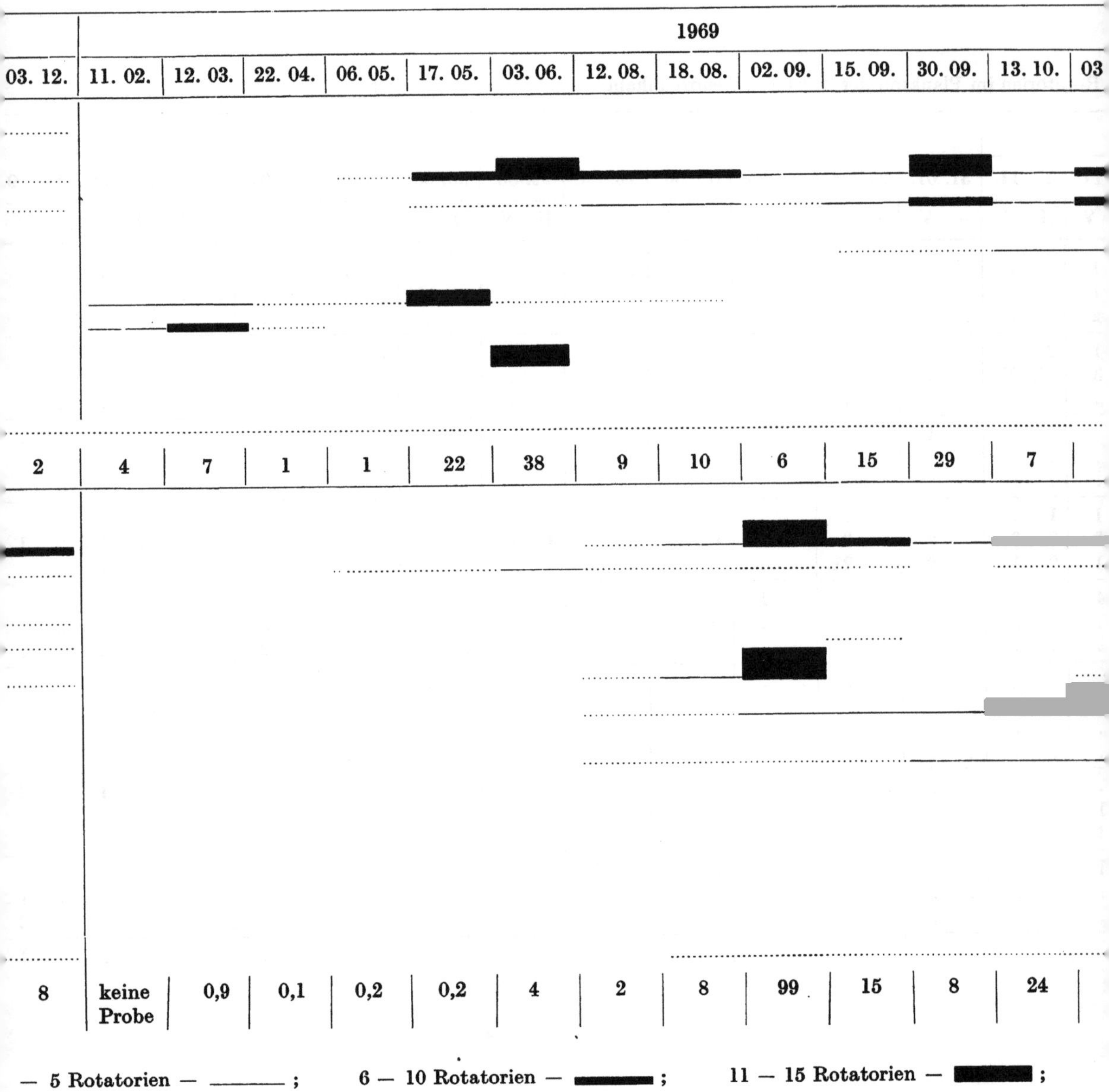

otatorien in den Talsperren Saidenbach und Neunzehnhain

		1970											
11.	02. 12.	31. 01.	10. 02.	24. 03.	09. 05.	02. 06.	16. 06.	22. 06.	30. 06.	06. 07.	22. 07.	13. 08.	08. 09.

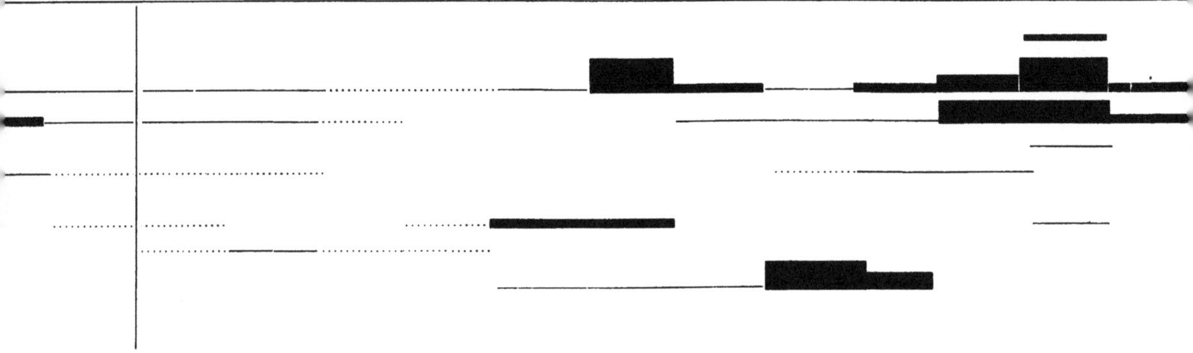

3	8	3	5	1	1	8	29	12	6	34	62	87	17

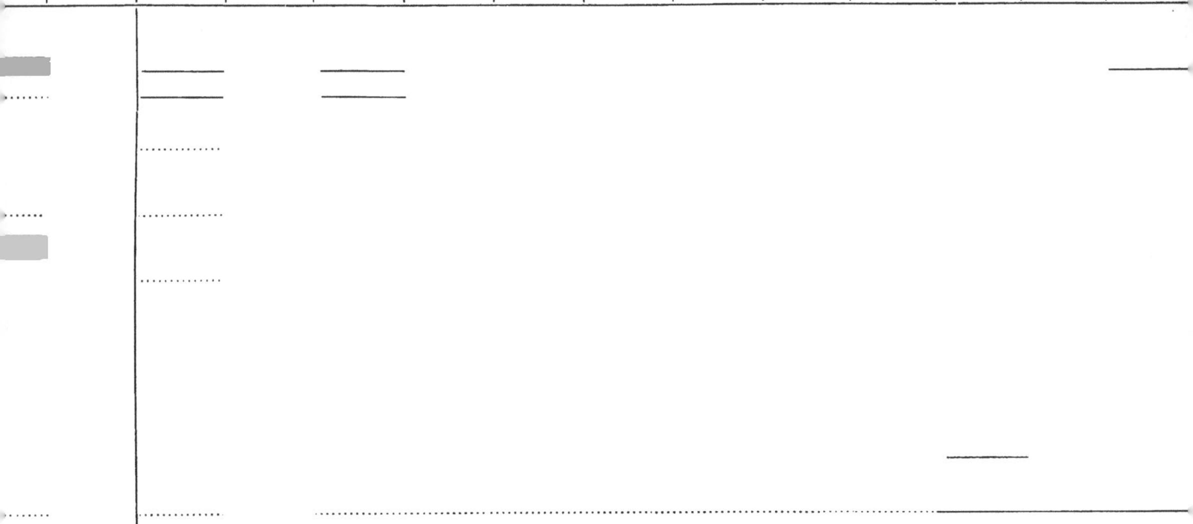

3	keine Probe	6	keine Probe	5	0,2	0,3	0,1	0,4	0,2	0,7	4	3	2

otatorien — ; 21 — 30 Rotatorien — ; mehr als 30 Rotatorien — ;

12. *Collotheca mutabilis* (HUDSON, 1885)

13. *Conochiloides natans* (SELIGO, 1900)

14. *Conochilus unicornis* ROUSSELET, 1892)

15. *Dicranophorus biastis* HARRING & MYERS,1928 (Abb. 12)

Verbreitung: Zwischen Wasserpflanzen in Teichen und Seen; verbreitet.
Maße: Lg. total 250—280 µm; Zehenlg. 35 µm; Kauerlg. 24 µm; Manubria 23 µm; Rami 17 µm; Uncus rechts 10 µm; Fulcrum 6 µm
Literatur: JARRING & MYERS (1928)

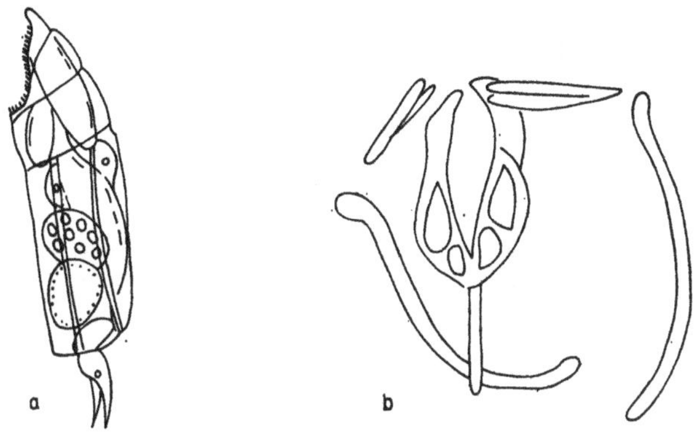

Abb. 12. *Dicranophorus biastis.*
a — Seitenansicht; b — Kauer

Bemerkungen: Nach HARRING & MYERS betragen die Maße: Lg. total 160 µm; Zehenlg. 20 µm und der Kauer 20 µm. Der *Dicranophorus*-Fund aus der Neunzehnhainer Talsperre hat Ähnlichkeit mit *D. biastis,* der in Amerika (Wisconsin) gefunden und beschrieben wurde. Der hiesige Fund ist fast doppelt so groß, die Kauermaße stimmen fast überein. Es handelt sich um einen augenlosen *D.* 8 Dotterstockkerne sind vorhanden. Die Blase ist groß. Auffällig sind die großen, im rechten Winkel abknickenden Fußdrüsen. Diese werden von HARRING & MYERS nicht gezeichnet. Die Zehenform entspricht gut der amerikanischen Art, am Zehengrund sind die Zehen bulbusartig verdickt, sie laufen in eine feine Spitze aus. Der Kauer weicht etwas von der Originalbeschreibung ab: er ist leicht asymmetrisch, der linke Uncus ist kleiner als der rechte. Die Unci sind gabelig. An den Rami wurden kleine Zähne gesehen, HARRING & MYERS bilden keine ab. Möglicherweise stellt der Fund aus Neunzehnhain eine besondere Form der amerikanischen Art dar, die bisher in Mitteleüropa noch nicht gefunden wurde.

Verbreitung: Bisher nur in Amerika bekannt.

16. *Dicranophorus robustus* var. *europaeus* WULFERT, 1936 (Abb. 13)

Maße: Gesamtlg. 420 µm; Zehenlg. 60 µm; Kauerlg. 40 µm; Manubria 32 µm; Fulcrum 20 µm
Literatur: VOIGT (1957), WULFERT (1936)

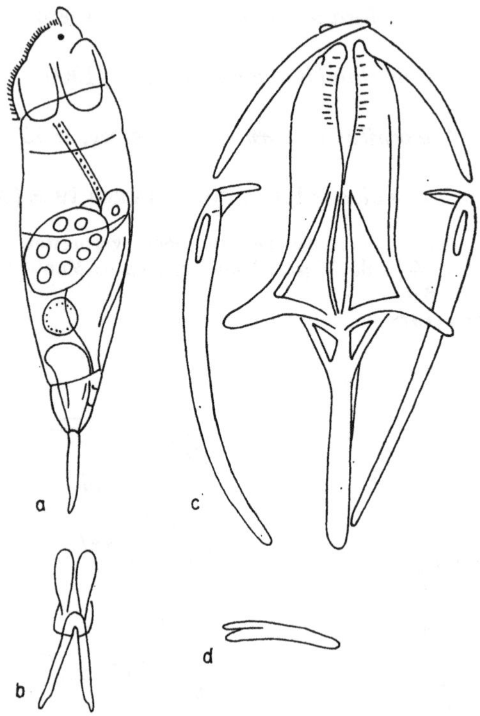

Abb. 13. *Dicranophorus robustus europaeus*

a – Seitenansicht; b – Zehen, Aufsicht; c – Kauer; d Uncus

Bemerkungen: Die vorliegende Varietät ist leicht an den charakteristischen Zehen (s. Abb. 13) und am unsymmetrischen Kauer zu erkennen.

Verbreitung: In Seen und Teichen.

17. *Dissotrocha macrostyla* (EHRENBERG, 1838)

Maße: Lg. 230 µm; Sporenlg. 44 µm; Dorsaltasterlg. 38 µm
Literatur: VOIGT (1957)
Verbreitung: Zwischen Wasserpflanzen; verbreitet.

18. *Gastropus hyptopus* (EHRENBERG, 1838)

19. *Gastropus stylifer* IMHOF, 1891

20. *Kellicottia longispina* (KELLICOTT, 1879)

21. *Keratella cochlearis* (GOSSE, 1851)

22. *Keratella quadrata* (O. F. MÜLLER, 1886)

23. *Lecane closterocerca* (SCHMARDA, 1895)

24. *Lecane flexilis* (GOSSE, 1889)

25. *Lecane luna* (O. F. MÜLLER,1776)

26. *Lecane lunaris* (EHRENBERG, 1832)

27. *Lecane scutata* (HARRING & MYERS, 1926)

28. *Lecane tenuiseta* HARRING, 1914

29. *Lepadella acuminata* (EHRENBERG, 1834)

30. *Lepadella patella* (O. F. MÜLLER, 1786)

31. *Macrochaetus subquadratus* PERTY, 1850

Maße: Gesamtlg. 162 µm; Panzerlg. kontrahiert 70 µm; Br. 66 µm; Zehenlg. 18 µm
Literatur: VOIGT (1957), WULFERT (1961)
Verbreitung: In stehenden Gewässern zwischen Wasserpflanzen; verbreitet.

32. *Microcodon clavus* EHRENBERG, 1830

Maße: Körperlg. 100 µm
Literatur: VOIGT (1957)
Verbreitung: In Teichen und Altwässern; verbreitet.

33. *Notommata aurita* (O. F. MÜLLER, 1786)

Maße: Gesamtlg. 290 µm; Zehenlg. 17 µm; Kauerlg. 58 µm; Manubria 44 µm; Fulcrum 24 µm
Literatur: VOIGT (1957)
Verbreitung: Zwischen Wasserpflanzen; verbreitet.

34. *Philodina megalotrocha* EHRENBERG, 1832

Maße: Gesamtlg. 200 µm; Sporenlg. 15 µm; Dorsaltaster 35 µm
Literatur: VOIGT (1957), WULFERT (1956, 1960a, 1960b)
Verbreitung: Zwischen Wasserpflanzen in Teichen und Seen, auch im Plankton.

35. *Pleurotrocha petromyzon* EHRENBERG, 1830

Maße: Gesamtlg. 200 µm; Zehenlg. 13 µm; Kauerlg. 20 µm; Manubria 15 µm
Literatur: VOIGT (1957), WULFERT (1935, 1956, 1960b, 1961)

Bemerkungen: Der zur Bestimmung dieser Art wichtige ventrale Zapfen am Manubrium wurde gut gesehen. Die Größenangaben für *P. petromyzon* schwanken sehr, der vorliegende Fund zählt zu den kleinen „Formen“.

Verbreitung: Zwischen Pflanzen größerer und kleinerer Gewässer; verbreitet.

36. *Ploesoma hudsoni* (IMHOF, 1891)

37. *Polyarthra dolichoptera* IDELSON, 1925

38. *Polyarthra vulgaris* CARLIN, 1943

39. *Rotaria haptica* (GOSSE, 1886)

40. *Rotaria rotatoria* (PALLAS, 1766)

41. *Synchaeta pectinata* EHRENBERG, 1832; zwei Einzelfunde!

42. *Taphrocampa annulosa* (GOSSE, 1851)

Maße: Gesamtlg. 160 µm; Zehenlg. 15 µm
Literatur: VOIGT (1967), WULFERT (1956, 1960a, 1960b, 1861)
Verbreitung: Zwischen Pflanzen kleinerer Gewässer.

43. *Trichocerca porcellus* (GOSSE, 1886)

44. *Trichotria pocillum* (O. F. MÜLLER, 1776)

45. *Trichotria tetractis* (EHRENBERG, 1830)

2.4. Graphische Darstellung des quantitativen Vorkommens der Rotatorien der Talsperren Saidenbach und Neunzehnhain

Im vorangegangenen systematischen Teil wurden die geschätzten Häufigkeiten angegeben. Im folgenden werden die Ergebnisse der Zählungen vorgestellt (Tab. 4).

Rotatorienvorkommen, die unter dem Wert 0,1 Tier/l lagen, wurden zusammengefaßt und unter der Rubrik „Rotatorien" geführt. Folgende Plankter erreichten den angegebenen Wert in der Saidenbachtalsperre nicht: *Ploesoma hudsoni, Ascomorpha ecaudis, Filinia maior, Notholca acuminata* var. *extensa, Notholca squamula* und *Gastropus stylifer*.
Dazu kommen die Tychoplankter, die *Lecane*-Arten usw. Unter der Rubrik „Zusammenfassung" wurden alle ermittelten Werte zusammengezählt und die Stückzahl/l eingesetzt.

In der Talsperre Neunzehnhain erreichen alle Plankter wenigstens einmal den Wert 0,1. Unter „Rotatorien" wurden die *Lecane*-Vorkommen, die Bdelloidea, Notommatidae und die Plankter, die in der betreffenden Probe unter dem Wert 0,1 lagen, zusammengefaßt.

Fiel beim Betrachten der Tab. 2 u. 3 vor allem die Abnahme der Artenzahlen auf, so zeigt die quantitative Tabelle für Saidenbach die massenhafte Zunahme einzelner Arten im Jahre 1970. Es sind: *Keratella cochlearis, Polyarthra dolichoptera, vulgaris* und *Conochilus unicornis*. Im Juli/August kommt es bei diesen Arten zu starken Populationsbildungen, wobei *Conochilus unicornis* beginnt, die Polyarthren sich anschließen und dann *Keratella cochlearis* folgt. Am 22. 7. werden viele Diatomeen beobachtet. Nach Abklingen der massenhaften Rotatorienvorkommen, Anfang September 1970, werden im Protokoll massenhaft Diatomeen genannt. Ein Zusammenhang zwischen der Abnahme der Rotatorien und der starken Zunahme der Diatomeen ist durchaus denkbar. Zwei Theorien wären möglich: Verdrängen die relativ großen Kieselalgen die Rotatorien oder wächst die Kieselalgenproduktion an, weil die Verbraucher (die Rotatorien) verschwinden? (Wobei noch geklärt werden muß, ob alle angeführten Arten Diatomeenkonsumenten sind.)

Die Tab. 4 (Talsperre Neunzehnhain) entspricht etwa der Tab. 3. Anfang September kommt es zur massenhaften Entwicklung von *Conochilus unicornis*, *Keratella cochlearis* ist häufig. Nach Abklingen dieses Maximums wird auch für Neunzehnhain reichlich Phytoplanktonbiomasse registriert. *Kellicottia longispina* hat im November 1969 ein Maximum. Keines wiederholt sich im Jahre 1970. Die Proben sind im zweiten Untersuchungsjahr öde. In den Protokollen wird häufig der Vermerk „Detritus" angetroffen, der vom Zufluß mit in die Sperre gebracht wird. 1970 ist Neunzehnhain sehr planktonarm.

Vergleichen wir die extremen Sichttiefenverhältnisse und Chlorophyllwerte[2]) mit dem quantitativen Auftreten der Rotatorien, so ergibt sich folgendes:

Für die Talsperre Neunzehnhain besteht im Jahre 1969 zwischen der Sichttiefe und dem Rotatorienvorkommen eine Korrelation: Die Sichttiefenmaxima im September und Oktober/November fallen mit einem Rotatorienmaximum zusammen. Ähnlich liegen die Verhältnisse für Saidenbach im Juni 1969 und im September 1969. Im August 1970 ist das Rotatorienmaximum nur mit einem unklaren Sichttiefenmaximum verbunden.

Werden die Chlorophyllwerte mit dem quantitativen Auftreten der Rotatorien verglichen, so ergibt sich: Im Mai 1969 setzt in Saidenbach etwa parallel mit dem Erscheinen hoher Chlorophyllwerte eine kräftige Entwicklung einiger Rotatorienarten

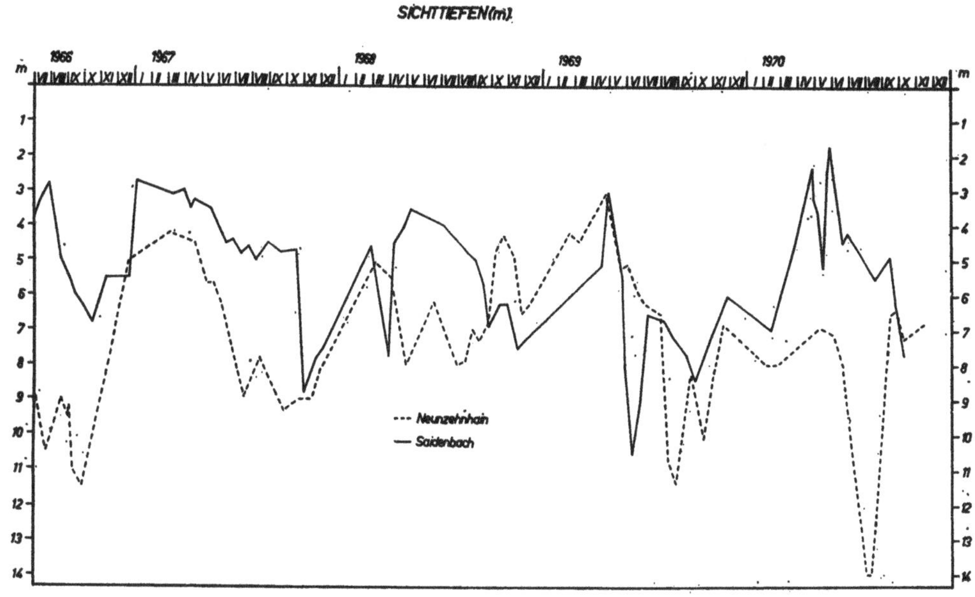

Abb. 14. Sichttiefenverhältnisse in den Talsperren Saidenbach und Neunzehnhain in den Jahren 1966—1970 (nach Höhne, unveröff.)

---

[2]) Die Sichttiefen- und Chlorophyllwerte wurden freundlicherweise von Höhne u. a. (1970) zur Verfügung gestellt.

ein: *Synchaeta pectinata, Polyarthra dolichoptera* (in 5 m Tiefe 270 Tiere/l), die Anfang Juni noch zunimmt und von *Conochilus unicornis* unterstützt wird. Im Juni 1970 nimmt mit einer zeitlichen Verzögerung gegenüber den Chlorophyllmaxima von Ende Mai die Rotatorienfauna zu. Es handelt sich um Polyarthren und um *Asplanchna.*

Für Neunzehnhain ist eine Korrelation zwischen Chlorophyllgehalt und maximalem Rotatorienauftreten nicht deutlich.

Ob es sich überhaupt um eine echte Wechselwirkung handelt, müßte durch weitere Beobachtungen geprüft werden.

3. *Auswertung*

3.1. Vergleich mit Ergebnissen früherer Untersuchungen an den Talsperren Saidenbach und Neunzehnhain und Vergleich mit der Talsperre Sedlice bei Zeliv

BEUSCHOLD (1961) nennt für die Saidenbachtalsperre 16 Rotatorienarten. Er gibt die Zeit ihres Vorkommens an und von einigen Arten den Monat des Maximums. Von diesen Rotatorienarten werden in der vorliegenden Arbeit bestätigt:

	zeitl. Verteilg. Monate	Maximum nach BEUSCHOLD Monate
Synchaeta pectinata	3–6, 9–2	1
Polyarthra vulgaris	3–2	7, 10
Keratella quadrata	4–10	8
Keratella cochlearis	3–2	8, 9
Notholca acuminata[3])	10	
Kellicottia longispina	3–2	6
Gastropus stylifer	9–11	9
Asplanchna priodonta	4–6, 9–2	11
Conochilus unicornis	5, 9, 12	

Die Arten: *Ascomorpha saltans, Trichocerca rousseleti, Brachionus urceus, Anuraeopsis fissa, Notholca labis, Filinia longiseta, Collotheca libera* wurden nicht gefunden.

Die von BEUSCHOLD angegebenen Zeiten stimmen nicht in allen Fällen mit den vorliegenden zeitlichen Angaben überein. Für *Polyarthra vulgaris, Keratella cochlearis, Kellicottia longispina, Gastropus stylifer, Asplanchna priodonta* ist der Unterschied gering. Die übrigen Angaben weichen stark voneinander ab.

In einer Arbeit von RUDOLF (1966) werden ebenfalls für Saidenbach Rotatorien mit ihrem Maximum angegeben. Es handelt sich um die Arten:

	Maximum
1. *Keratella quadrata,*	Juli
2. *Kellicottia longispina,*	Juli (November)*)
3. *Keratella cochlearis,*	September (Juli, August)*)
4. *Gastropus stylifer,*	September
5. *Polyarthra vulgaris,*	Oktober
6. *Asplanchna priodonta,*	November (Juni)*)
7. *Synchaeta pectinata*	Dezember

[3]) Möglicherweise liegt hier die Varietät *extensa* vor.

Filinia longiseta und *Trichocerca rousseleti*, die auch in dieser Faunenliste genannt werden, können nicht bestätigt werden. Mit kleinen Verschiebungen stimmen die Zeiten für 1, 2, 3, 6 überein. Bei den mit *) versehenen Angaben handelt es sich um Ergebnisse aus der vorliegenden Arbeit, die zugefügt wurden.

RUDOLF stellte fest, daß sich das Rotatorienplankton in den oberen 10 m konzentriert. Im allgemeinen kann das bestätigt werden, eine Ausnahme bildet *Polyarthra dolichoptera-vulgaris*.

Vergleichen wir die Rotatorienliste der untersuchten Talsperren mit den qualitativen Aufzeichnungen von ŠTĚPÁNEK & CHALUPA für die Talsperre Sedlice, so sind die Hauptvertreter für alle drei Sperren die gleichen. Für Sedlice werden genannt:

Asplanchna priodonta
Keratella stipitata-Formen, gehören zum Formenkreis der *K. cochlearis*
Keratella quadrata
Polyarthra („Unter dieser Gattung verbargen sich verschiedene Arten, gewöhnlich handelte es sich um *P. dolichoptera*“. Übers. v. Verf.)

Eine untergeordnete Rolle spielen die Arten:

Filinia longiseta		
Sychaeta oblonga	×	
Polyarthra longiremis		
Polyarthra vulgaris	×	○
Notholca squamula	×	
Notholca acuminata		
Brachionus angularis		○
Brachionus urceolaris		
Brachionus calyciflorus		
Kellicottia longispina	×	○
Euchlanis deflexa		
Trichocerca cynlindrica.		

Die mit × versehenen Spezies kommen auch in Saidenbach vor, die mit ○ gekennzeichneten finden wir in Neunzehnhain.

Vergleicht man die Darstellungen des quantitativen Vorkommens der oben genannten Hauptvertreter der drei Talsperren, so ergibt sich für Sedlice und Saidenbach einige Übereinstimmung. Einzelne Hauptvorkommen liegen zeitlich zusammen, wie das Auftreten der *Keratella stipitata tecta* (= *cochlearis*-Formenkreis) im August, *Keratella quadrata* im April, *Polyarthra* im Juni, September, Juli und August. Von weiteren ausführlichen Vergleichen soll hier abgesehen werden (näheres s. ŠTĚPÁNEK & CHALUPA). Der Versuch, die Rotatorienfaunen von Sedlice, Saidenbach und Neunzehnhain vergleichsweise nebeneinander zu sehen, zeigt gewisse Übereinstimmungen der beiden Talsperren Sedlice und Saidenbach.

3.2. Vergleiche mit den Ergebnissen anderer Autoren

3.2.1. Das Saprobiensystem von KOLKWITZ & MARSSON und seine Anwendung

KOLKWITZ & MARSSON benutzten auch die Rotatorien in ihrem Saprobiensystem (1909) als Indikatoren. Sie führten für die einzelnen Saprobiegrade verschiedene

Vertreter (11 Arten) an. LIEBMANN (1962) lehnt die Rotatorien als Verunreinigungsanzeiger ab. In seinem revidierten KOLKWITZschen Saprobiensystem sind nur noch wenige Rotatorienarten genannt (3). Dennoch ist es nicht ganz von der Hand zu weisen, die Rotatoriengemeinschaften einzelner Talsperren zu untersuchen und zu vergleichen, um daraus eventuell eine Eutrophierung abzulesen. Wie die Untersuchungen an den Taslperren Saidenbach und Neunzehnhain gezeigt haben, besitzt jede Sperre trotz zahlreicher gemeinsamer Arten eine charakteristische Rotatorienzusammensetzung, die eine Unterscheidung beider Talsperren zu jeder Jahreszeit ermöglicht. Von den im alten KOLKWITZSschen System genannten Vertretern wurden in der vorliegenden Arbeit nur wenige gefunden. Die Bevorzugung des oligotrophen Gewässertyps durch *Kellicottia longispina* kann bestätigt werden. *Rotaria rotatoria* war in beiden Taslperren zu selten, um eine sichere Aussage zu erlauben. Der Wert der *Keratella*-Arten als Trophiegradanzeiger muß für den vorliegenden Fall bezweifelt werden, da die Planktonfauna der oligotrophen Neunzehnhainer Talsperre gerade in den Zeiten der Minimalbesiedlung hauptsächlich aus *Keratella*-Arten bestand. Das gleiche gilt für *Lecane lunaris*. Keratellen und *Rotaria rotatoria* werden als Trophieanzeiger wahrscheinlich erst durch ihr massenhaftes Vorkommen nutzbar, ihr geringfügiges Auftreten besagt nichts.

3.2.2. Vergleich mit der Arbeit von JÄRNEFELT (1952)

JÄRNEFELT (1952) sucht in seiner Arbeit über die finnischen Seen nach Gesetzmäßigkeiten im Auftreten des Planktons, um diese als Indikatoren für den Trophiegrad der Seen einsetzen zu können. Hauptsächlich befaßt er sich mit dem Phytoplankton, er nennt in seinen Planktonlisten, die er für die Trophiegruppen der Seen anführt, aber auch Rotatorien. Vergleichen wir seine Ergebnisse mit den hier vorliegenden, so zeigt sich Übereinstimmung im Vorkommen von *Kellicottia longispina* im oligotrophen Bereich. *Polyarthra dolichoptera* kommt in oligotrophen Seen vor, ist aber am stärksten in meso- und polyhumosen Seen vertreten.

Polyarthra vulgaris und *Keratella stipitata* (*K.-cochlearis*-Kreis) gehören zu den indifferenten Arten, JÄRNEFELT findet sie weniger in oligotrophen und mehr in eutrophen Gewässern.

In die Mittelgruppe gehören auch, mit stärkerer Neigung zur Eutrophie: *Synchaeta* sp., *Trichocerca porcellus* und *Asplanchna priodonta. Asplanchna priodonta* kommt zwar in Neunzehnhain vor, ist aber in Saidenbach stärker vertreten. Diese Ergebnisse JÄRNEFELTS stehen in Übereinstimmung mit einigen Resultaten der vorliegenden Arbeit.

Cephalodellen und Collothecaceen werden ohne Artenbenennung hauptsächlich für eutrophe Gewässer aufgeführt, beide Gattungen kamen aber auch in oligotrophen Seen vor. Diese Angabe ist nicht verwendbar, sie sagt zu wenig aus. Ausnahmslos in nahrungsreichen Seen fand JÄRNEFELT *Trichocerca pusilla.*

Keratella stipitata, Typ *hispida*, } Formenkreis *K. cochlearis*
Keratella stipitata, Typ *ecauda*,
Keratella quadrata

Tab. 5. Qualitative Tiefenserien, Zählergebnisse (jeweils Anzahl der Tiere/l), Talsperre Saidenbach

Art	0 m			5 m			10 m			15 m			25 m			35 m		
	17. V. 1969	20. IX. 1969	19. VI. 1970	17. V. 1969	20. IX. 1969	19. VI. 1970	17. V. 1969	20. IX. 1969	19. VI. 1970	17. V. 1969	20. IX. 1969	19. VI. 1970	17. V. 1969	20. IX. 1969	19. VI. 1970	17. V. 1969	20. IX. 1969	19. VI. 1970
Trichocerca porcellus	—	3	—	—	3	—	—	—	—	—	—	—	—	6	—	—	—	—
Polyarthra dolichoptera u. *vulgaris*	111	213	174	270	57	468	48	24	96	42	42	102	57	39	36	15	—	24
Keratella cochlearis	6	138	30	9	30	102	6	45	21	3	39	12	3	15	9	—	—	3
Keratella quadrata	—	3	—	3	3	—	3	—	6	—	—	—	—	6	—	—	—	3
Kellicottia longispina	6	9	9	—	6	3	—	—	3	—	6	—	6	—	—	—	—	—
Synchaeta pectinata u. a.	27	3	18	204	6	60	165	—	6	129	—	12	69	—	—	21	—	—
Conochiloides natans	21	—	—	6	—	—	3	—	—	—	—	—	—	—	—	—	—	—
Conochilus unicornis	—	—	3	—	—	45	—	—	3	—	—	—	—	—	—	—	—	3
Collotheca pelagica	—	—	—	—	—	—	—	—	3	—	—	9	—	—	—	—	—	—
Collotheca mutabilis	—	3	—	—	—	—	—	—	—	—	—	—	—	—	—	—	—	—
Filinia maior	—	—	—	—	—	—	—	—	3	—	—	—	—	—	—	—	—	—
Rotatorien	42	12	3	12	—	—	3	—	3	—	—	3	3	—	—	3	—	—
Gesamtanzahl der Tiere/l	213	384	237	504	105	678	228	69	144	174	87	138	138	66	45	39		33

Tab. 6. Quantitative Tiefenserien, Zählergebnisse (jeweils der Anzahl Tiere/l), Talsperre Neunzehnhain

Name	0 m			5 m			10 m			15 m			25 m		
	18, V. 1969	22. IX. 1969	30. VI. 1970	18. V . 1969	22. IX. 1969	30. VI. 1970	18. V. 1969	22. IX. 1969	30. VI. 1970	18. V. 1969	22. IX. 1969	30. VI. 1970	18. V. 1969	22. IX. 1969	30. VI. 1970
Keratella cochlearis	—	930	—	—	249	—	—	210	—	—	15	—	—	6	—
Keratella quadrata	—	6	—	5	—	—	25	—	—	5	—	—	—	15	—
Kellicottia longispina	—	258	—	—	126	—	—	114	—	—	3	—	—	6	—
Ascomorpha saltans	—	12	—	—	—	—	—	—	—	—	—	—	—	—	—
Ascomorpha ecaudis	—	3	—	—	—	—	—		—	—	—	—	—	—	—
Gastropus stylifer	—	6	—	—	—	—	—	—	—	—	—	—	—	—	—
Ploesoma hudsoni	—	9	—	—	—	—	—	—	—		—	—	—	—	—
Lecane flexilis	—	3	—	—	—	—	—	—	—	—	—	—	—	—	—
Brachionus angularis	—	—	—	—	—	—	—	—	3	—	—	3	—	—	—
Polyarthra dolichoptera u. *vulgaris*	—	—	—	—	—	—	—	—	3	—	—	3	—	—	—
Rotatorien	—	3	—	—	6	—	—	3	—	—	—	—	—	3	—
Gesamtanzahl der Tiere/l		1230		5	381		25	327	6	5	18	6		30	

Über den Wert der *Keratella*-Arten als Trophiegradanzeiger wurde oben bereits berichtet.

Trichocerca pusilla kam als Einzelfund in Saidenbach vor.

Untersuchungen über Trophiegradindikatoren, vor allem phytoplanktischer Art, wurden außer von JÄRNEFELT auch von anderen Autoren unternommen (s. HÖHNE & KLOSE 1966). Unter anderem wird hier das Ergebnis zitiert, daß nicht allein das Vorkommen einer Art etwas aussagt, sondern daß „der zahlenmäßige Umfang" der Artbestände Rückschlüsse über die Beschaffenheit des Wassers zuläßt. Diese Feststellung entspricht den Ergebnissen der vorliegenden Arbeit.

THUNMARK (1945) und NYGARD (1949) stellten Phytoplanktonquotienten auf, die „das gegenseitige Verhältnis von Artenzahlen der für die Charakterisierung des Trophiegrades wichigen taxonomischen Gruppen" beinhalten. Diese Methode wurde von HÖHNE & KLOSE (1966) für die Talsperren Saidenbach und Neunzehnhain angewandt. Nicht alle vorgeschlagenen Phytoplanktonquotienten führten zum Ziel, einige erwiesen sich aber als brauchbar und zeigten den oligotrophen Charakter der Talsperre Neunzehnhain und den eutrophen der Talsperre Saidenbach an.

3.3. Die Rotatoriengemeinschaften der untersuchten Talsperren und ihre Unterschiede

In Saidenbach wurden 42, in Neunzehnhain 45 Arten registriert. Davon sind 28 Arten gemeinsam. Für beide Talsperren tritt 1970 in der Faunenliste eine Veränderung ein, die bei Neunzehnhain zu einer Verarmung führt. Auch bei Saidenbach fallen einige Arten aus (z. B. *Gastropus*).

Offensichtlich handelt es sich um Schwankungserscheinungen um ein langjähriges Populationsmittel. Die Schwankungen betreffen die einzelnen Arten in verschieden starkem Maße. Ob die Zuordnung der Schwankungsgröße zu den einzelnen Arten gesichert ist, muß in Anbetracht der kurzen Untersuchungsdauer offen gelassen werden. Außerdem ist die Anzahl der Proben zu gering, um über die räumliche Verteilung des Rotatorienplanktons mit Sicherheit Aussagen machen zu können.

Zu den Hauptvertretern beider Talsperren gehören: *Keratella cochlearis, Keratella quadrata, Kellicottia longispina, Polyarthra dolichoptera, Polyarthra vulgaris, Asplanchna priodonta, Collotheca mutabilis* und *Conochilus unicornis*. In Saidenbach kommt in entscheidendem Maße die Gattung *Synchaeta* hinzu, die in Neunzehnhain, bis auf zwei Einzelfunde, fehlt. Die Kombination *Polyarthra — Synchaeta* und das zahlreiche Vorhandensein dieser zwei Gattungen scheint für das nahrungsreichere Gewässer charakteristisch zu sein, vgl. Sedlice. *Polyarthra* wird in Neunzehnhain zwar „häufig" beobachtet, es kommt aber nie zu erheblichen Zahlenbefunden. Nur 1968 werden bis zu 5 Tiere/l gezählt, während 1969 noch nicht einmal dieser Wert erreicht wird. Das zeigt, daß auch bei gleichen Artvorkommen in beiden Sperren wohl normalerweise ein quantitativer Unterschied besteht. Dabei überragen in Saidenbach die zwei angeführten Gattungen und einige andere Arten, z. B. *Trichocerca porcellus, Asplanchna priodonta, Conochiloides natans*, während in Neunzehnhain *Kellicottia longispina* stärker vertreten ist. Nur in Neunzehnhain kommt der 1970 erstmals beobachtete *Microcodon clavus* vor. Erwähnenswert ist für Saidenbach das Auftreten der Gattung

Notholca, die in Neunzehnhain fehlt. Auffallend sind die relativ häufigen Funde der aus dem „Litoral“ stammenden *Lecane*-Arten im Plankton beider Sperren. In Neunzehnhain sind sie auch in den artenärmsten Planktonproben des Jahres 1970 vorhanden, ebenso die *Rotaria*-Arten. Von den Einzelfunden sind die Cephalodellen erwähnenswert, da sie sich auf Neunzehnhain beschränken. Die Faunenliste jeder Talsperre enthält außerdem ihre eigenen Arten. Sie treten aber so begrenzt auf, daß sie nicht als typische Unterschiede bewertet werden können.

Offenbar sind die Rotatorienarten wegen ihrer weiten Verbreitung und ihrer Anpassungsfähigkeit zur Gewässercharakteristik in qualitativem Sinne wenig geeignet. Sehr wohl sollte sich aber auf Grund der offensichtlichen Bevorzugung bestimmter Gewässertypen an Hand von Zählergebnissen eine quantitative Charakterisierungsmöglichkeit ergeben. Im folgenden sei ein möglicher Anhaltspunkt zur quantitativen Kennzeichnung eines Gewässers vorgeschlagen.

Aus den bereits angegebenen quantitativen Daten ist zu erkennen, daß Unterschiede in den Besiedlungsdichten in beiden Sperren, vor allem bei den Gattungen *Kellicottia*, *Polyarthra* und *Synchaeta* vorkommen. Die erste ist in der oligotrophen Neunzehnhainer Talsperre, die beiden letzteren sind in der nährstoffreicheren Saidenbachsperre häufiger. Zur Charakterisierung wurde der Wert

$$n = \frac{a + b}{c}$$

berechnet. Dabei bedeuten:

a = maximal gefundene Individuenzahl der Gattung *Polyarthra*
b = maximal gefundene Individuenzahl der Gattung *Synchaeta*
c = maximal gefundene Individuenzahl der Gattung *Kellicottia*

Für die Saidenbachtalsperre erhält man $n = \sim 10$, für Neunzehnhain $n = 0{,}1$.

Allerdings sind methodische Bedenken am Platze, da die Individuenzahlen meistens nicht zur gleichen Zeit gezählt wurden. Trotzdem sind die Unterschiede in den Populationsdichten so deutlich, daß die allgemeine Verwendbarkeit des angegebenen Quotienten naheliegt und in anderen Untersuchungen ähnlicher Zielstellung geprüft werden sollte.

4. *Zusammenfassung*

1. Bei Untersuchungen in den Jahren 1969 und 1970 wurden in der Talsperre Saidenbach 42, in der oberen Talsperre Neunzehnhain 45 Rotatorienarten gefunden. In den beiden Sperren gemeinsam kamen 28 Arten vor.

2. Die Unterart *Synchaeta kitina monodactyla* ist neu. *Dicranophorus biastis* wird erstmalig in Europa und *Notholca acuminata* var. *extensa* erstmals im Süden der DDR gefunden.

3. Der Artenreichtum ist in beiden Sperren etwa gleich, nicht aber das quantitative Vorkommen. Letzteres ist in Saidenbach bedeutend größer, das spricht für den größeren Trophiegrad dieses Gewässers.

4. In Tabellen ist das qualitative und quantitative Vorkommen der Rotatorien zusammengestellt und ausgewertet. Die Artenzahlen nehmen in beiden Talsperren von 1969 zu 1970 ab. Besonders auffällig ist die Verödung in Neunzehnhain.

5. Das Saprobiensystem von KOLKWITZ & MARSSON kann nur begrenzt angewendet werden. *Kellicottia longispina*, charakteristisch für oligosaprobe Gewässer, kommt vorzugsweise in der oberen Talsperre Neunzehnhain vor. Die erzielten Ergebnisse werden mit der Arbeit von JÄRNEFELT (1952) verglichen. Auf gewisse Übereinstimmungen wird hingewiesen.

6. Trotz der gemeinsamen Arten besitzt jede Talsperre eine charakteristische Rotatorienzusammensetzung, die eine Unterscheidung beider Talsperren zu jeder Jahreszeit ermöglicht. Es wird versucht, die untersuchten Gewässer quantitativ zu kennzeichnen.

5. *Literatur*

BEUSCHOLD, E., 1961: Limnologische Untersuchungen am Hauptbecken der Saidenbachtalsperre. — Internat. Rev. ges. Hydrobiol. **46**, 18—42.

DONNER, J., 1954: Zur Rotatorienfauna Südmährens. — Österr. Zool. Z. **5** (1/2), 30—117.

HARRING, H. K., & F. J. MYERS, 1924: The Rotifer Fauna of Wisconsin. II. A Revision of the Notommatid Rotifers, exclus. of the Dicranophorinae. — Trans. Wis. Acad. Sci. **21**, 415—549.

— 1926: The Rotifer Fauna of Wisconsin. III. A Revision of the genera *Lecane* and *Monostyla*. — Trans. Wis. Acad. Sci. **22**, 315—423.

HAUER, J., 1952: Rädertiere aus dem Naturschutzgebiet „Wutachtal". — Beitr. naturkdl. Forsch. SW-Deutsch. **11** (1), 45—53.

HEDLICH, R., 1959: Über den Einfluß von Kahlschlägen auf den Trophiegrad einer Trinkwassertalsperre. — Dipl.-Arb. Zool. Inst. K.-M.-Univ. Leipzig.

— 1966: Über die Auswaschung von Pflanzennährstoffen aus Waldböden und ihr Einfluß auf die Wasserqualität von Talsperren. — Wiss. Z. K.-M.-Univ. Leipzig **15** (1), 217—227.

HÖHNE, E., 1965: Die chemischen und biologischen Verhältnisse in den Trinkwassertalsperren und -teichen des mittleren Erzgebirges. — Wiss. Z. K.-M.-Univ. Leipzig. **14** (2), 231—237.

HÖHNE, E., & H. KLOSE, 1966: Soziologische Methoden zur Erfassung des Trophiegrades. — Limnologica (Berlin) **4** (4), 201—214.

HOHMANN, H. D., 1963: Zur Kenntnis des Stoffhaushaltes von Mittelgebirgsbächen (Einzugsgebiet Obere Talsperre Neunzehnhain). — Dipl.-Arb. Zool. Inst. Leipzig.

JÄRNEFELT, H., 1952: Plankton als Indikator der Trophiegruppen der Seen. — Ann. Sci. Fennicae Biol. **18**, 2—29.

KLAPPER, H., 1957/58: Biologische Untersuchungen an den Einläufen und Vorbecken der Saidenbach-Talsperre (Erzgeb.). — Wiss. Z. K.-M.-Univ. Leipzig, Math.-Nat. R. **7** (1), 11—47.

— 1961: Die Eutrophierung von Trinkwassertalsperren und ihre möglichen Auswirkungen dargestellt am Beispiel der Saidenbachtalsperre. — Wiss. Z. K.-M.-Univ. Leipzig **10** (1), 81—87.

KOCH-ALTHAUS, B., 1963: Systematische und ökologische Studien an Rotatorien des Stechlinsees. — Limnologica (Berlin) **1** (5), 375—456.

KOLKWITZ, R., & M. MARSSON, 1909: Ökologie der tierischen Saprobien. — Internat. Rev. Hydrobiol. **2**, 126—152.

LIEBMANN, H., 1962: Handbuch der Frischwasser- und Abwasserbiologie. Bd. 1. 2. Aufl. — VEB G. FISCHER Verlag Jena.

NYGARD, G., 1949: Hydrobiological studies on some Danish ponds and lakes. Part II. The Quotient Hypothesis and some new or little known phytoplancton organisms. — Biol. Skrifter **VII** (1), 1—293.

ROUSSELET, CH. F., 1902: The genus *Synchaeta*: A Monographic Study with Descriptions of Five New Species. — J. Micr. Soc. London **1902**, 269—290, 393—411.

RUDOLF, K., 1966: Untersuchungen über die Ernährungsbeziehungen in der Freiwasserregion der Saidenbachtalsperre. — Wiss. Z. K.-M.-Univ. Leipzig. Math. Nat. R. **15** (1), 229—245.

RUTTNER, A., 1959: Sitzung der mathematisch-naturwissenschaftlichen Klasse vom 8. Jänner 1959. — Österr. Akad. Wiss., Anz. math.-naturwiss. Kl., 1—6.

SLÁDEČEK, V., 1950: Studies of the Zooplancton of the Ponds of Padr (Bohemia) with Special Reference of the Cladoceran *Holopedium gibberum*. — Bull. internat. Acad. tcheque Sci. No. 22, 1—28.

ŠTĚPÁNEK, M., & J. CHALUPA, 1958: Limnological Study of the reservoir Sedlice near Zeliv. II. Biological Part. — Scientific Papers from Institute of Chemical Technology, Prague, Faculty of Technology of Fuel and Water 2. Part 2, 313—565.

SUDZUKI, M., 1955: Life history of some japanese Rotifers. I. *Polyarthra trigla* EHRENB. — Sci. Rep. Tokyo kvoiku Daigaku Nr. 118.

THUNMARK, S., 1945: Zur Soziologie des Süßwasserplanktons. — Folia Limnol. Scand. **3**, 5—65.

UHLMANN, D., 1961: Über den Einfluß von Planktonorganismen auf ihr Milieu. — Internat. Rev. ges. Hydrobiol. **46** (1), 115—129.

— 1966: Artendichte. — Limnologica (Berlin) **4** (2), 221—233.

VOIGT, M., 1957: Rotatoria. Die Rädertiere Mitteleuropas. — Berlin-Nikolassee.

WULFERT, K., 1935: Beiträge zur Kenntnis der Rädertierfauna Deutschlands, Teil I: Notommatidae. — Arch. Hydrobiol. **28**, 583—602.

WULFERT, K., 1936: Beiträge zur Kenntnis der Rädertierfauna Deutschlands Teil II: Dicranophorinae. — Arch. Hydrobiol. **30**, 401—437.

— 1937: Beiträge zur Kenntnis der Rädertierfauna Deutschlands. Teil III: Cephalodellae. — Arch. Hydrobiol. **31**, 592—635.

— 1939: Beiträge zur Kenntnis der Rädertierfauna Deutschlands. Teil IV: Die Rädertiere der Saale—Elster-Niederung bei Merseburg in ökologisch-faunistischer Beziehung. — Arch. Hydrobiol. **35**, 563—624.

— 1956: Die Rädertiere des Teufelssees bei Friedrichshagen. — Arch. Hydrobiol. **51** (4), 457 bis 495.

— 1960a: Die Rädertiere saurer Gewässer der Dübener Heide. Teil I: Die Rotatorien des Zadlitzmoors und des Wildenhainer Bruchs. — Arch. Hydrobiol. **56** (3), 261—298.

— 1960b: Die Rädertiere saurer Gewässer der Dübener Heide. Teil II: Die Rotatorien des Krebsscherentümpels bei Winkelmühle. — Arch. Hydrobiol. **56** (4), 311—333.

— 1961: Die Rädertiere saurer Gewässer der Dübener Heide. Teil III: Die Rotatorien des Presseler und des Winkelmühler Teiches. — Arch. Hydrobiol. **58** (1), 72—102.

Eingegangen am 2. 5. 1972.

Anschrift des Verfassers:
Dr. BRIGITTE KOCH-ALTHAUS
DDR-801 Dresden A 1
Pillnitzer Str. 24

Limnologica (Berlin)	11	1	1976	47—55

LUIS A. BENEDETTO

Annual and diurnal periodicity of stonefly-drift at the arctic circle

With 9 Figures and 1 Table

1. Introduction

After TANAKA (1960), WATERS (1962), and MÜLLER (1963a,b) downstream-drift of invertebrate fauna is known as a periodic phenomenon with characteristic patterns. In most cases drift is a result of the locomotoric activity of the members of the bottom fauna (MÜLLER 1970a, b, MÜLLER & BENEDETTO 1970) which was called behavioral drift (WATERS 1965). Plecoptera nymphs are night-active (SCHWARZ 1967, MÜLLER 1970a, b, c); in this circadian rhythm light acts as a phase-setting mechanism "Zeitgeber" (ASCHOFF 1960).

Analysing the annual and diurnal activity of several organisms MÜLLER (op. cit.), ERKINARO (1969), and MÜLLER-HAECKEL (1967, 1970a) showed, that under the extreme conditions of light-dark changes at high latitudes there are unusual particularities in the periodicity.

Little information is available on the drift pattern of several species from northern Europe. The most important studies are those of MÜLLER (1970c) and THOMAS (1970a, b). Taking into account that stonefly nymphs are negatively phototactic (BRINCK 1949), and that every species must be looked upon as being an ecological unit, it was considered to be necessary to anâlyse the drift pattern of every stonefly-species separately.

2. Material, methods, aud study area

From July 1965 until June 1966 the stonefly-nymphs drifting down the Jaltisjokk were sampled continuously with an automatic Stream Drift Sampler described by MÜLLER (1965). This was done near Messaure, some 20 km. north of the arctic circle at 66°42′ N. lat. and 20°25′ E. long. A description of the study area and of the biotic and abiotic conditions of the Kaltisjokk has been summarized by MÜLLER (1970 d, e) and by MÜLLER-HAECKEL (1967, 1970a, b). Only a few points need be repeated here. The Kaltisjokk is ice-covered from the middle of October until the end of April, and its temperature is approximately constant near 0 °C. There are no diurnal changes of temperature from the middle of November until the time of ice melting. The temperature rises and diurnal variations appear in May; in summer the amplitude reaches 5° to 6 °C. between the minima and maxima. Maximum temperatures in summer may be as high as 20 °C. The light conditions vary from a light-dark change of approximately nLD 4:20 in winter until nLL in summer when the midnight-sun is visible for some 3 weeks. The material was preserved in 70% alcohol and determinated under a binocular microscope. For every month the diel variation of the drift is calculated as the percental deviation of two hours figures from the 24 hours mean value.

3. Results

a) Annual periodicity of drift

As pointed out by BENEDETTO (1973) the stonefly nymphs drifting in Kaltisjokk showed an annual periodicity which was parallel to the annual periodicity of the

Diatoms drift: A winter maximum occurred in November after which the monthly drift-rate decreased continuously until April, when a minimum was found. Finally in May the major maximum of stonefly drift was observed during the time of emergence. Approximately the same tendency was observed in the Diatoms drift (MÜLLER-HAECKEL 1967, 1970b, MÜLLER & MÜLLER-HAECKEL 1970).

The investigation of the separate specific drift patterns however brought to light that the cited facts have their roots in the complex combination of several patterns in a mixed population.

Patterns of *Brachiptera risi*, *Protonemura meyeri*, *Amphinemura sulcicollis*, *Amphinemura borealis*, *Nemoura flexuosa*, *Leuctra hippopus*, *Leuctra fusca*, and *Siphonoperla burmeisteri* are plotted against the total stonefly drift (Fig. 1) as examples for five

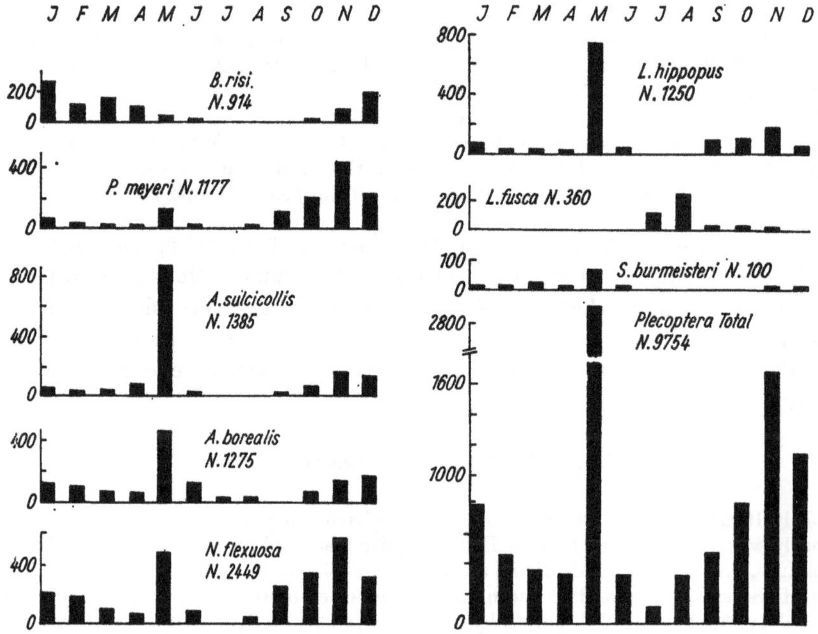

Fig. 1. Specific and total annual patterns of drifting Plecoptera in Kaltisjokk

Tab. 1. Annual drift pattern of stonefly nymphs in northern Europe

Classe	Definition	Drift pattern	Typic species
A	Winter nymphs	only a maximum in winter	*B. risi*
B	Spring nymphs	only a maximum in spring	*S. burmeisteri*
C	Summer nymphs	only a maximum in summer	*L. fusca*
D	Winter–Spring nymphs	a major maximum in winter and a secondary one in spring	*P. meyeri* and *N. flexuosa*
E	Spring–Winter nymphs	a major maximum in spring and a secondary one in winter	*A. sulcicollis*, *A. borealis* and *L. hippopus*

types of annual patterns found for stonefly nymphs drifting in Kaltisjokk. So the studies stonefly nymphs may be grouped into the five classes (Table 1).

From the thirteen species studied, one belongs to classe A, two to classe B and C respectively, three to classe E, and five to classe D.

b) Diurnal periodicity of stonefly nymphs

Brachiptera risi MORT. (914 nymphs)

B. risi nymphs drifted in the course of this investigation from October until June. October month was not analysed since only 6 specimens were catched. Fig. 2 shows the monthly diel variation of drift activity from November until June. The nymphs

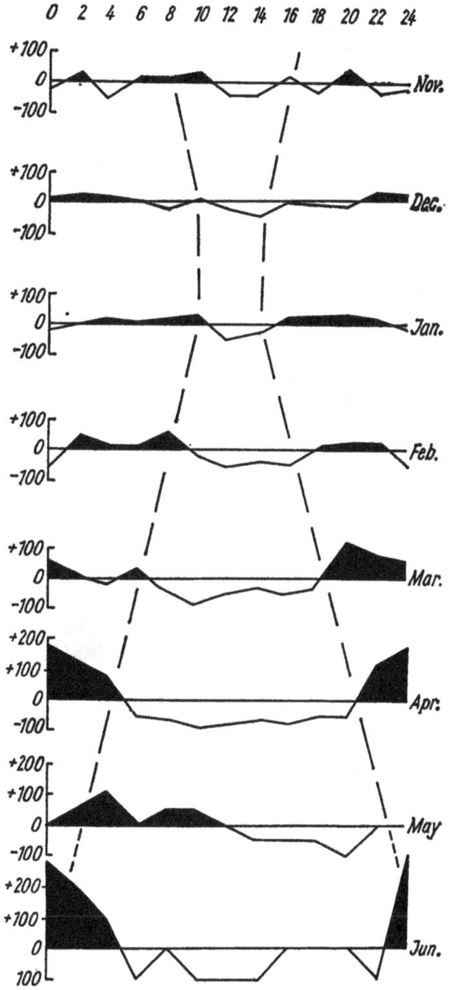

Fig. 2. *Brachiptera risi*, diurnal pattern of drifting nymphs along the year (Broken lines show time for Sunrise and Sunset.)

of this species were always synchronized on the 24 hours period. They were night-active until April. A phase shift took place in May, when most of the nymphs were full-grown (BENEDETTO 1973) and the nymphs became dawn-active. This phase lag persisted in June, when the nymphs drifted on the time of less light intensity (approximately 500 Lux).

Amphinemura sulcicollis STEPH. (1385 nymphs)

From the results of this investigation and from logical speculations about the life cycles of stoneflies it was stated that during their aquatic life stonefly nymphs have different activities which may result in drift. A first hatching-activity was studied (Fig. 3); the drift of the egg-larvae of *A. sulcicollis* shows a day-active, alternans pattern which seems to be a last result of the embryonic rhythm.

Following the development of the drift of the nymphs of *A. sulcicollis* it was observed that after hatching a 180° phase shift occurred, after which the nymphs followed

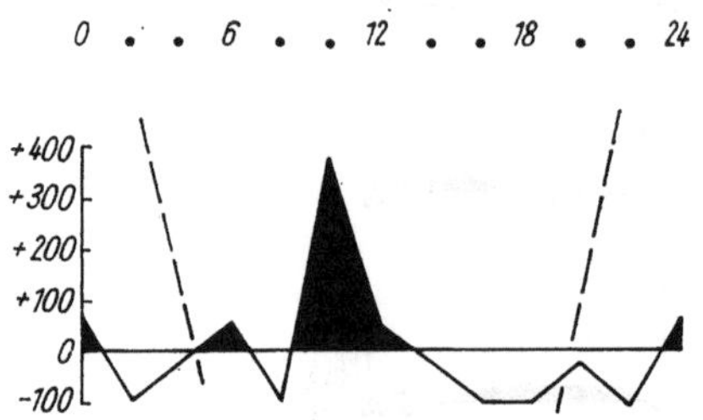

Fig. 3. *Amphinemura sulciollis*, egg-larvae, diurnal pattern of drifting individuals during September

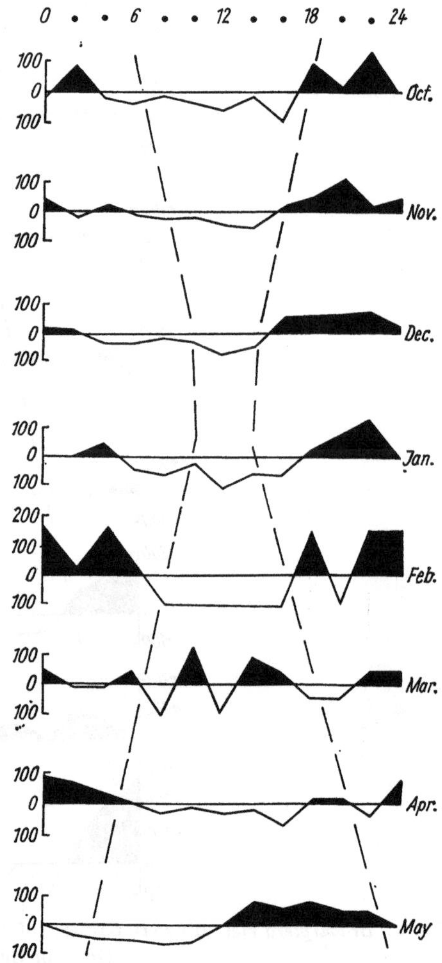

Fig. 4. *Amphinemura sulcicollis*, diurnal pattern of drifting nymphs along the year

the typical night-active rhythm (Fig. 4) which was mentioned before by MÜLLER (op. cit.) and by MÜLLER & BENEDETTO (1970). From October until April the nymphs were night-active. A new phase shift occurred in May simultaneously with the emergence of the species. So the nymphs became once more day-active, which remains the characteristic rhythm of the sexual phase of the stonefly species studied hitherto. Fig. 5 contains a schematic representation of the synchronizing agent light as a continuous function, a sinus wave generated by a rotating vector. In this the theory of the phase angle and the phase difference were plotted against the three activities of *A. sulcicollis* nymphs.

Day-active nymphs were synchronized on the 24 hours period, and the synchronizer seems to be water temperature.

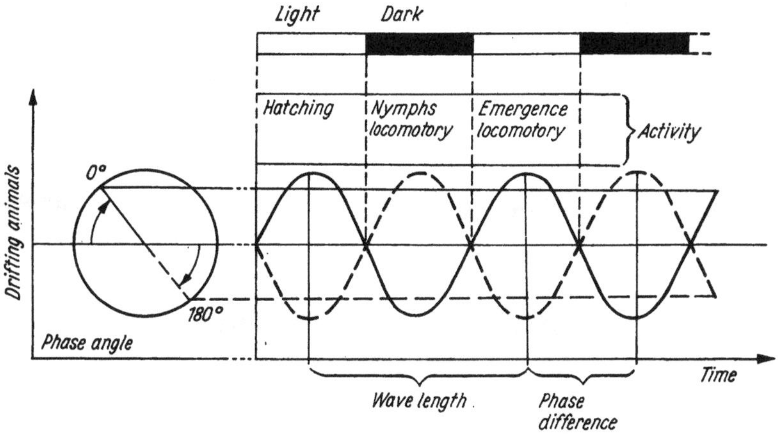

Fig. 5. Diurnal rhythm of different activities of *Amphinemura sulcicollis* larvae and nymphs plotted again the light-dark cycle, to show the phase shifts occurring between the different activities

Nemoura cinerea RETZ. (351 nymphs)

The nymphs of this species began to drift in September, when only six larvae were catched in the sampler. In Fig. 6 the results obtained from October until June are summarized. The values of March and April are offered only to preserve the continuity of the figure, but the number of specimens (8 per month) was considered to be insufficient for an analysis of the daily pattern. Nymphs of this species were night-active until they reached full-growth (BENEDETTO op. cit.); at this time they shifted phase and became day-active. In June the nymphs were perfectly synchronized on the 24 hours period and the phase setting mechanism was water temperature.

Capnia atra MORT. (135 nymphs)

C. atra is a species of particular interest because of its early flight-period which begin in April, more than a month before the midnight-sun period. Five specimens were catched in October. From November until March the nymphs were night-active (Fig. 7), but in April under light conditions of approximately nLD 16:8, 50% of the

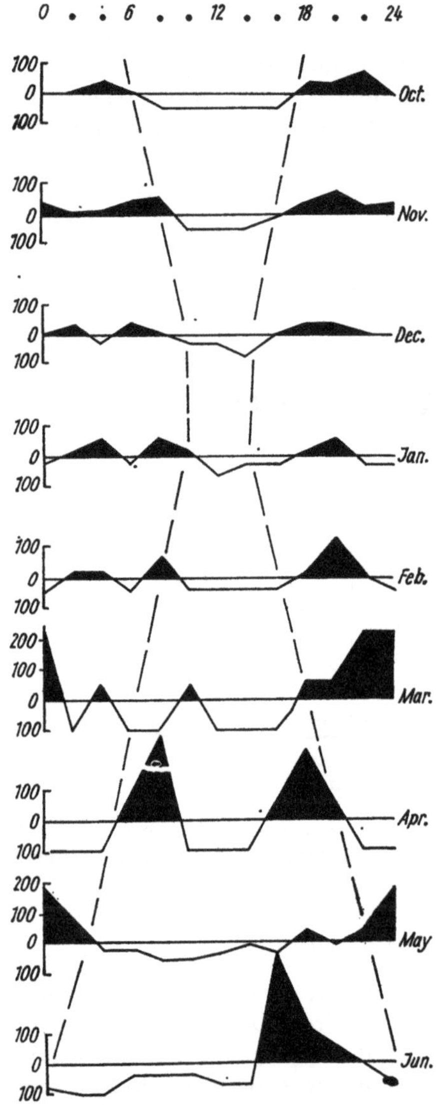

Fig. 6. *Nemoura cinerea*, diurnal pattern of drifting individuals along the year

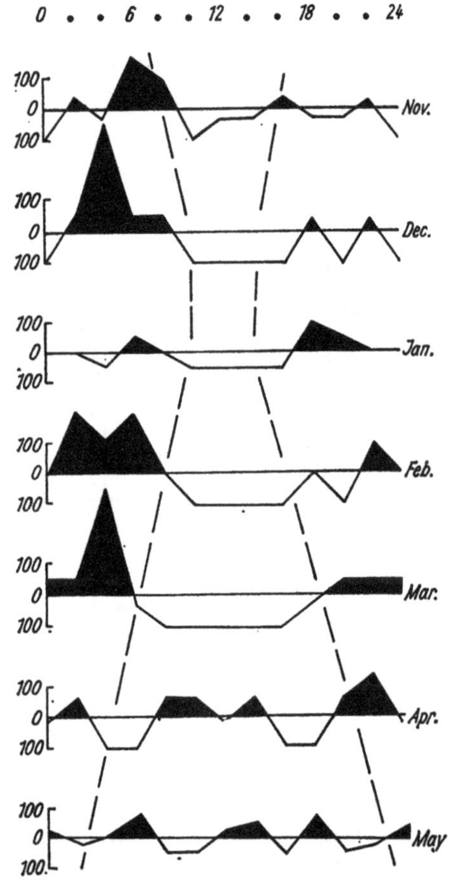

Fig. 7. *Capnia atra*, diurnal pattern of drifting nymphs along the year

nymphs were drifting in sunshine and so it was in May. This drift behaviour seems to coincide with the desynchronization-effect described by Müller (1970a, b, c) with only one difference, that it occurs more than a month before. But, a closer consideration of this behaviour during May brings to light, that most of the nymphs drifting in the first decade of May were night-active, while those drifting in the second decade were day-active (Fig. 8). No records of *C. atra* were registered on the thirth decade of May.

The results from other species of the same material generally agree with those described above. Fig. 9 shows the diel pattern of the full-grown nymphs of *Nemoura avicularis* Mort. during May, when they were day-active. *Amphinemura borealis* Mort., *Nemoura flexuosa* Aub., *Protonemura meyery* Pict., *Leuctra hippopus* Kemp., and *Siphonoperla burmeisteri* Pict. were night-active until they were full-grown and became day-active afterwards. Since the period of emergence starts at different times in different species, mixed populations undergo a transition phase of both day and night-activities. The only exception are nymphs of *Leuctra fusca* L. which are always night-active.

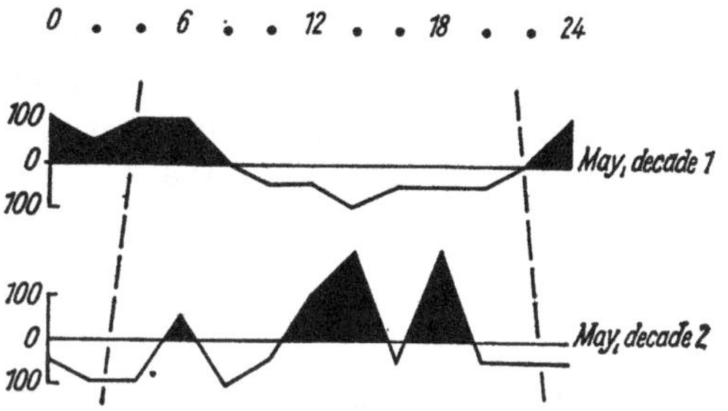

Fig. 8. *Capnia atra*, deeper analysis of the diurnal pattern of the nymphs drifting in May

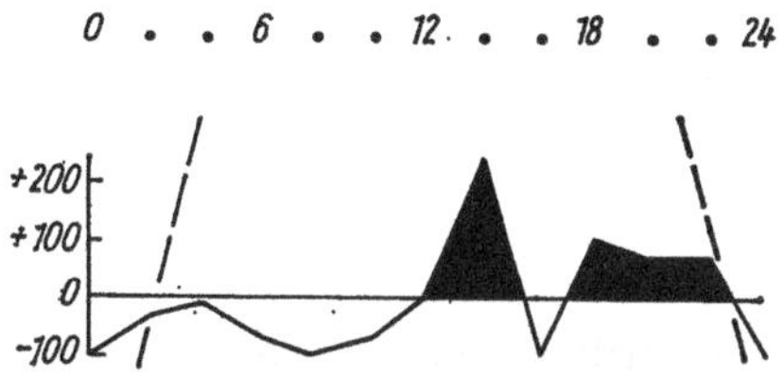

Fig. 9. *Nemoura avicularis*, full-grown nymphs, diurnal pattern of drifting individuals during May

4. *Discussion*

a) Annual periodicity of drift

Müller (1970a, b) regards the annual drift pattern of stonefly nymphs with one winter-peak as typical for central Europe, and those with two maxima: a major one in summer and a secondary one in winter, as typical for northern Europe. Schwarz (1970), on the other hand, observed that the maximum drift rate of *Diura bicaudata* L. in central Europe occurs in summer; so the results of this investigation stated that the annual periodicity of stonefly-drift is an specific phenomenon and that there are clear ecological differences between different species. That means that no regional stonefly-drift patterns exist. Mixed stonefly populations will show annual patterns which result from the interaction of several drift patterns of the species which settle at a stream locality. So there are a lot of possibilities of variations just depending on the species occurring in one biotope or another.

b) Diurnal periodicity of stonefly nymphs

Though very few is known hitherto about the physiology of stoneflies, we know that they have a seasonal, sexual cycle, and that spermatogenesis and ovogenesis occur during the last nymphal stage, which must produce important hormonal changes.

As observed by Brinck (1949) many species of Filipalpia emerge in the morning; and Thomas (1969, 1970b) demonstrated that *N. cinerea*, *A. sulcicollis*, *A. standfussi*, *A. borealis*, *L. hippopus*, and *S. burmeisteri* are day-active during emergence.

From my results I conclude that nymphs of *A. sulcicollis* and *C. atra* become positive-phototactic during April and May. Nymphs of *N. avicularis* during May, and those of *B. risi* and *N. cinerea* during May and June; this in some contradiction to the opinion of BRINCK (1949).

The work of MÜLLER & BEDEDETTO (1970) was based on the locomotoric activity of different larval stages of *Dinocras cephalothes,* in which the possibilities about hatching and emergence were not considered. MÜLLER (1970a, b, c) affirmed that within the Arctic Circle stonefly nymphs are desynchronized during midsummer time. But, since biological systems contain innumerable oscillators with different frequencies which may interact (SOLLBERGER 1965), and since the desynchronization described by MÜLLER (op. cit) is based on a study of mixed populations, containing different species as well as different age-size subpopulations, this can no longer be supported. The taxonomical separation of species and the careful analysis of the different age-size groups demonstrated that every species was synchronized with the 24 hours period.

5. *Summary*

The drift activity of stonefly nymphs was investigated in the Kaltisjokk, in northern Sweden.

Different diurnal rhythms were observed in different age-size groups within the same species, which demonstrated that phase shifts occur simultaneously with some hormonal changes of the nymphs.

All the species mentioned in this paper were always synchronized on the 24 hours period.

Hatching larvae of *Amphinemura sulcicollis are* day-active. After hatching nymphs of all species were night-active until they became full-grown. Shortly before emergence full-grown nymphs became once more day-active.

6. *References*

ASCHOFF, J., 1960: Exogenous and endogenous components in circadian rhythms. — Cold Spring Harbor Symp. Quant. Biol. **25**, 11—28.

BENEDETTO, L., 1973: Growth of stoneflies nymphs in swedish Lappland. — Ent. Tidskr. **94**: 15—19.

BRINCK, P., 1949: Studies on swedish Stoneflies (Plecoptera). — Op. Ent. Suppl. **11**, 1—250.

ERKINARO, E., 1969: Der Phasenwechsel der lokomotorischen Aktivität bei *Microtus agrestis* (L.), *M. arvalis* (PALL.) und *M. oeconomus* (PALL.). — Aquilo, Ser. Zool. **8**, 1—31.

MÜLLER, K., 1963a: Tag-Nachtrhythmus von Baetidenlarven in der Organischen Drift. — Naturwissenschaften **50**, 161.

— 1963b: Temperatur und Tagesperiodik der „Organischen Drift" von *Gammarus pulex*. — Naturwissenschaften **50**, 410—411.

— 1965: An automatic Stream Drift Sampler. — Limnology and Oceanography **10** (3), 483—485.

— 1970a: Die Drift von Insektenlarven in Nord- und Mitteleuropa. — In: Der Kaltisjokk, Die Biologie eines Lappländischen Waldbaches. — Österreichs Fischerei **23** (5/6), 111—117.

— 1970b: Tages- und Jahresperiodik der Drift in Fließgewässern in verschiedenen geographischen Breiten. — Oikos Suppl. **13**, 21—44.

— 1970c: Die Tagesperiodik der Driftaktivität von Plecopterenlarven am Polarkreis. — Ent. Scand. 1, 55—62.

— 1970d: Das Untersuchungsgebiet der Ökologischen Station Messaure. — Oikos Suppl. **13**, 9—13.

— 1970e: Die Hydrographie des Kaltisjokk. — In: Der Kaltisjokk. Die Biologie eines Lappländischen Waldbaches. — Österreichs Fischerei **23** (5/6), 77—82.

MÜLLER, K., & L. BENEDETTO, 1970: Die lokomotorische Aktivität der Larven von *Dinocras cephalotes* CURTIS (Plecoptera). — Oikos Suppl. **13**, 69—74.

MÜLLER-HAECKEL, A., 1967: Tages- und Jahresperiodicität von *Ceratoneis arcus* KÜTZ. (Diatomaea). — Oikos **18**, 351—356.

— 1970a: Tages und Jahresperiodik einzelliger Algen in fließenden Gewässern. — In: Der Kaltisjokk. Die Biologie eines Lappländischen Waldbaches. — Österreichs Fischerei **23** (5/6), 97—100.

— 1970b: Die Hydrochemie des Kaltisjokk. — In: Der Kaltisjokk. Die Biologie eines Lappländischen Waldbaches. — Österreichs Fischerei **23** (5/6), 83—90.

MÜLLER-HAECKEL, A., & K. MÜLLER, 1970: Chronobiologie in Fließgewässern. — In: Der Kaltisjokk. Die Biologie eines Lappländischen Waldbaches. — Österreichs Fischerei **23** (5/6), 90—96.

SCHWARZ, P., 1967: Untersuchungen zum Besiedlungskreislauf von *Isoperla goertzi* (Insecta Plecoptera). — Arch. Hydrobiol. **64** (1), 75—87.

— 1970: Autökologische Untersuchungen zum Lebenszyklus von *Setipalpia*-Arten (Plecoptera). — Arch. Hydrobiol. **67** (1), 103—140.

SOLLBERGER, A., 1965: Biological rhythm research. — Elsevier Publishing Company, Amsterdam—London—New York, 461 pp.

TANAKA, H., 1960: On the daily change of the drifting of benthic animals in stream, specially on the taxonomic groups of insects. — Bull. Freshwater Fish. Res. Lab. Tokyo **9**, 13—24.

THOMAS, E., 1969: Zur Tagesperiodik des Schlüpfens von Ephemeropteren und Plecopteren. — Oecologia **3**, 230—239.

— 1970a: Die Oberflächendrift im Kaltisjokk. — In: Der Kaltisjokk. Die Biologie eines Lappländischen Waldbaches. — Österreichs Fischerei **23** (5/6), 101—110.

— 1970b: Die Oberflächendrift eines lappländischen Fließgewässers. — Oikos Suppl. **13**, 45—64.

WATERS, T. F., 1962: Diurnal periodicity in the drift of stream invertebrates. — Ecology **43**, 316—320.

— 1965: Interpretation of invertebrate drift in streams. — Ecology **46**, 327—334.

Received February 5, 1975.

Address of the author:

Dr. LUIS A. BENEDETTO
Limnologische Flußstation des
Max-Planck-Instituts für Limnologie
D-6407 Schlitz
Postfach 102

Limnologica (Berlin)	11	1	1976	57—100

HEINZ SCHNEE

Beitrag zur Limnologie von Braunwässern unterschiedlicher Typen

Mit 12 Abbildungen und 9 Tabellen

Inhalt

1. Einleitung

Die Braunwasserseen wurden im limnologischen Seetypensystem anfangs als dystropher Typus neben die oligotrophen und eutrophen Seen gestellt. Dabei wurde die Dystrophie von NAUMANN als Sonderfall der Oligotrophie betrachtet. Doch erwies sich diese Einteilung als unbefriedigend, weil es sowohl oligotrophe als auch eutrophe Braunwässer gibt. Das THIENEMANNsche Einteilungsprinzip, das Braun- und Klarwasserseen unterscheidet, wird den Tatsachen besser gerecht. Die Braunfärbung der hier zu behandelnden Gewässer beruht auf dem Vorhandensein von im Wasser gelösten Huminstoffen, die allochthoner Herkunft sind. Das ist der wesentliche Unterschied zwischen Klarwässern und Huminwässern mit allen seinen Konsequenzen für den Stoffhaushalt und die Lebewelt der letzteren. Dagegen sind die häufig saure Reaktion und die bisweilen „besondere" Nährstoffarmut der Braunwässer keine hinreichenden Kriterien für ihre Charakterisierung, weil sie nicht immer zutreffen und weil sie auch Klarwässern zukommen können.

Um die hydrobiologische Erforschung der Huminwässer haben sich insbesondere nordeuropäische und amerikanische Autoren verdient gemacht. Aus Mitteleuropa liegen hingegen relativ wenige Untersuchungen vor. Dabei kommt der Erforschung der genannten Gewässer keineswegs nur theoretisches Interesse zu. Da die Wasserversorgung immer größere Probleme aufwirft, wird in Zukunft verstärkt auch huminbraunes Oberflächenwasser zur Gewinnung von Trink- und Brauchwasser verwendet werden müssen. Es ist daher notwendig, die Limnologie der Braunwässer besser kennenzulernen.

Im Rahmen der vorliegenden Arbeit wurden von Mai 1965 bis Juli 1966 die unterschiedlichsten natürlichen Humingewässer von der Moorschlenke bis zum See und Fließgewässer untersucht. Dabei fand neben einer biologischen Bestandsaufnahme die vergleichende Erfassung der Nährstoffverhältnisse besondere Berücksichtigung. Es wurde ferner versucht, durch biologische Tests Aufschlüsse über die Bioaktivität der Untersuchungsobjekte zu erhalten.

Dieser Beitrag ist die stark gekürzte Fassung einer am ehemaligen Zoologischen Institut der Karl-Marx-Universität Leipzig angefertigten Diplomarbeit. Herrn Prof. Dr. UHLMANN bin ich für die Themenstellung sowie für zahlreiche wertvolle Ratschläge sehr zu Dank verplichtet.

2. *Methodik*

Entnahme, Transport und Aufbewahrung der Wasserproben wurden nach den in der Hydrochemie üblichen Methoden vorgenommen. Die Analysen auf Sauerstoff (WINKLER), BSB, Farbgrad (HAZEN), Nitrat (Na-Salizylat), Silizium (WINKLER) und Mangan (MARSCHALL) erfolgten nach den in den Deutschen Einheitsverfahren (1960) vorgeschriebenen Methoden. Sauerstoff wurde ferner nach dem Joddifferenzverfahren von OHLE (1953), die freie Kohlensäure nach OHLE (1952), das SBV nach SCHÄPERCLAUS, der pH-Wert elektrometrisch und die elektrolytische Leitfähigkeit konduktometrisch bestimmt. Aus HÖLL (1960) wurden folgende Analysenmethoden entnommen: $KMnO_4$-Verbrauch (KUBEL), Chlorid (MOHR), Sulfat (WINKLER) und Eisen (Rhodanid). Die Bestimmung der Huminstoffe (HS) erfolgte nach CHALUPA (1963), des Ortho- und Gesamtphosphats nach VOGLER (1965) sowie des Nitrits (Indol) nach CZENSNY (1960). Ammoniak und organisch gebundener Stickstoff wurden nach den Einheitsverfahren der RGW-Länder (1965) analysiert, ebenso Kalzium und Magnesium. Die flammenphotometrischen Natrium- und Kaliumbestimmungen schließlich wurden freundlicherweise in der OFM Karl-Marx-Stadt vorgenommen.

Erhebliche Schwierigkeiten verursachte bei allen kolorimetrischen und einigen titrimetrischen Analysen die starke Eigenfärbung der Braunwässer (vgl. BERG 1962). Es wurden deshalb die Blindproben stets aus der zu untersuchenden Probe genommen. Zur Nitrat-, Chlorid-, Siliziumsowie zur Ca- und Mg-Bestimmung mußten die Proben mit $KAl(SO_4)_2$ entfärbt werden. Ca und Mg sowie Sulfat wurden auch im Glührückstand ermittelt.

Die zur Planktonzählung bestimmten Proben wurden sofort mit Jodjodkaliumazetat fixiert. Das Phytoplankton wurde mit dem umgekehrten Mikroskop nach UTERMÖHL ausgezählt, wobei nur die dominanten Formen Berücksichtigung fanden. Zur qualitativen Planktonuntersuchung dienten Netzplankton- (Nr. 25) und zentrifugierte Proben.

Die Erfassung des Benthos war rein qualitativ. Die Häufigkeit wurde nur grob geschätzt.

Zur Testung der autotrophen Bioaktivität der Braunwässer dienten Reinkulturen von *Haematococcus pluvialis* FLOT. (s. HEDLICH 1965). Je 10 ml bakterienfiltriertes Probenwasser wurden in Kulturröhrchen mit der Testalge beimpft und die Kulturen täglich 16 h bei 20 °C und 10 cm Abstand von einer Leuchtstoffröhre beleuchtet. Mit dem PULFRICH-Nephelometer erfolgte dann alle 2—4 Tage die Messung der dem Wachstum proportionalen Trübung.

In Anlehnung an KNÖPP (1964) wurden BSB-(= ZZ-)Tests durchgeführt. Das auf 20 °C temperierte, 30—45 min belüftete Versuchswasser wurde folgendermaßen weiterbehandelt:

A: sofortige O_2-Bestimmung
B: normaler BSB-Ansatz
C: wie B, aber bei Ansatz Zugabe von Glukoselösung (3 h bei 150 °C sterilisiert) — 5, 10, 20 und 100 mg/l.

B und C wurden 2 oder 5 Tage bei 20 °C im Dunkeln bebrütet und dann der Restsauerstoff bestimmt (4 Parallelen).

Es gelten folgende Beziehungen:

A — B = BSB
B — C = ZZ (zusätzliche Zehrung)

3. *Charakteristik der untersuchten Gewässer*

3.1. Großes und Kleines Kranichseehochmoor

Diese beiden ombrogenen Hochmoore (Abkürzungen GK und KK) liegen auf dem Kamm des westlichen Erzgebirges in 930 m Höhe über NN. Ihre Größe beträgt 37 ha (GK) und 25 ha (KK) (KÄSTNER & FLÖSSNER 1933). Der Untergrund der Moore besteht aus Eibenstocker Turmalingranit (PIETSCH 1962). Die Mächtigkeit der Torfschicht beträgt nach FRENZEL (1930) 3,5 m (GK) und 2,5 m (KK). Beide Hochmoore sind Wasserscheidemoore. Der südliche Abfluß des GK hat insofern große Bedeutung, als er in die Wiltzsch mündet, die der Hautpzuleiter der Trinkwassertalsperre Carlsfeld ist.

Die Jahresmitteltemperatur beträgt im Gebiet 4,2 °C. Die jährliche Niederschlagsmenge übersteigt 1200 mm (GOLDSCHMIDT 1950). Die Proben wurden an je einer Erweiterung der beiden Bachbetten (GKA bzw. KKA) unmittelbar nach Verlassen der Torfschicht und an einigen Schlenken (GKS bzw. KKS) gewonnen. Die Tiefe der letzteren betrug nur wenige dm.

3.2. Wildenhainer und Zadlitzbruch

Die Zwischenmoore Wildenhainer (W) und Zadlitzbruch (Z) liegen etwa 2 km voneinander entfernt am Südrand der Dübener Heide. Das erstere ist 180 ha groß und ruht auf kalk- und eisenhaltigem Talsand. Es bildet die Wasserscheide zwischen Elbe und Mulde. Der 84 ha große Zadlitzbruch liegt in einer pleistozänen Senke auf Plateausand (v. LINSTOW 1919). Anfang des 19. Jahrhunderts sind beide Moore fast völlig ausgetorft worden und gegenwärtig in Regeneration begriffen. Die jährliche Niederschlagsmenge beträgt im Gebiet der beiden Brüche nach FUESS (1932) 550 mm.

Die Untersuchungsstellen lagen im W an einem flachen Teich am „Flügeldamm“ und am Abfluß an der Westgrenze des Moores. Im Z wurden die Proben aus einer Schlenkenreihe (ZS) im Südteil und einmal aus einer nördlich gelegenen Schlenke (ZN) sowie aus dem Abflußbach (ZG) entnommen.

3.3. Die Moorgewässer um Serrahn

Der „Strelitzer Bogen“ der baltischen Endmoräne des Würmglazials prägt das Landschaftsbild um die Biologische Station Serrahn in Mecklenburg, in deren Nähe die kleinen Hochmoore Klockenbruch (KB, 1 ha groß), Beilbruch (BB, 6 a) sowie der teichartige „Carpiner Bruch“ (CB, 50 a) liegen.

Diese Moore sind im Gegensatz zu denen des Erzgebirges durch Verlandung kleiner Seen (Toteisbildungen) entstanden. Ihr Untergrund besteht aus pleistozänen Ablagerungen. Beim KB und BB ist die Hochmoorfläche von einem 1—2 m breiten Lagg umgeben, aus dem die Proben entnommen wurden. Im CB erfolgte die Probenentnahme vom Teichufer aus.

Die Jahresmitteltemperatur beträgt im Serrahner Gebiet 8 °C, die jährliche Niederschlagsmenge 560 mm (SCAMONI 1959).

3.4. Herthasee

Der 1,5 ha große, maximal 9,5 m tiefe Herthasee (H) auf Rügen liegt im Kreidegebiet der Halbinsel Jasmund. Das Gelände ist durch Endmoränen gekennzeichnet, deren Geschiebemergel den Untergrund des Sees bildet. Das Gewässer ist der Rest einer sonst verlandeten moorigen Talsenke (OHLE 1934). Die Jahresmitteltemperatur der Umgebung liegt bei 7,5 °C, die Niederschlagssumme erreicht im Jahr 625 mm (Atlas 1962).

Der Herthasee wurde am 26. 4. 1966 kurz nach der Schneeschmelze untersucht.

3.5. Straminke

Die 1 ha große „Straminke (S)“ bei Zingst liegt nur 100 m vom Ostseestrand entfernt auf holozänem Sand (Atlas 1962) (Rest einer alten Flutrinne). Das Gewässer ist 1,5 m tief und sehr windexponiert. Die Untersuchung fand am 27. 4. 1966 statt.

4. Chemische Untersuchungen

4.1. Sauerstoff

Vergleichende O_2-Bestimmungen nach dem Joddifferenzverfahren und nach der herkömmlichen WINKLER-Methode ergaben beträchtliche Unterschiede, weil die Jodzehrung durch die organischen Stoffe bei der letzteren viel zu niedrige Werte vortäuscht. Das OHLEsche Joddifferenzverfahren lieferte 6—26% höhere O_2-Konzentrationen als die unmodifizierte WINKLER-Methode. Im folgenden wird nur auf die mit dem ersteren ermittelten Werte eingegangen.

Die Schlenken der beiden Kranichseemoore waren außer im Winter stets recht O_2-reich. Dank ihrer exponierten Lage wird ihr Wasser sehr leicht durchmischt. In der Vegetationszeit kommt eine erhebliche phytoplanktogene O_2-Produktion dazu. Übersättigungen bis zu 147% waren in dieser Zeit keine Seltenheit. Im Herbst stellen sich bereits leichte Defizite ein, da das Phytoplankton keine Rolle mehr spielt. Allerdings können stärkere Regenfälle wieder O_2-Sättigung bewirken.

Während des langen Winters verhindern dicke Eis- und Schneedecken das Wachstum von Algen und den Austausch mit der Luft. Die Folge sind starke O_2-Defizite (bis 51% der Sättigung). In den Abflußbächen ist der O_2-Gehalt deutlich von Niederschlägen abhängig. Nur bei starker Wasserführung wird die Sättigungskonzentration erreicht.

Im Wildenhainer Bruch lagen die O_2-Konzentrationen fast stets um die Sättigung. Da die O_2-Eigenproduktion hier völlig unbedeutend ist, kann das nur durch die Aufnahme atmosphärischen Sauerstoffs gewährleistet werden. Der stets sehr niedrige Huminstoffgehalt dürfte keinen Einfluß haben.

In den Zadlitzschlenken herrschten dagegen meist stark O_2-zehrende Prozesse, deren Ursachen hohe- HS- und Fe-Gehalte, Laubeinfall und windgeschützte Lage sind. Im Winter kommt es unter Eis zu anaeroben Verhältnissen, wobei der H_2S-Gehalt 10 mg/l erreicht. Bei längerer Trockenheit kann aber auch im Sommer O_2-Schwund eintreten. Selbst eine üppige *Tribonema*-Entwicklung vermag das nicht zu verhindern. Dagegen trug eine im Anschluß an starke Regenfälle im Juni auftretende Phytoplanktonblüte zu einer erstaunlich hohen O_2-Übersättigung bei (153%).

Der O_2-Haushalt der Serrahner Moorgewässer hat eine gewisse Ähnlichkeit mit dem mancher Abwasserteiche (s. UHLMANN 1964). Im Sommer ist die das Wasser erreichende Lichtintensität gering, weil der umgebende Buchen-Traubeneichenwald sehr dicht und eine *Lemna-minor*-Sperrschicht (Bedeckung etwa 50%) ausgebildet ist. Das Phytoplankton kann sich schon deshalb nicht entwickeln. Außerdem tritt eine Massenentfaltung von *Daphnia pulex* auf. Der O_2-Gehalt wird folglich im Sommer sehr niedrig (bis 7,2% der Sättigung), weil die Zehrung durch das Fallaub stark ist und eine nennenswerte Sauerstoffproduktion ja nicht zustande kommt. Im Frühjahr ist das Licht nicht mehr begrenzender Faktor für das Phytoplankton. Es können sich Vegetationsfärbungen ausbilden. Die O_2-Spannung des Wassers ist deshalb erhöht. Doch wurde — zumindest zum Zeitpunkt der Untersuchung (März) — die Sättigungskonzentration nicht erreicht oder gar überschritten.

Widriger Umstände wegen liegen vom Herthasee nur folgende O_2-Werte vor:

0 m 20,45 mg/l (182% der Sättigung)
1 m 21,0 mg/l (180% der Sättigung)
8 m 7,6 mg/l (60% der Sättigung)

Die starke O_2-Übersättigung in der Oberflächenschicht beruht auf planktogener Produktion. Das Defizit bei 8 m zeigt das Vorhandensein einer Schichtung an, obwohl etwa 2 Wochen nach dem Eisbruch Vollzirkulation zu erwarten wäre. OHLE (1934) stellte im Sommer an der Oberfläche ein leichtes Defizit und O_2-Schwund im Tiefenwasser fest.

Im Februar fand v. PETERSON (mdl. Mitt.) in der Straminke unter 15 cm starkem Eis 86% der O_2-Sättigung. Die eigene Frühjahrsuntersuchung ergab in 0 und 1 m Tiefe Werte, die 113% der Sättigungskonzentration entsprechen und die ebenfalls nur auf Assimilations-Sauerstoff zurückzuführen sind.

4.2. Freie Kohlensäure, Säurebindungsvermögen und pH-Wert

In den Schlenken der beiden Erzgebirgshochmoore ist der CO_2-Gehalt im Frühjahr und Sommer im Gegensatz zur O_2-Konzentration auffallend niedrig (bis 1,8 mg/l). Das ist sicher eine Folge der intensiven Assimilation des Phytoplanktons. Es ist denkbar, daß CO_2 als Minimumfaktor auftritt, zumal Bikarbonat nie zur Verfügung steht. Allerdings kann mit einem ständigen Nachschub durch bei der Zersetzung von HS entstehendes Kohlendioxyd gerechnet werden.

Im Herbst und Winter wurden höhere CO_2-Konzentrationen festgestellt, z. B. im März in Schlenke KK_1 unter Eis und Schnee am Grunde (in 50 cm Tiefe) 16,2 mg/l. Es überrascht, daß die CO_2-Akkumulation (und das O_2-Defizit, s. S. 61) nicht noch viel höher war, da die Eisdecke bereits 3 Monate bestanden hatte.

Wie die Sauerstoffverhältnisse sind auch die Kohlensäure-Konzentrationen in den Abflußbächen witterungsabhängig, insbesondere im GKA. Nach starken Niederschlägen dominiert der Hochmooreinfluß, d. h., der CO_2-Gehalt ist relativ gering, das SBV fehlt völlig und der pH-Wert ist sehr niedrig (bis 3,5). Bei Trockenheit überwiegt hier das Grundwasser. Das freie CO_2 erreicht sehr hohe Werte (120 mg/l), der pH-Wert steigt bis auf 5,4, und es ist etwas (Fe(II))-Bikarbonat nachweisbar.

Erwartungsgemäß sind die pH-Werte in den Hochmoorgewässern sehr niedrig. Mit Ausnahme des GKA schwanken sie zwischen 3,3 und 4,25, meist liegen sie jedoch unter 4.

In den untersuchten Gewässern des W variieren die CO_2-Menge und des SBV weniger stark, ohne daß diese Unterschiede jahreszeitlich bedingt sind: 9,2—22,6 mg/l bzw. 0,4—1,35 mval/l. Hohen O_2-Sättigungen kommen niedrigere CO_2-Konzentrationen zu und umgekehrt. Es waren Beziehungen zu den Niederschlägen angedeutet.

Die pH-Werte bewegen sich mit 6,5—7,1 um den Neutralpunkt.

Während in den ZS Bikarbonate völlig fehlen, ist der CO_2-Gehalt stets sehr hoch: 26,4—137 mg/l. Bei der Mineralisation großer Fallaubmengen und abgestorbener *Tribonema*-Bestände sowie beim Abbau der Huminstoffe entstehen erhebliche CO_2-Mengen. Durch die Gegenwart starker Mineralsäuren (H_2SO_4) ist aber keine Karbonatbildung möglich. Aus demselben Grunde erreicht die Nicht-Kohlensäureazidität sehr hohe Werte, weshalb das Wasser stets stark sauer ist (pH 3,2—3,9).

In den vom Fallaub beeinflußten Serrahner Moorgewässern ist die CO_2-Produktion hoch. Sowohl im Sommer (BB: 37 mg/l) als auch im Frühjahr (BB: 26,5 mg/l) wurden beträchtliche Konzentrationen gefunden. Auch eine *Glenodinium*-Massenentwicklung im Frühjahr konnte diese Mengen nicht aufbrauchen, doch war in den dichtesten Planktonwolken im BB eine geringere Konzentration (18,5 mg CO_2) festzustellen als an planktonärmeren Stellen.

Im Gegensatz zu den Erzgebirgshochmooren weist das Wasser einen, wenn auch geringen Bikarbonatgehalt auf (0,02—0,09 mval/l). Die pH-Werte liegen mit 5,9 (BB), 5,7 (KB) und 5,25 (CB) im mäßig sauren Bereich.

Freie Kohlensäure, Bikarbonat und pH-Wert zeigten im Herthasee eine deutliche, biogen bedingte Schichtung, obwohl man zum Untersuchungszeitpunkt (Ende April) doch Vollzirkulation erwarten sollte (s. Abb. 1a). Bis in eine Tiefe von 1 m war das Wasser völlig CO_2-frei, eine Folge der überaus starken Entwicklung planktischer Diatomeen. Der geringere Bikarbonatgehalt an der Oberfläche ist auf biogene Entkalkung zurückzuführen, eine Schlußfolgerung, die auch durch die Stratifikation des Ca^{++} (Abb. 1b) bewiesen wird. In den Oberflächenschichten war sogar etwas Monokarbonat vorhanden. Der pH-Wert verringert sich mit der Zunahme des CO_2 ständig und ist im interstitiellen Wasser mit 6,4 am niedrigsten.

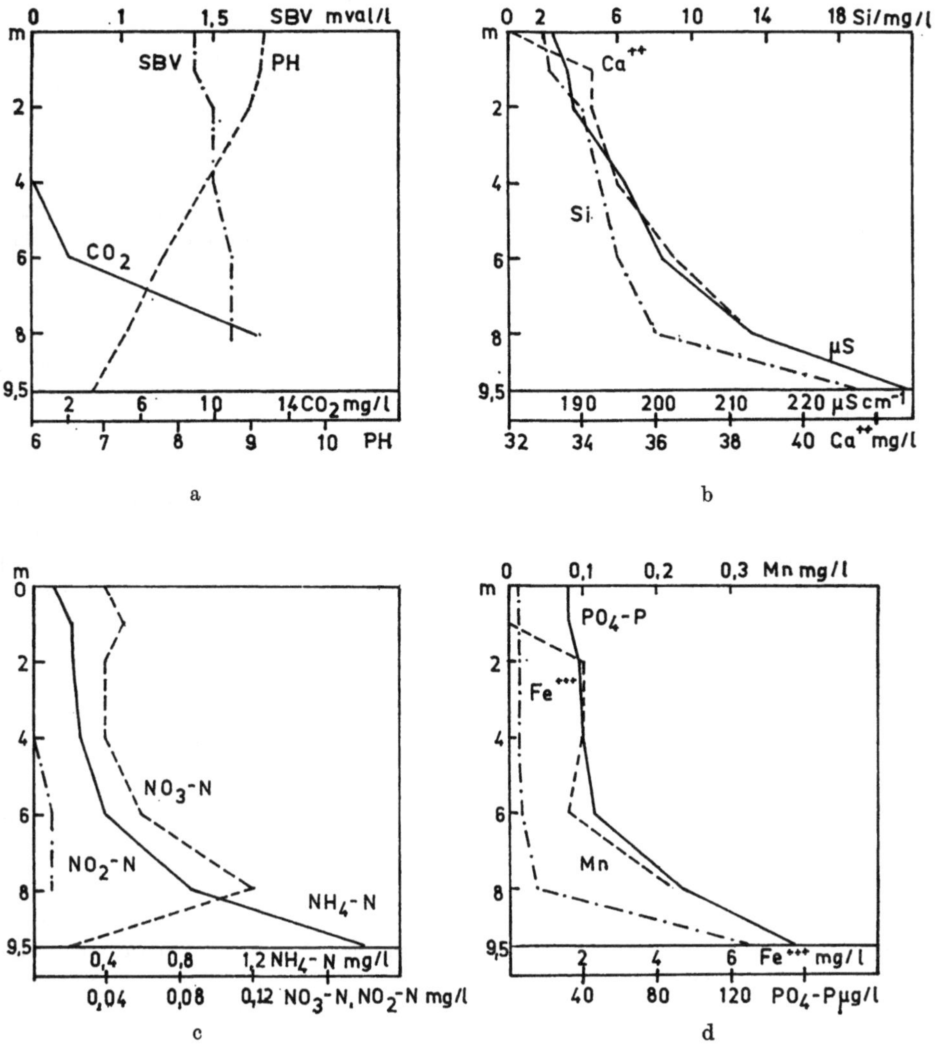

Abb. 1a—d. Schichtungsverhältnisse im Herthasee am 26. 4. 1966

An der Oberfläche der Straminke fand sich ein CO_2-Gehalt von 12,3 mg/l und ein pH-Wert von 8,1. In 1 m Tiefe war die CO_2-Konzentration etwas höher (15,8 mg/l) und der pH mit 8,0 etwas niedriger. Das SBV von 2,6 bzw. 2,5 weist auf eine gute Fruchtbarkeit dieses Fischzuchtweihers hin.

4.3. Huminstoffe

4.3.1. Allgemeines und Diskussion der angewandten Huminstoff-Bestimmungsmethoden

Ausführliche Angaben über Entstehung, Nomenklatur, Struktur und Eigenschaften der Huminstoffe (HS) im Wasser bringen u. a. ASCHAN (1908), ODÉN (1919), SHAPIRO (1957, 1964, 1966a), CHALUPA (1963, 1966, 1968), BLACK & CHRISTMAN (1963b), GHASSEMI & CHRISTMAN (1968). Danach stellen die Huminsubstanzen des Wassers eine heterogene Stoffgruppe mit sehr unterschiedlichen Molekulargewichten dar. Sie haben aromatischen Charakter (Polyphenolkondensate), negative Ladung und saure Reaktion. Für den Nährstoffhaushalt der Braunwässer sehr wichtige Eigenschaften der HS sind ferner ihre starke Lichtabsorption, die Kationenaustauschfähigkeit und die Chelataktivität gegenüber Metallionen, die ihnen zusammen mit ihrer Adsorptionsfähigkeit eine wichtige Rolle als Nährstoffregulatoren im Sinne OHLES (1937) geben.

In den meisten Fällen gelangen die HS durch Auswaschung von Torfböden in die Gewässer. Die die Moore entwässernden Bäche führen besonders nach starken Niederschlägen häufig große Mengen dieser Stoffe mit.

Die Huminsubstanzen können sowohl molekular als auch kolloidal im Wasser gelöst sein, wie aus Befunden von SHAPIRO (1957), OHLE (1933, 1945, 1940), BLACK & CHRISTMAN (1963a) und SIMM (1969) hervorgeht.

Eine Schätzung des HS-Gehalts der Braunwässer ist schon durch Bestimmung des Platin-Farbgrades (FG) möglich. Doch sind Fe(III) und Mn (IV) an der Farbbildung beteiligt (s. ÅBERG & ROHDE 1942 und Abb. 7). Daß die Farbe ferner indikatorartig vom pH-Wert abhängt, ist seit OHLE (1934) eine bekannte Tatsache, die offenbar für alle Humusgewässer zutrifft (ELSTER 1955, BAYLY 1964 u. v. a.). Das konnte auch bei den eigenen Untersuchungen stets beobachtet werden. Aus diesem Grunde und weil die Farbtönungen der untersuchten Gewässer voneinander abweichen, ist die Farbe nur innerhalb eines Gewässers zur Charakterisierung des HS-Gehalts brauchbar und für Vergleiche kaum geeignet. Genauigkeiten, wie sie z. B. von SIMM (1969) angegeben werden, erscheinen fiktiv.

Die Bestimmung der Oxydabilität der Braunwässer mittels $KMnO_4$ (PV) als Maß der HS-Belastung weist ebenfalls Fehlerquellen auf (u. a. Vorhandensein farbloser organischer Stoffe und nicht vollständige Oxydation).

Das direkte HS-Bestimmungsverfahren nach CHALUPA (1963) beruht auf doppelter Extraktion der HS mit Amylalkohol und verdünnter NaOH und photometrischer Messung des Extraktes. Mit ihm wurden in dieser Arbeit recht gute Erfahrungen gesammelt. Das Verfahren von OBENAUS (1962, 1963, 1964) beachtet die Heterogenität der Huminstoffe durch Ermittlung des für jeden HS spezifischen PV. Eigene Analysen konnten mit dieser Methode nicht durchgeführt werden.

4.3.2. Ergebnisse der Huminstoff-Untersuchungen

Die größten Unterschiede im HS-Gehalt der Kranichseemoore wurden im GKA festgestellt (s. Tab. 1). Diese Schwankungen sind witterungsbedingt. Die hohen Konzentrationen fanden sich im niederschlagsreichen Sommer 1965. Durch die starke Wasserführung des Baches konnte das eisenhaltige Grundwasser soweit verdünnt werden, daß eine Ausfällung der HS nicht möglich war (s. S. 71). Der FG ist im GKA verglichen mit den HS- und PV-Werten sehr hoch. Die Ursache ist kolloidales $Fe(OH)_3$ (und $Mn(OH)_4$), das dem Wasser auch bei geringer Huminstoffbelastung (bei Trockenheit) eine braune Farbe verleiht.

Nach ihrem Gehalt an gelösten organischen Stoffen sind die Kranichseegewässer als polyhumos zu bezeichnen.

Tab. 1. Huminstoffgehalt in den Kranichseehochmooren[1])

	HS [mg/l]	PV [mg/l]	FG [mg/l]
GKS	39–55 (48)	104–200 (147)	175–215 (195)
GKA	8–50 (25)	45–200 (121)	160–350 (221)
KKS	22–29 (25)	63–135 (94)	85–150 (114)
KKA	20–55 (36)	79–220 (137)	120–250 (170)

[1]) Die Durchschnittswerte sind hier – wie auch in den nächsten Tabellen – in Klammern angegeben.

Abb. 2 veranschaulicht die Beziehungen der HS-Konzentration zum PV der untersuchten Hochmoorgewässer des Erzgebirges. Zwischen beiden Größen besteht eine signifikante Korrelation. Die Tatsache, daß die Gerade die y-Achse bei 2,3 schneidet, zeigt das fast ausschließliche Vorhandensein huminoider Substanzen an. An dieser Stelle sei Tab. 2 angeführt, die die Ergebnisse einiger von Klöcking durchgeführter HS-Bestimmungen nach ihrer Methode enthält.

Tab. 2. HS-Bestimmungen von Klöcking (pers. Mitt.)

Probe	PV Gesamt [mg/l]	PV_2 [mg/l]	PV_1 [mg/l]	spez. PV	spez. KHS	HS [mg/l]
BB	167,4	5,1	162,3	3,225	0,310	50,3
KKA	167,5	1,9	165,5	3,238	0,309	51,1
ZS	169,4	12,6	156,7	3,370	0,297	46,5
WA	28,8	3,4	25,3	—	—	—

Der auf farblose organische Stoffe zurückzuführende PV_2 von KKA ist auch hier vernachlässigbar niedrig, was gut mit Abb. 2 übereinstimmt. Bemerkenswert ist noch, daß der Regressionskoeffizient aus Abb. 2 ($b = 3{,}5$) dem allerdings einen Einzelwert repräsentierenden spez. PV von KKA (3,238) sehr nahe kommt.

Die Abb. 3 und 4 zeigen statistisch signifikante Zusammenhänge zwischen HS bzw. PV und FG. Aus beiden Darstellungen ist ferner ersichtlich, daß Eisen an der Farbbildung Anteil haben muß, da jeweils bei Null HS bzw. PV noch ein Rest-FG vorhanden ist.

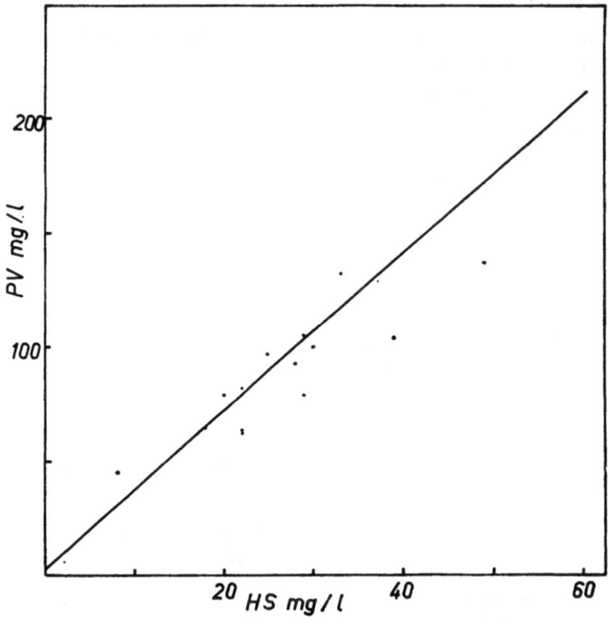

Abb. 2. Zusammenhang zwischen Huminstoffgehalt und Oxydabilität in den Kranichseehochmooren

$$y = 2{,}3 + 3{,}5x \; ; \qquad r = 0{,}91$$

Aus den geschilderten Beziehungen zwischen HS, PV und FG kann man schlußfolgern, daß alle drei angewandten Bestimmungsmethoden zumindest in den Kranichseemooren reale Aussagen über die HS-Konzentration im Wasser ermöglichen.

Im Wildenhainer Bruch war der HS-Gehalt nur gering (s. Tab. 3). Die Ursache dürfte in der vollständigen Austorfung des Moores zu sehen sein. Eindeutige Beziehungen zur Witterung ließen sich nicht erkennen. Die Gewässer sind hier oligohumos.

Aus Abb. 5 ersieht man eine klare, lineare Beziehung zwischen HS und PV im W. Ungefärbte organische Stoffe sind hier stärker vertreten.

Da auch der Z fast völlig ausgetorft ist, ist das in den Torfstichen stehende Wasser so gut wie ungefärbt. In den Schlenken wurden dagegen stets hohe HS-Konzentrationen festgestellt, während das Wasser des Zadlitzgrabens sehr wenig HS enthielt (s. Tab. 3).

Aus Abb. 6 geht hervor, daß auch im Z ein signifikanter Zusammenhang zwischen HS und PV besteht, wobei der Anteil farbloser organischer Substanzen erheblich

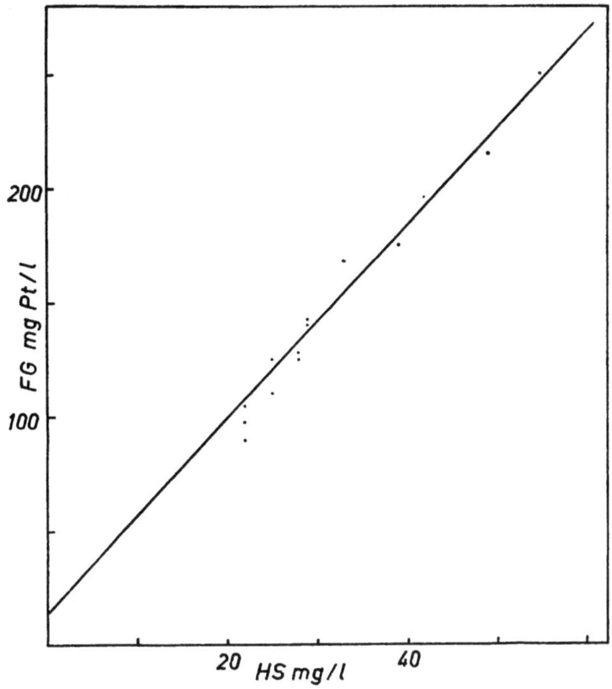

Abb. 3. Zusammenhang zwischen Huminstoffgehalt und Farbgrad in den Kranichseehochmooren

$y = 14{,}6 + 4{,}2x$; $r = 0{,}97$

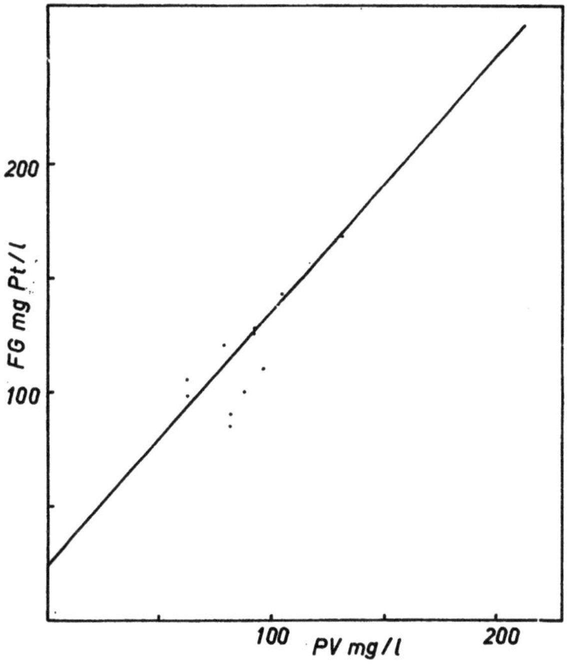

Abb. 4. Zusammenhang zwischen Oxydabilität und Farbgrad in den Kranichseehochmooren

$y = 22{,}5 + 1{,}1x$; $r = 0{,}89$

Tab. 3. Huminstoffgehalt im Wildenhainer und Zadlitzbruch

	HS [mg/l]	PV [mg/l]	FG [mg/l]
WF	4 und 7	30–73 (46)	20 und 26
WA	2,5–14 (7,5)	21–70 (40)	20–60 (38)
ZS + ZN	25–60 (34)	163–300 (203)	150–285 (206)
ZG	3–7 (4)	26–50 (33)	20–40 (30)

ist. Der Regressionskoeffizient ($b = 3{,}8$) unterscheidet sich nicht wesentlich vom spez. PV (ZS in Tab. 2) nach KLÖCKING.

Daß der hohe Eisengehalt (s. S. 71) in den ZS bedeutenden Einfluß auf die Farbe des Wassers nimmt, geht aus Abb. 7 hervor.

Auf Grund der wenigen verfügbaren Werte über die HS-Konzentration in den Braunwässern bei Serrahn muß eine Darstellung möglicher Beziehungen zwischen HS, PV und FG unterbleiben.

Die Analysen ergaben folgende Werte: HS: 32–52 mg/l, PV: 121–490 mg/l und FG: 137–200 mg Pt/l. Es konnte beobachtet werden, daß in den Serrahner Moorgewässern die HS nicht nur dem Torf entstammen, sondern daß sie auch durch Abbau von Laub entstanden sein müssen, denn die höchsten Werte wurden in den stark laubexponierten Wasserteilen der Moore gefunden.

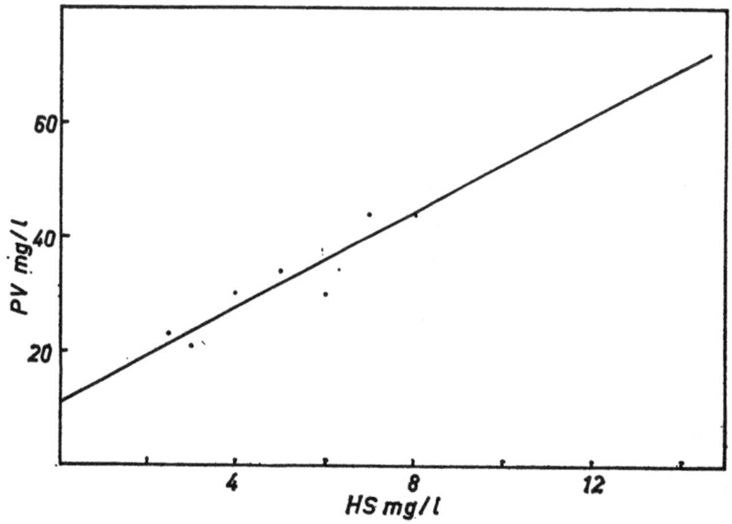

Abb. 5. Zusammenhang zwischen Huminstoffgehalt und Oxydabilität im Wildenhainer Bruch
$y = 10{,}8 + 4{,}2x$; $r = 0{,}99$

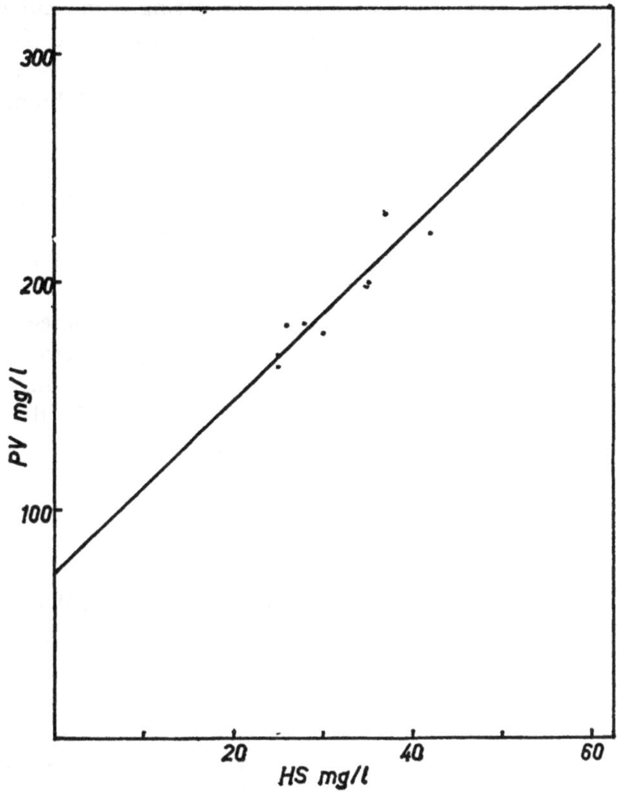

Abb. 6. Zusammenhang zwischen Huminstoffgehalt und Oxydabilität im Zadlitzbruch
$y = 72{,}7 + 3{,}8x$; $r = 0{,}98$

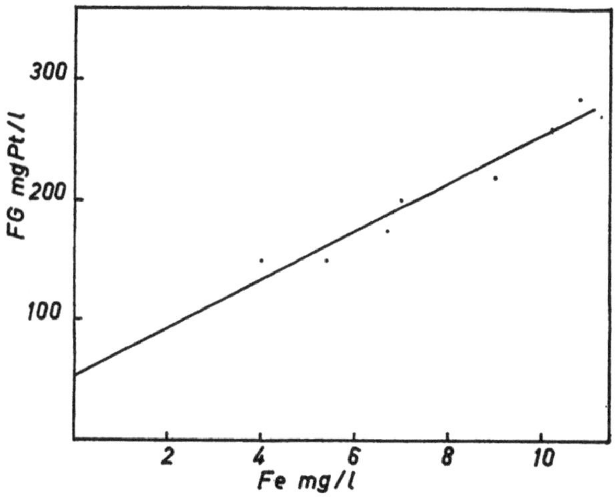

Abb. 7. Beziehungen zwischen Eisengehalt und Farbgrad im Zadlitzbruch
$y = 52 + 20{,}3x$; $r = 0{,}97$

Im Herthasee war die Sichttiefe mit 90 cm sehr gering. Eine Zunahme der Farbe bzw. des PV mit der Tiefe konnte nicht festgestellt werden. Die Konzentration der organischen Stoffe nahm im Gegenteil von 4 m an etwas ab (von 85 auf 72 mg $KMnO_4$ je l). Der hohe FG bei 9,5 m (130 mg/l) ist auf Fe(III) zurückzuführen. Leider sind die Werte nicht sehr zuverlässig, da die Proben erst eine Woche nach der Entnahme auf ihren HS-Gehalt untersucht werden konnten, so daß sie nur ungefähre Anhaltspunkte bieten. Das scheint auch der Grund dafür zu sein, daß hier im Gegensatz zu vielen anderen Wasserinhaltsstoffen keine everse Schichtung nachgewiesen werden konnte. Allerdings war schon beim Schöpfen der Proben die gleichmäßige Farbintensität von 0—8 m aufgefallen.

Wie schon OHLE (1934) festgestellt hat, ist auch im Herthasee der Laubeintrag durch den umgebenden Buchenwald neben dem Zufluß von Moorwasser für die allochthone Zufuhr der braunen organischen Stoffe verantwortlich. OHLE fand bei seiner Untersuchung eine starke Farbschichtung und schon an der Oberfläche mit 150 mg/l PV einen höheren HS-Wert als den, der bei der eigenen Untersuchung ermittelt wurde (83 mg $KMnO_4$/l $\triangleq$ 24 mg HS/l).

Die Sichttiefe in der Straminke betrug 60 cm, war also noch geringer als im Herthasee. In Übereinstimmung damit war der HS-Gehalt höher. In 0 und 1 m Tiefe wurden 33 bzw. 34 mg/l HS gefunden.

Es gibt an der Ostseeküste viele Gewässer mit stark braunem Wasser. Die Herkunft der oft in ungeheuren Mengen gelösten HS ist nicht immer leicht zu erklären. MÜNCH & PETZOLD (1955/56) beobachteten bei der Untersuchung von Grundwasser der Bessiner Sandbank (Hiddensee) eine deutliche Farbvertiefung vom Strand nach dem Landesinneren zu. Sie vermuten, daß die starke Färbung des Grundwassers in den strandferneren Isotelen auf faulende Anwurfstreifen im Sand zurückzuführen ist. Man könnte Ähnliches auch für die Straminke annehmen.

Ganz extrem hohe HS-Konzentrationen wurden in zwei Braunwassertümpeln gefunden. B1 befindet sich zwischen Ahrenshoop und Born, etwa 200 m vom Strand. B2 liegt mitten im Darßwald.

	B 1	B 2
HS	470 mg/l	3300 mg/l
PV	835 mg/l	4150 mg/l
FG	1709 mg/l	8000 mg/l

Im Tümpel B1 entstammen die HS sicher dem Abbau des dort massenhaft gedeihenden Adlerfarns.

Auffallend niedrig ist das Verhältnis von PV:HS, wie es sonst nie gefunden wurde. OBENAUS' Werte (1962) von Wässern der küstennahen Rostocker Heide sind ebenso niedrig.

4.4. Eisen und Mangan

Diese beiden Schwermetalle kommen in den Oberflächenschichten der Klarwässer nur in Spuren vor. In Braunwässern sind dagegen selbst an der Oberfläche stets höhere Fe-Konzentrationen zu finden, weil das Eisen durch die HS offenbar in Lösung gehalten wird, was auch für das Mangan zutrifft.

Mittlere Fe-Konzentrationen wurden in den Schlenken der Kranichseen festgestellt:

im KK 0,25—0,55 mg/l und im GK 0,4—1,05 mg/l.

Das Eisen entstammt hier natürlich ausschließlich dem *Sphagnum*-Torf, der ziemlich reich an diesem Element ist. HARNISCH (1929) und WEBER (1902; zit. n. KIVINEN 1935) bringen ähnliche Fe-Werte von Hochmoorgewässern.

Obwohl sich Fe(II) und Fe(III) mit der Rhodanidmethode nicht sicher trennen lassen (s. VERBEEK 1961), kann doch gesagt werden, daß sowohl Fe(II) als auch Fe (III) nachgewiesen werden konnten und daß die O_2-Spannung des Wassers wenig Einfluß auf die Verteilung der beiden Wertigkeitsstufen hatte. Selbst bei O_2-Übersättigung bis 144% wurden höhere Fe(II)- als Fe(III)-Gehalte gefunden.

Nach GJESSING (1964) befand sich in den von ihm untersuchten Braunwässern (Phenanthrolin-Methode) fast sämtliches Fe im 2wertigen Zustand, doch konnte SHAPIRO (1966b) wahrscheinlich machen, daß es sich hierbei um analytische Fehler GJESSINGS handelte. Anderseits stellte auch KJENSMO (1968) hohe Fe(II)-Konzentrationen in einem Braunwassersee fest, so daß SHAPIROS (1966b) Ansicht vom fast ausschließlichen Vorkommen von Fe(III) in Humusgewässern kaum zutreffen dürfte.

Durch Grundwassereinfluß bedingt sind die vor allem bei Trockenheit hohen Fe-Werte im GKA: 1,25—16,6 mg/l. Für die unterhalb des Großen Kranichseemoors gelegene Talsperre Carlsfeld ist es bedeutungsvoll, daß im GKA Eisen in größeren Mengen auftritt. Bei nicht zu hoher Wasserführung werden die aus dem Moor stammenden HS früher oder später noch im Bach zusammen mit dem Eisen ausgefällt. Das Resultat ist ein sehr klares Talsperrenwasser mit Sichttiefen bis 5,25 m (HÖHNE 1966 — persönl. Mitt.). In niederschlagsreichen Perioden funktioniert diese natürliche „Fällanlage" nicht mehr, weil das Fe soweit verdünnt wird, daß es keine Entfärbung bewirken kann; dann gelangt allerdings stark braunes Wasser in die Sperre.

Mangan konnte in den Schlenken nur einmal nachgewiesen werden: KK_1 0,1 mg/l. Die höchste Konzentration fand sich im GKA mit 0,44 mg/l (neben 12,4 mg Fe/l). Die Probe hatte mit 350 mg Pt/l eine starke Färbung, an der das Mn als Mn(IV) maßgeblich beteiligt war.

Im Wildenhainer Bruch waren die Fe-Konzentrationen auf Grund der Oligohumosität des Wassers niedrig (0,05—0,9 mg/l). Die dominierende Form war hier das Fe(III).

In den Zadlitzschlenken schwankten die Fe-Werte zwischen 4,0 und 10,8 mg/l. Das Verhältnis Fe(II):Fe(III) war in der überwiegenden Zahl der Fälle stark zugunsten des Fe(II) verschoben, da durch das schwefelsaure Milieu und oft erhebliche O_2-Defizite eine Oxydation sehr erschwert ist. Das Fe(II) dürfte hier auch als Sulfat vorliegen.

Wurde das Wasser einige Wochen stehengelassen, so zeigte es eine milchige Trübung, die nach Säurezusatz wieder verschwand. Demnach war das gelöste Eisen hier nur bei den sehr niedrigen natürlichen pH-Werten stabil. Ähnliches konnte auch im GKA beobachtet werden.

Mangan war im W nicht feststellbar. In ZS ergab eine Analyse 0,22 mg/l.

In den Serrahner Braunwässern fanden sich Fe-Gehalte um 0,5 mg/l. Mangan war in keinem Fall nachweisbar.

Wie aus Abb. 1d hervorgeht, zeigen Fe(III) (Ges.-Fe wurde nicht bestimmt) und Mn im Herthasee eine deutliche Schichtung. Besonders in der bodennahen Wasserschicht nimmt der Fe-Gehalt rapide zu. Da dasselbe vom CO_2 zu konstatieren ist (Abb. 1a), dürfte es sich zum Teil um $Fe(HCO_3)_2$ handeln.

Die Eisenkonzentrationen in der Straminke lagen bei 0,3 mg/l.

4.5. Phosphor- und Stickstoffverbindungen

In Tab. 4 sind die während der Untersuchungsperiode in den Kranichseemooren vorgefundenen P- und N-Konzentrationen zusammengefaßt.

Tab. 4. Ergebnisse der P- und N-Analysen in der Kranichseemooren

	PO_4-P [µg/l]	Ges.-P [µg/l]	NO_3-N [mg/l]	NH_4-N [mg/l]	org. N [mg/l]
GKS	0–22 (8)	14–44 (36)	0 –0,14 (0,05)	0,22–0,9 (0,45)	1,2–2,5 (2,0)
GKA	4–30 (16)	12–68 (34)	0,03–0,16 (0,09)	0,4 –0,9 (0,55)	1,3–2,5 (1,7)
KKS	0–10 (3)	10–45 (25)	0 –0,10 (0,03)	0 –1,0 (0,35)	1,3–2,6 (1,9)
KKA	3–27 (11)	16–56 (30)	0,01–0,12 (0,06)	0,16–0,5 (0,34)	1,2–2,7 (2,1)

Was das Verhältnis des PO_4-P zum Ges.-P anbetrifft, so fällt auf, daß es in den Schlenken jeweils viel kleiner ist als in den Abflußbächen. Im GK beträgt der Anteil des PO_4-P am Ges.-P 22% (Schlenke) und 50% (Abfluß). Im KK sind die entsprechenden Zahlen 12% und 37%. Diese Unterschiede lassen sich damit erklären, daß das Bachwasser einerseits durch seine Berührung mit dem mineralischen Untergrund Orthophosphat aufnehmen kann und andererseits in den Schlenken eine starke biogene Zehrung in der Vegetationsperiode den PO_4-Gehalt herabsetzt.

Nach Untersuchungen von OHLE (1924), ÅBERG & RODHE (1942) und PEUKERT (1964) ist es wahrscheinlich, daß zwischen dem Gehalt an HS und P eine Proportionalität besteht.

Wie Abb. 8 zeigt, existiert auch in den Braunwässern des GK und KK ein statistisch gesicherter Zusammenhang zwischen HS und Ges.-P, mit allerdings erheblicher Streuung der Einzelwerte. Da bewiesen ist, daß die Huminstoffe des Wassers P-haltig sind (B. SCHNEE 1968), war eine solche Beziehung zu erwarten. Die Näherungsgerade schneidet die y-Achse bei 9,1. Dieser Wert entspricht dem durchschnittlichen Orthophosphat-Gehalt und ist fast identisch mit dem aus Tab. 4 erhaltenen Gesamtdurchschnittswert (9,5 µg PO_4-P/l).

Der Jahresgang des PO_4-P-Gehaltes in den Kranichseemooren (Abb. 9) weist den üblichen Verlauf auf. Im Frühjahr und Sommer waren die Werte auf Grund organismischen (insbesondere planktischen) Verbrauchs niedrig, in der vegetationslosen Zeit dagegen relativ hoch (s. auch GESSNER 1933).

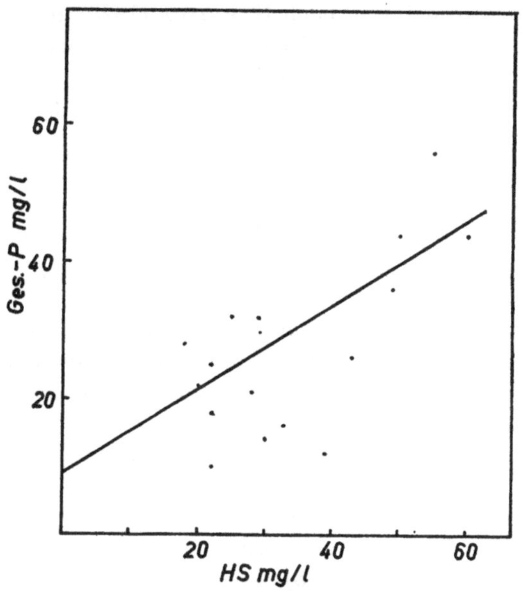

Abb. 8. Beziehungen zwischen Huminstoff- und Gesamt-P-Gehalt in den Kranichseehochmooren
$y = 9{,}2 + 0{,}62x$; $r = 0{,}60$

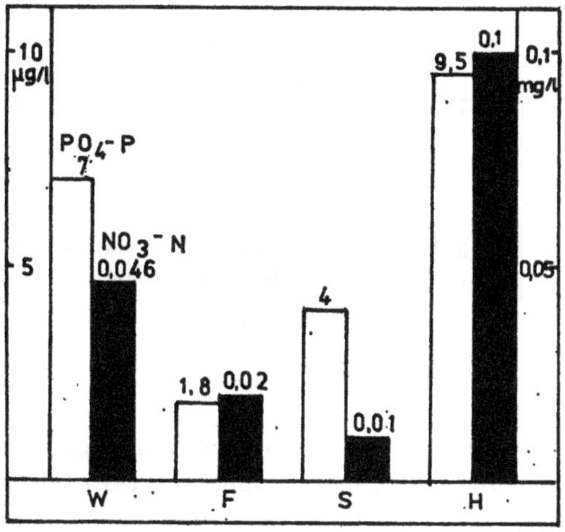

Abb. 9. Jahresgang des Orthophosphat-P und des Nitrat-N in den Kranichseehochmooren

Bei der Betrachtung des Stickstoff-Haushaltes im GK und KK fällt der sehr niedrige Anteil des Nitrat-N auf (Abb. 10a). Wie die Phosphat- sind auch die Nitratwerte in den Schlenken niedriger als in den Bächen (Tab. 4). Wiederum ist biogene Zehrung dafür verantwortlich. Die höchsten NO_3-N-Mengen wurden im niederschlagsreichen Jahr 1965 festgestellt. Das Regenwasser ist in Hochmooren die einzige Nitratquelle, da nitrifizierende Bakterien völlig fehlen. Zwei Analysen im Frühjahr 1966 ergaben 0,3 und 0,48 mg/l. Der Jahresgang des NO_3-N (Abb. 9) ist in erster Linie vom

Phytoplankton abhängig. Doch sind nach HARNISCH (1929) und BENADE (1954) auch denitrifizierende Bakterien an der Beseitigung des Nitrats beteiligt. Der relativ hohe NH_4-N-Gehalt des Hochmoorwassers (Abb. 10a) entsteht bei der Zersetzung der Torfbildner durch eiweißabbauende, ammoniakbildende Mikroben (BENADE 1954). Das NH_4^+-Ion scheint die einzige N-Quelle von Bedeutung für die Pflanzenwelt der Hochmoore zu sein. Während der Phytoplanktonblüten im Juni war in zwei Schlenken des KK aber auch kein Ammoniak mehr nachweisbar.

Der organisch gebundene Stickstoff ist die bei weitem überwiegende N-Form der Hochmoorgewässer (Abb. 10a). Über die Umwandlung in NH_4—N könnte er für die Pflanzenernährung von Nutzen sein. Sofern der organisch gebundene N gelösten Proteinen oder Aminosäuren angehört (letztere konnten in Braunwässern von POVOLEDO 1964 nachgewiesen werden), kann er leicht zu Ammoniak reduziert werden. Als eventueller Bestandteil von HS ist er sicher sehr fest gebunden (SCHREIBER 1927) und für Bakterien schwerer angreifbar.

Nitrit ist in Hochmoorgewässern offenbar ohne Bedeutung, denn die bei Stichproben gefundenen Konzentrationen bewegten sich um die Nachweisgrenze.

Entsprechend dem oligotrophen Charakter von WF und WA sind die P-Werte hier niedrig. Jahreszeitliche Schwankungen wurden deshalb nicht beobachtet. Dieselben Verhältnisse weist der ZG auf (s. Tab. 5).

Tab. 5. Ergebnisse der P- und N-Analysen in Wildenhainer- und Zadlitzbruch

	PO_4-P [μg/l]	Ges.-P [μg/l]	NO_3-N [mg/l]	NO_2-N [mg/l]	NH_4-N [mg/l]	org. N [mg/l]
WA	3—14 (7,7)	7—30 (17)	0,02—0,12 (0,06)	0	0 —0,35 (0,09)	1,65—1,75 (1,7)
WF	0—14 (6)	15—25 (20)	0,02—0,05 (0,03)	0	0; Sp.	1,4
ZS + ZN	4—10 (6,5)	12—58 (27)	0,09—0,19 (0,09)	0,02—0,09 (0,04)	0,28—3,0 (1,9)	2,4 —4,5 (3,4)
ZG	0—7 (6,5)	2—12 (7)	0,03—0,12 (0,07)	0	0,3—0,65 (0,47)	1,6

In den ZS wäre wohl eine größere Orthophosphatmenge zu erwarten gewesen, da Fallaub ins Wasser gelangt. Aber einmal erfolgt die P-Freisetzung aus dem Laub in dem stark schwefelsauren Milieu nur langsam und zum anderen ist der Verbrauch durch permanente *Tribonema*-Bestände sicher erheblich.

Der weitaus größte Teil des P ist organisch gebunden (Tab. 5). Eine Parallelität zwischen dem Ges.-P und den HS war angedeutet, doch auf Grund zu geringen Materials statistisch nicht ganz signifikant.

Die Nitratkonzentrationen im W und Z liegen in etwa der gleichen Größenordnung (s. Tab. 5). Eine Regenwasseranalyse aus dem Z ergab 0,57 mg NO_3–N/l. Dementsprechend wurden die Höchstwerte während oder nach Niederschlägen gefunden. Es war zu erwarten, daß in den extrem sauren Braungewässern des Z nur minimale Nitratmengen vorliegen können (s. auch Abb. 10b). Der ebenfalls sehr niedrige

NO_3–N-Gehalt in den zirkumneutralen WF und WA ist dagegen etwas überraschend, da hier eine Nitrifikation durchaus möglich sein müßte. Offenbar ist die bakterielle Denitrifikation im Schlamm stark (hohes Verhältnis von N-verbrauchender Schlammoberfläche zum Wasservolumen), während planktogener Verbrauch unbedeutend ist. Abb. 10b veranschaulicht den geringen Anteil der anorganischen N-Verbindungen am Ges.-N im W.

Im Gegensatz zum W spielen die reduzierten anorganischen N-Verbindungen im Z eine wichtige Rolle. Die hohe NH_4–N-Menge in den ZS entstammt absterbenden *Tribonema*-Watten. Der Anteil des Ammoniaks am Ges.-N ist mit 35% sehr hoch. Nitrit konnte hier bei jeder Untersuchung nachgewiesen werden.

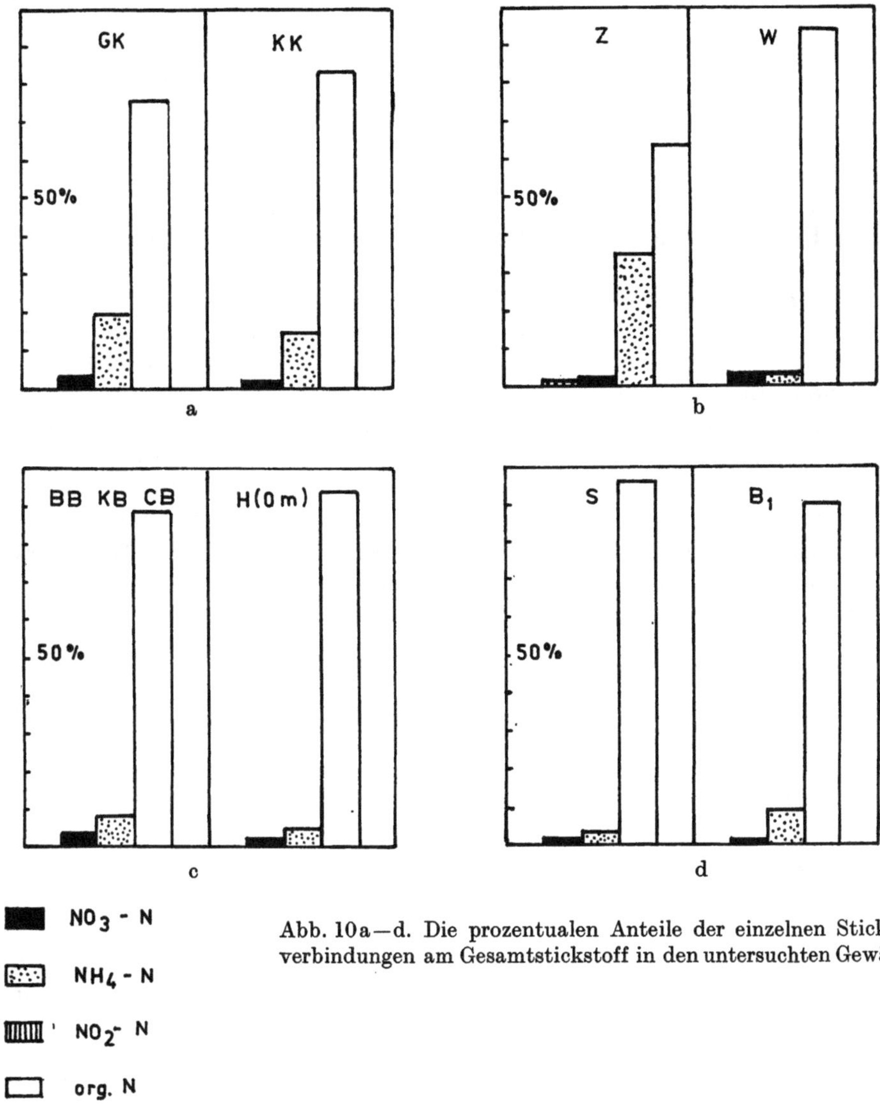

Abb. 10a—d. Die prozentualen Anteile der einzelnen Stickstoffverbindungen am Gesamtstickstoff in den untersuchten Gewässern

Der organisch gebundene N ist auch in den beiden Zwischenmooren die vorherrschende N-Form (s. Abb. 10b). Die erheblichen Mengen dürften auf den Abbau der Fadenalgenbestände zurückzuführen sein.

Die Serrahner Moorgewässer unterscheiden sich grundlegend durch ihren sehr hohen P-Standard von den bisher behandelten (s. Tab. 6).

Tab. 6. Ergebnisse der P- und N-Analysen in den Serrahner Gewässern

	PO_4-P [µg/l]	Ges.-P [µg/l]	NO_3-N [mg/l]	NH_4-N [mg/l]	org. N [mg/l]	PV [mg/l]
BB	620; 560	800; 710	0,09; 0,09	0,4; 0,26	3,4; 2,5	220; 165
KB	230	300	0,09	0,32	2,2	121
CB	225	280	0,1	Sp.	2,4	167

Die hohen P-Werte sind durchaus verständlich, wenn man weiß, welche enormen P-Mengen im Laub inkarniert sind (s. WARNKE 1964). Der Anteil des PO_4–P am Ges.-P ist hoch und in allen Proben etwa gleich (77—80%). Daß gerade der BB die höchsten P-Konzentrationen aufzuweisen hat, verdankt er der relativ stärksten Laubzufuhr.

Der gleichmäßig niedrige NO_3–N-Gehalt ist im Sommer vorwiegend auf bakterielle und im Frühjahr vor allem auf planktogene Zehrung zurückzuführen. Am Gesamt-N hat das Nitrat auch hier nur einen minimalen Anteil. Nitrifikation dürfte in dem mäßig sauren Milieu noch keine Bedeutung haben.

Es wurde für Fallaubgewässer erstaunlich wenig NH_4–N gefunden, während organisch gebundener N die wichtigste N-Form darstellt (vgl. Abb. 10c).

Abb. 1d zeigt die Schichtung des PO_4–P und des Fe(III) im Herthasee. Von einem Antagonismus der beiden Stoffe kann hier keine Rede sein. Die Konzentrationen nehmen vielmehr sowohl beim PO_4–P als auch beim Fe(III) mit der Tiefe stark zu. Die HS halten das Fe in Lösung und verhindern die Fällung des Phosphats.

Trotz starker Zehrung durch eine Phytoplanktonblüte kommt das Orthophosphat auch an der Oberfläche in reichlichem Maße vor. Die Ursache des hohen P-Standards des Sees ist starker Laubeintrag im Herbst (s. OHLE 1934).

Der Ges.-P nimmt ebenfalls mit der Tiefe zu: 56—176 µg/l. Der PO_4–P ist mit über 50% wesentlich beteiligt.

Die oberen Schichten des Sees sind infolge planktogenen Verbrauchs stark an Nitrat verarmt (s. Abb. 1c). Von 4—8 m Tiefe nimmt es jedoch zu, da die Zehrung durch das Plankton aufgehört und die bakterielle noch nicht ihr Höchstmaß erreicht hat. Am Grund dagegen ist das Nitrat fast völlig verschwunden. Einer Nitrifikation dürfte im Herthasee nichts im Wege stehen, so daß der Verbrauch ständig kompensiert werden kann.

Die reduzierten anorganischen N-Verbindungen nehmen mit der Tiefe zu, doch ist NO_2–N erst ab 6 m nachweisbar (s. Abb. 1c). OHLE fand während der Sommerstagnation ebenfalls eine NH_4–N-Stratifikation.

Beim organisch gebundenen N ist eine Schichtung nicht klar zu erkennen. Die Werte sind, wenn man den relativ niedrigen PV betrachtet, recht hoch (2,4—2,6 mg/l).

In 9,5 m Tiefe wurde mit 2,0 mg N/l die niedrigste Konzentration gefunden, aber die größte NH_4–N-Menge. Es scheint hier also ein erheblicher Teil des org. geb. N zu NH_4–N reduziert zu werden. Auch KARCHER (1939) beobachtete eine solche Abnahme des org. geb. N in Sedimentnähe, die bei O_2-Schwund besonders stark war (vgl. NÜMANN 1941).

Abb. 10c orientiert über die Anteile der einzelnen N-Verbindungen am Ges.-N bei 0 m.

Die Straminke wird als Fischzuchtweiher gedüngt, so daß die hohe PO_4–P-Konzentration (280 µg/l) verständlich ist. Der Ges.-P erreicht bei 0 m 405 µg/l.

An N-Verbindungen ergab die Stichprobe (0 m):

NO_3–N	0,02 mg/l
NH_4–N	0,1 mg/l
org. N	3,3 mg/l (vgl. Abb. 10d).

Auch im Sommer herrscht akuter Mangel an NO_3–N (0,03 mg/l), wie anläßlich einer Exkursion im Juli 1964 festgestellt werden konnte. Diese Armut an assimilierbarem Stickstoff ist das Resultat eines erheblichen biogenen Verbrauchs.

Abb. 10d zeigt, daß auch im B_1 ähnliche Relationen der N-Verbindungen bestehen wie in den anderen Braunwässern.

4.6. Silizium

Wie bei den meisten anderen mineralischen Komponenten wurden die kleinsten Si-Mengen in den Hochmoorgewässern gefunden:

BB	2,5 mg/l
KB	4,0 mg/l
GKA	4,4 mg/l.

Im W ergaben zwei Stichproben 9 und 15 mg Si/l.

Im Herthasee war eine ausgeprägte Schichtung der Kieselsäure zu verzeichnen (Abb. 1b). Eine Massenentfaltung von *Stephanodiscus hantzschii* GRUN. ließ das Wasser der oberen Schichten stark an Si verarmen. Im Hypolimnion ging die Kieselsäure aus den absinkenden zarten Schalen wieder in Lösung, wie der starke Anstieg bei 9,5 m beweist. Die Untersuchung des Sediments ergab dementsprechend im Gegensatz zum Plankton nur ganz vereinzelt *Stephanodiscus*-Zellen.

4.7. Sulfat

Da das Sulfat in den starksauren Erzgebirgshochmooren (außer durch Absorption von atmosphärischem SO_2) nur durch bakterielle und Luftoxydation des beim Eiweißabbau gebildeten H_2S in geringen Mengen entstehen kann, wurden hier die niedrigsten Sulfatkonzentrationen festgestellt. Schwefelwasserstoff war nur im Schlamm nachweisbar, da die O_2-Spannungen im Wasser nie so tief sanken, daß es hier auftreten konnte. Ein Teil des Sulfats tritt als freie Schwefelsäure in Erscheinung und bedingt die hohe H^+-Ionenkonzentration des Moorwassers.

In den Serrahner Mooren wurden etwas höhere Sulfatgehalte registriert (um 20 mg/l.) Sehr hohe Werte wurden mit 132—193 mg/l im Zadlitzbruch ermittelt. Sie lassen auf einen gipshaltigen Untergrund schließen, da das Kalzium ebenfalls reichlich vertreten war (s. Abb. 11c). Der stets sehr niedrige pH-Wert ist auch hier auf freie Schwefelsäure zurückzuführen. Die H_2S-Bildung war häufig intensiv.

Bis in 8 m Tiefe pendelte der Sulfatgehalt im Herthasee um 20 mg/l, um am Grund mit 45 mg/l ein Maximum zu erreichen. Schwefelwasserstoff fehlte sowohl im Wasser als auch im Schlamm.

Die Sulfatkonzentration von 117 mg/l in der Straminke entspricht den Brackwassergegebenheiten.

Auf die Chlorid-, Natrium- und Kaliumuntersuchungen soll hier nicht näher eingegangen werden. Erwähnt sei nur, daß sich eine Reihe steigender Konzentrationen von den Kranichseehochmooren über die Serrahner Moore, die beiden Zwischenmoore und den Herthasee bis zur halinen Straminke ergab. Die Relationen auch dieser Ionen sind aus Abb. 11a—g ersichtlich.

4.8. Kalzium und Magnesium

Die Gewässer der ombrogenen Erzgebirgshochmoore sind verständlicherweise sehr arm an Erdalkalien (0,9—6,4 mg Ca/l und 0,6—3,0 mg Mg/l).

Der Erdalkalistandard der Serrahner Moore ist mit 1,4—10,7 mg Ca/l und 1,7 bis 5,4 mg Mg/l etwas höher.

Die hohen Werte im Zadlitzbruch (38—59 mg Ca/l und 4,8—12,6 mg Mg/l) sind auf den gipshaltigen Untergrund zurückzuführen. Im Wildenhainer Bruch sind die entsprechenden Konzentrationen um etwa die Hälfte niedriger.

Das Kalzium ist im Herthasee deutlich geschichtet (Abb. 1b). Diese Schichtung ist das Ergebnis der starken Assimilationstätigkeit des Phytoplanktons, wodurch an der Oberfläche biogene Entkalkung erfolgt und das absinkende $CaCO_3$ sich im Hypolimnion wieder als $Ca(HCO_3)_2$ löst.

In der Straminke übertrifft der Mg-Gehalt mit 52 mg/l die Ca-Konzentration mit 49 mg/l. Dieselbe Tatsache wurde im B 1 festgestellt, nur sind die absoluten Konzentrationen viel geringer.

4.9. Gesamtelektrolytgehalt und Ionenrelationen

Als Maß für den Elektrolytgehalt der Humusgewässer diente die elektrolytische Leitfähigkeit, wobei natürlich zu beachten ist, daß bei niedrigem pH auch H^+-Ionen erfaßt werden (Tab. 7).

Die regelmäßig untersuchten Gewässer zeigen meist eine deutliche Abhängigkeit des Elektrolytgehalts von den Niederschlägen. Bei Trockenheit werden entgegengesetzt dem HS-Gehalt die höchsten Werte gefunden. Besonders große Schwankungen zeigen die sehr elektrolytarmen Hochmoorgewässer, da sie auf Grund ihrer Flachheit und exponierten Lage der Austrocknung oder Regengüssen am stärksten ausgesetzt sind.

Tab. 7. Elektrolytische Leifähigkeit der untersuchten Gewässer [μScm⁻¹]

GKS	43– 91	ZN	824
GKA	52– 87,5	ZG	268 u. 316
KKS	50–118	KB	32
KKA	68–112	BB	68,5 u. 74
		CB	120
WF	206 u. 251	H	186–236 (0–9,5 m)
WA	181–243	S	2110
ZS	455–531	B1	222

Die Abhängigkeit der Elektrolytkonzentration des Hochmoorwassers von der Witterung ist in der Literatur oft herausgestellt worden (z. B. KIVINEN 1935, GORHAM 1956b).

Im Z und W finden sich viel höhere Leitfähigkeitswerte. In beiden Fällen, besonders stark aber im Z, sind sie durch gipshaltigen Untergrund bedingt (s. Abb. 11c).

Es erscheint wegen des elektrolytreichen Wassers kaum möglich, daß die beiden Moore wieder zu echten Hochmooren regenerieren können, zumal ja die Niederschlagsmenge gering ist.

Im Herthasee zeigt die Leitfähigkeit eine Schichtung, die vor allem durch die Stratifikationen des Ca^{++} und der Alkalimetalle verursacht wird (s. Abb. 1b). Wie OHLE (1934) feststellen konnte, ist die Elektrolytkonzentration während der Sommerstagnation noch stärker geschichtet.

Entsprechend ihrem Brachkwassercharakter hat die Straminke die weitaus höchste Leitfähigkeit.

Die Ionenfelddiagramme (Abb. 11a–g) zeigen, daß von allen untersuchten Braunwässern nur der Herthasee die RODHEsche Standardionenkomposition repräsentiert, also ein typisches Bikarbonatgewässer darstellt (Abb. 11e). Er ist auch das einzige Gewässer, in dem alle acht erfaßten Hauptionen vertreten sind.

Im W erreicht und übertrifft das Sulfat in seinem Äquivalentanteil das HCO_3^-. Es handelt sich also um gemischte $Ca(HCO_3)_2$- und $CaSO_4$-Gewässer (Abb. 11b). Ähnliche Ionenrelationen kommen im BB vor, obwohl die absoluten Konzentrationen viel geringer sind (Abb. 11d).

Die Braunwässer der beiden Kranichseemoore und die des Z sind als $CaSO_4$-Typen anzusprechen (Abb. 11a, 11c). Ihre Ionenverhältnisse sind praktisch dieselben, obwohl in den absoluten Konzentrationen sehr große Unterschiede bestehen. In den Hochmooren ist allerdings auch $MgSO_4$ stärker vertreten. Die Straminke weicht in ihrer Ionenkomposition vom Meerwasser ab, denn Na^+ und Mg^{++} sind etwa zu gleichen Teilen vorhanden (Abb. 11f).

Ein interessantes Verteilungsbild stellt das Diagramm des B1 dar (Abb. 11g). Es überwiegt hier Na_2SO_4, erst an zweiter Stelle folgt NaCl.

5. *Diskussion zum Nährstoffhaushalt der Braunwässer*

Die Untersuchungen über den O_2-Haushalt verschiedener humoser Gewässer ergeben, daß eine große Mannigfaltigkeit herrscht, von Anaerobie bis zu hochgradigen O_2-Übersättigungen, wobei

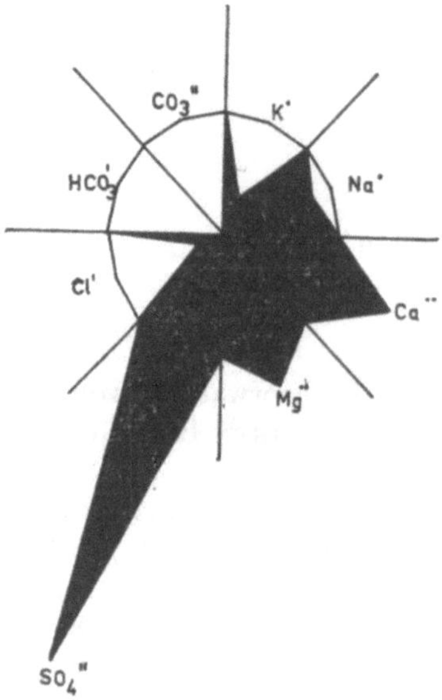

a

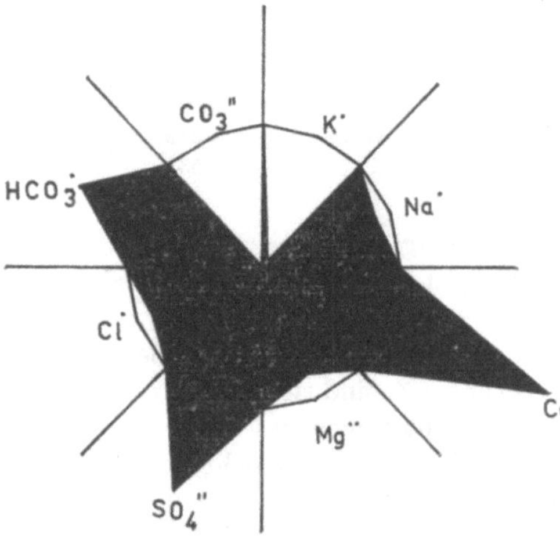

b

Abb. 11a—g. Ionenfelddiagramme. a — Kleines Kranichseehochmoor; b — Wildenhainer Bruch; c — Zadlitzbruch; d — Beilbruch; e — Herthasee; — f — Straminke; g — Braunwassertümpel 1

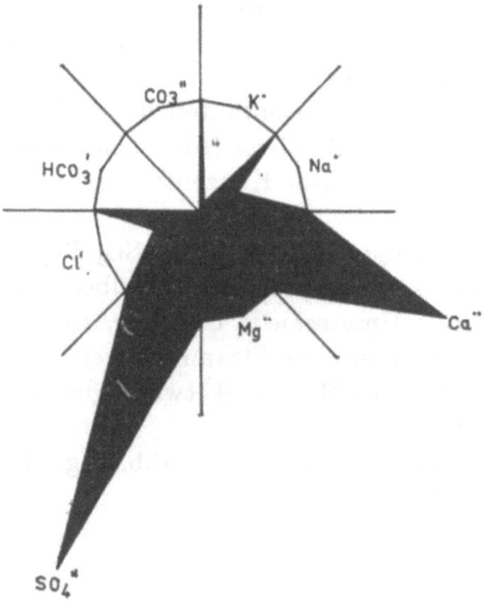

c

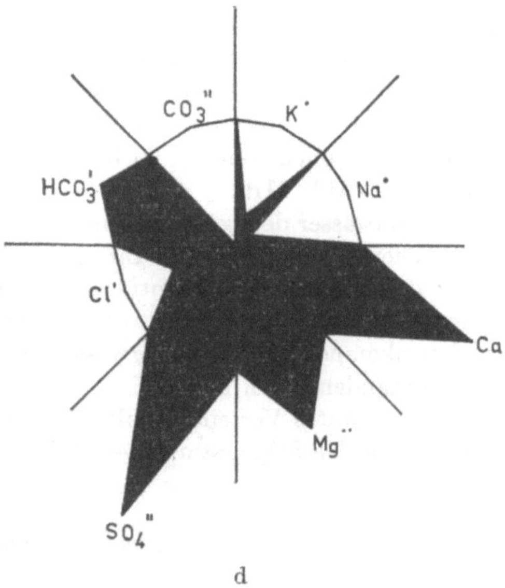

d

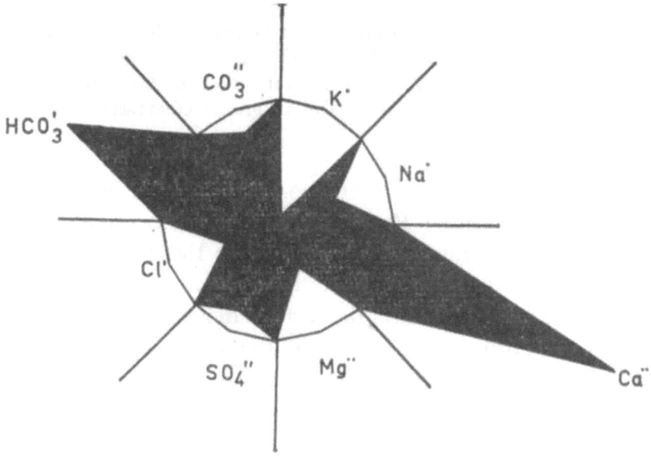
CO3''
K·
HCO3'
Na·
Cl'
SO4''
Mg··
Ca··

e

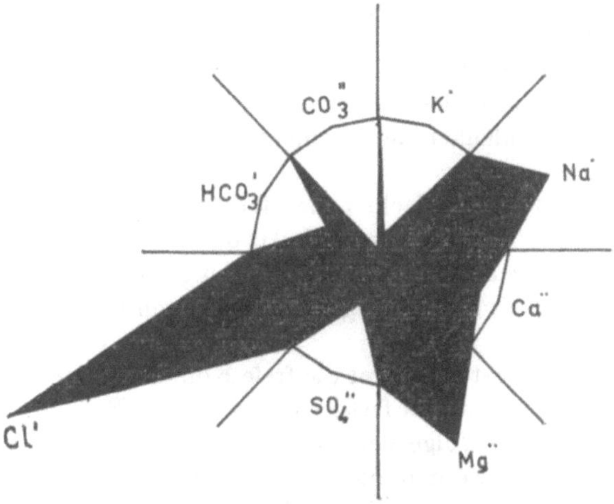
CO3''
K·
HCO3'
Na·
Ca··
Cl'
SO4''
Mg··

f

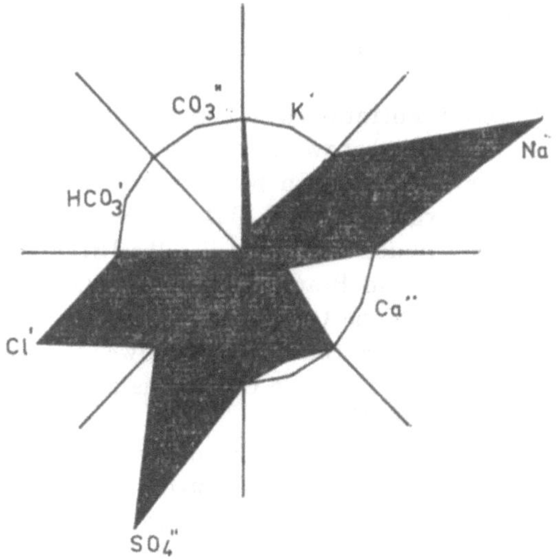
CO3''
K·
HCO3'
Na··
Ca··
Cl'
SO4''

g

die Abweichungen von der Sättigung nach unten überwiegen. In der Literatur findet man fast nur Angaben über mehr oder weniger starke O_2-Defizite (WILLER, zit. n. HARNISCH 1929, in Hochmoorblänken; GESSNER 1934 sowie ÅBERG & RODHE 1942 in südschwedischen und OHLE 1934 in norddeutschen dystrophen Seen). Dabei nimmt das Defizit an der Oberfläche mit steigendem Humusgehalt zu.

Größere Planktonentwicklungen haben die oben genannten Autoren nicht vorgefunden, so daß die O_2-zehrenden Prozesse die O_2-produzierenden stets überwogen. Während der Stagnationsperioden war die durch die Destruktion der HS verursachte Zehrung so stark, daß Sauerstoffschwund im Hypolimnion eintrat. In eutrophen humosen Seen, wie im Herthasee, kam dazu noch der erhebliche Sauerstoffverbrauch durch das absinkende Phytoplankton. Fast alle untersuchten Seen in Schweden, Deutschland und Wisconsin (JUDAY & BIRGE, zit. n. ÅBERG & RODHE) wiesen in den Stagnationsperioden eine klinograde O_2-Kurve auf. Doch haben LUNDBERG (1929), JÄRNEFELT (1932) und JUDAY & BIRGE (1932) (alle zitiert n. ÅBERG & RODHE 1942) einige wenige polyhumose Seen vorgefunden, die am Ende der Sommerstagnation noch reichlich Sauerstoff im Hypolimnion enthielten. Es wurde eine „Inagilität" der HS in diesen Seen vermutet, d. h., der Humus soll besonders schwer und langsam abbaubar sein. Die Ursache dieser Erscheinung könnte Nährstoffmangel sein. Ähnliches wurde ja in den Schlenken der Kranichseemoore beobachtet, und hier fehlte es vor allem an Nitrat.

Beim Herthasee liegt der Verdacht nahe, daß er zumindest im Untersuchungsjahr frühjahrsmeromiktisch war. Von einer Zirkulation konnte keine Rede sein, denn zwei Wochen nach der Eisschmelze müßte sie längst eingetreten sein. Es wurde aber eine deutliche Schichtung vieler Stoffe festgestellt, was auch trotz der wenigen Werte für den Sauerstoff zutraf.

Im Gegensatz zum Sauerstoff nimmt die freie Kohlensäure mit steigendem Humusgehalt zu, wie von vielen Autoren festgestellt wurde (z. B. GESSNER 1933, ÅBERG & RODHE 1942). Das ist eine Folge der Entstehung des CO_2 bei der Destruktion der organischen Stoffe. Doch werden diese Verhältnisse vom Plankton und von der Atmosphäre beeinflußt. So verschwindet die freie CO_2 im Frühjahr in den oberen Schichten des Herthasees durch den Verbrauch planktischer Diatomeen völlig. Aus demselben Grunde wird sie in der gleichen Jahreszeit in den Schlenken der Kranichseemoore stark dezimiert.

Die extrem hohen CO_2-Konzentrationen in den ZS und im GKA werden von verfaulenden Fadenalgen und Fallaub bzw. vom Grundwasser zusammen mit den HS verursacht. Daß in den Fallaubgewässern bei Serrahn ebenfalls viel CO_2 gelöst ist, erscheint selbstverständlich (vgl. WARNKE 1964). Dagegen überrascht der auch im Winter relativ geringe Gehalt in den KKS, der dieselbe Ursache haben dürfte wie das relativ niedrige O_2-Defizit. Auch in Braunwässern sind O_2- und CO_2-Gehalt Indikatoren von Abbauvorgängen im Sinne OHLES (1952). Doch kann über ihre Quantität nichts ausgesagt werden, weil andere Einflüsse (Grundwasser, organische Produktion, anorganische Oxydationsprozesse u. a.) den Gashaushalt beeinflussen.

In den meisten Fällen dürften Mineralsäuren für die niedrigen pH-Werte vieler humoser Gewässer verantwortlich sein, z. B. in den Kranichseemooren und im Zadlitzbruch (s. GORHAM 1956a, b, GORHAM & CRAGG 1960, TOLPA & GORHAM 1961). SHAPIRO (1957), BERG (1962) und BAYLY (1964) konnten jedoch nachweisen, daß auch HS allein pH-Werte unter 4 zu bewirken vermögen. Die eigenen Untersuchungsergebnisse haben gezeigt, daß HS bei pH-Werten von 3,2—9,2 im Wasser gelöst sein können. Das Vorkommen von Huminstoffen ist also nicht an bestimmte pH-Werte gebunden. Sowohl im stark sauren als auch im alkalischen Bereich werden z. T. hohe HS-Konzentrationen festgestellt. Nach BERG kommen im Kongo ebenfalls Braunwässer im pH-Bereich von 3,5—8 vor.

Es hat sich ferner ergeben, daß Humingewässer die verschiedensten Ionenrelationen und Elektrolytgehalte haben können, mit anderen Worten, das Vorkommen von HS im Wasser ist nicht auf niedrige Salzkonzentrationen beschränkt, wie z. B. GESSNER (1959) meint. Er schließt das aus seinen Untersuchungen am Rio Negro, dessen „Schwarzwasser" eine außerordentlich geringe Leitfähigkeit hat (etwa 10 µS). Weiter zitiert er WILLÉN (1954), der in schwedischen Seen eine umgekehrte Proportionalität zwischen Humus- und Elektrolytgehalt mit allerding sehr starker Streuung festgestellt hatte. Für einen bestimmten Gewässertyp trifft das sicher zu, doch kann man diese Ergebnisse nicht verallgemeinern. Denn auch BERG (1962) unterscheidet im Kongo Humingewässer mit sehr differenten Elektrolytmengen.

Bezüglich des Verhältnisses der HS zum Fe wurde festgestellt, daß hohe Eisenkonzentrationen fast stets an hohe HS-Gehalte gebunden sind, während die umgekehrte Relation nicht gegeben war. Über den Mechanismus der Bindung zwischen den organischen Stoffen und dem Eisen siehe Diskussion bei SCHNEE (1968).

Der Schwankungsbereich, innerhalb dessen bei den vorliegenden Untersuchungen HS gefunden wurden, war mit 2,5—3300 mg/l ungeheuer groß. Die extrem hohen Konzentrationen kamen in in Küstennähe gelegenen Kleingewässern vor. Daß sie an der Ostseeküste durchaus keine Ausnahme sind, geht aus den Analysen von OBENAUS hervor. Die Autorin gibt in vielen Fällen mehrere Hundert mg/l HS an. In größeren Gewässern und im Binnenland scheinen solche hochkonzentrierten HS-Lösungen sehr selten zu sein. Hier wurden in der überwiegenden Zahl der Fälle HS-Gehalte von einigen 10 mg/l gefunden, was gut mit zahlreichen Literaturangaben übereinstimmt, die etwa äquivalente $KMnO_4$-Oxydabilitätswerte bringen. In den echten Hochmoorgewässern sind die HS praktisch die einzigen organischen Stoffe, soweit sie mit dem PV erfaßbar sind (s. Abb. 2). Die Torfbildner werden also weitgehend humifiziert. In anderen Fällen, z. B. im Wildenhainer und Zadlitzbruch kommen neben den HS noch farblose organische Substanzen vor, doch überwiegen stets die HS.

Die in den Kranichseemooren zwischen HS, PV und FG jeweils festgestellten signifikanten, linearen Korrelationen lassen den Schluß zu, daß hier alle drei angewandten Untersuchungsverfahren zur Ermittlung des HS-Gehalts geeignet sind. Die HS-Konzentration wird ziemlich genau durch die Oxydabilität erfaßt. Die Farbe weist einen durch Fe(III) bedingten „Blindwert" auf, der aber relativ gering ist. Im Gegensatz dazu gibt der PV die HS-Belastung in den Übergangsmooren auf Grund unvollständiger Humifizierung weit ungenauer wieder. Bei hohem Eisengehalt, wie im Zadlitzbruch, kann der FG zur Charakterisierung der HS-Menge überhaupt nicht verwendet werden.

Interessant ist ein Vergleich der Permanganat- mit der Bichromat-Oxydabilität, wie er von SIMM (1969) angestellt wurde. Danach beträgt der Anteil des PV am CSB in polyhumosen Gewässern auf Grund des hohen Gehalts stark gefärbter und leicht oxydierbarer Fulvosäuren 50%. In oligohumosen Seen sinkt er wegen der größeren Menge farbloser bzw. schwachgefärbter Substanzen, die schwerer oxydierbar sind, auf 30%.

In dieser Arbeit konnten Bichromatanalysen nicht durchgeführt werden. Aus Abb. 2 geht aber hervor, daß die Permanganat-Oxydabilität in den Kranichseemooren fast ausschließlich auf Huminstoffe (Fulvosäuren) zurückgeführt werden kann, d. h. also, daß der PV hier, wie schon erwähnt, durchaus repräsentativ für den HS-Gehalt ist.

Bezüglich der Beziehungen des Phosphats zum Eisen herrschen in Braunwässern andere Verhältnisse als in Klarwässern. Das Fe(III) wird selbst bei hohem Redoxpotential durch die HS an der Ausfällung gehindert, so daß auch das Phosphat gelöst bleibt. Es ist deshalb nicht verwunderlich, daß Huminwässer oft phosphatreicher sind als Klarwässer. So hat OHLE (1934) bei der Untersuchung norddeutscher humoser Seen stets recht hohe PO_4–P-Gehalte festgestellt. GESSNER (1933, 1934) und ÅBERG & RODHE (1942) konnten bei Untersuchungen von Hochmoorblänken des Isergebirges bzw. schwedischer Humusseen ähnliches konstatieren. Die eigenen Analysendaten entsprechen diesen Angaben. Insbesondere in den mecklenburgischen Braunwässern war der P-Standard sehr hoch, wobei der Orthophosphatanteil stets über 50% ausmachte.

Die Ursache der realtiv niedrigen, aber immer noch ausreichenden P-Konzentrationen (s. S. 72) der Hochmoorgewässer des Erzgebirges dürfte der geringe P-Gehalt des *Sphagnum*-Torfes sein, denn nach SCHREIBER (1927) geben mehrere Autoren 0,024–0,066% P im Torf an. Nach B. SCHNEE (1968) enthält eine aus Hochmoorwasser isolierte Huminsäure 0,055% P. Da die Mineralisation der Torfbildner nur langsam vonstatten geht und die HS recht stabile Verbindungen sind, resultieren daraus eben relativ geringe pflanzenverfügbare P-Mengen.

Von OHLE (1963) durchgeführte Versuche haben ergeben, daß das Orthophosphat durch die Bildung von Eisenhumat aus adsorptiver Bindung an $Fe(OH)_3$ befreit wird. Nach WETZEL (1968) kann durch Chelierungsprozesse zwischen HS und Eisen eine Phosphatfreisetzung aus fester Salzbildung erfolgen. Diese Ergebnisse bieten eine Erklärung dafür, daß im Abfluß des GK Phosphat überhaupt und dazu in recht ansehnlichen Mengen zu finden war, denn das ausfallende $Fe(OH)_3$ allein hätte es schnell adsorbiert.

Insgesamt kann festgestellt werden, daß der Phosphorhaushalt humoser Gewässer dadurch gekennzeichnet ist, daß Mangel an diesem wichtigen Element nur selten auftritt. Meist steht das Phosphat dem Phytoplankton und anderen Organismen auf Grund verschiedener allochthoner Einflüsse in genügenden Mengen zur Verfügung.

Auf die Stickstoffarmut von Hochmoorgewässern ist besonders von HARNISCH (1929) hingewiesen worden. Danach sind die hohe Azidität und die dadurch bedingte „außerordentlich geringe Fäulniskraft“ des Hochmoores, das Fehlen nitrifizierender und nur seltene Vorkommen Luft-N bindender Bakterien die Ursachen dieser ungünstigen Verhältnisse. Dementsprechend fanden GESSNER (1933) und GORHAM (1956b) in Hochmoorschlenken kaum über der Nachweisbarkeitsgrenze liegende Nitratmengen vor. Denselben Verhältnissen begegnet man in sauren humosen Seen, wie aus den Arbeiten von ASCHAN (1908), KARCHER (1939), ÅBERG & RODHE (1942) und HARKEWITSCH (1964) hervorgeht.

Die eigenen Ergebnisse bezüglich der N-Konzentrationen stehen im Einklang mit der zitierten Literatur. Der Stickstoff-Haushalt der verschiedenen Braunwässer weist eine bemerkenswerte Ähnlichkeit auf. Die anorganischen N-Verbindungen werden mengenmäßig in jedem Falle weit vom organischen Stickstoff übertroffen. Nur im Zadlitzbruch kommt dem NH_4–N größere Bedeutung zu.

Auffallend ist besonders der äußerst niedrige NO_3–N-Gehalt nicht nur in den sauren Gewässern. Trotzdem wurden sowohl im sauren als auch im alkalischen Milieu starke Algenproduktionen beobachtet, wobei das Nitrat völlig verschwinden konnte. Es ist klar, daß der NO_3–N am ehesten ins Minimum geraten kann (vgl. KARCHER 1939), jedenfalls eher als das Phosphat, an dem es kaum mangelte. Daß trotz der stets geringen Nitratmengen Massenproduktionen möglich waren, dürfte auf einen kurzgeschlossenen Kreislauf des Nährstoffs zurückzuführen sein. Doch spielt wahrscheinlich der oft in größeren Mengen vorhandene NH_4–N hierbei eine wesentliche Rolle. Allerdings kann auch er völlig aufgebraucht werden.

Trotz der ähnlich niedrigen Konzentrationen an assimilierbarem Stickstoff sowohl in den sauren als auch in den alkalischen Braunwässern ist die Annahme berechtigt, daß in den letzteren auf Grund günstigerer mikrobieller Verhältnisse bessere Bedingungen für die Remineralisierung, aber auch für die Luft-N-Bindung bestehen, daß der Verbrauch also viel schneller kompensiert werden kann als in den umsatzschwachen sauren Gewässern.

6. *Biologische Bestandsaufnahmen*

Die biologischen Untersuchungen mußten sich auf qualitative und quantitative Bestandsaufnahmen in den einzelnen Braunwässern beschränken. Eine eingehende Analyse jahreszeitlicher Veränderungen war auch in den regelmäßiger untersuchten Gewässern nicht möglich. Es kann angenommen werden, daß bei intensiverer Bearbeitung des Materials weitere Arten zu finden wären.

6.1. Kleines und Großes Kranichseehochmoor

In den Schlenken des KK wurden insgesamt 22 Pflanzen und 23 Tierarten festgestellt. Besonders bemerkenswert ist die geringe Bedeutung der Desmidiaceen, einer Algengruppe, die für saure Hochmoorgewässer typisch ist. LEHER (zit. nach GESSNER 1959) fand bei seinen umfangreichen Untersuchungen oberbayrischer Hochmoore sieben Arten, die nur bei pH-Werten unter 5 vorkamen, sehr viele (etwa 200) dagegen zwischen 4,5 und 6,0. WEHRLE (1927, zit. n. HARNISCH 1929) konnte zwischen pH 4,2 und 3,2 noch 23 Arten antreffen. Die meisten Desmidiaceen bevorzugen ein schwach saures Milieu.

Den 4 Zieralgen- stehen 5 Blaualgenarten gegenüber, von denen *Leptobasis spirulina* STEIN auf Hochmoorgewässer beschränkt ist (GESSNER 1959). Aber auch *Chroococcus turgidus* (KÜTZ) NÄG. bevorzugt diesen Biotop. Die anderen Cyanophyceenarten haben allgemeinere Verbreitung.

Von den 5 angetroffenen Diatomeenarten sind 3 acidophil (im Sinne von HUSTEDT 1956), nämlich *Frustulia rhomboides* var. *saxonica* (EHR.) DETONI, *Eutonia tenella* (GRUN.) A. CL. und *Pinnularia gibba* var. *linearis* EHR. Die erstgenannte Art war die bei weitem häufigste. Von Fadenalgen wurden nur wenige Arten gefunden. Am zahlreichsten war noch *Mougeotia* sp., die auch ins Plankton geht (FOTT 1959). Ebenfalls stark vertreten war im Benthos *Tribonema aequale* (?) PASCHER. Auf in KK 2 eingehängten Objektträgern bildete diese Heterokonte dichte Beläge.

Der compound-index nach NYGAARD (1949; zit. n. KLOTTER 1953) beträgt für die SCHLENKEN im KK 1,75. Das würde mäßige Eutrophie bedeuten. In der Tat werden sehr hohe Planktonzahlen gefunden. Doch stammen die die Massenentfaltung verursachenden Algen aus Gruppen, die im compound-index überhaupt nicht erfaßt sind, so daß er hier keinen Wert besitzt. Es sind drei Vertreter verschiedener Flagellatenreihen, die die überragende Rolle in den Hochmoorschlenken des KK spielen, und zwar *Cryptomonas ovata* EHR, *Dinobryon divergens* IMH. und *Chlamydomonas* sp. Nach HARNISCH (1929) wird das Plankton in Hochmoorgewässern in der Hauptsache von Flagellaten gebildet, während die Desmidiaceen mehr dem gebundenen Wasser angehören. Die übrigen Algengruppen spielen im Plankton keine Rolle. Die Ergebnisse der quantitativen Untersuchungen zeigen, daß in den relativ tiefen Schlenken KK 1 und KK 2 die Verteilung der drei dominierenden Flagellatenarten ähnlich ist. *Dinobryon* ist als wärmeliebende Form (GESSNER 1933) Anfang Mai noch ohne Bedeutung. Es herrscht *Chlamydomonas* vor (43260000 Ind./l). Im Juni tritt sie aber zurück, und *Cryptomonas* (8440000 Ind./l) sowie *Dinobryon* (11560000 Ind./l) erreichen Maxima. In der kleinen Schlenke KK 7 kommt *Chlamydomonas* zusammen mit *Cryptomonas* auch im Juni noch massenhaft vor. Merkwürdigerweise fehlt *Dinobryon* in den flachen Schlenken KK 3—7 entweder völlig oder ist ganz unbedeutend.

Da quantitative Planktonuntersuchungen von Hochmooren in der Literatur sehr spärlich sind, können nur wenige Vergleiche gezogen werden. GESSNER (1933) fand in Moorblänken im Isergebirge bei pH-Werten von 3,5—4,5 bis 10 Mill. Ind./l. *Dinobryon divergens*, Cryptomonaden und *Ankistrodesmus falcatus* (CORDA) RALFS. dominierten. In einzelnen Fällen konnte er sogar unter Eis Massenentwicklungen von Desmidiaceen (*Staurastrum*) beobachten (8 Mill. Ind./l). PEUKERT (1964) gibt ähnliche Planktonzahlen vom dystrophen Floßteich bei Reitzenhain (Erzgeb.) an.

Im GK traten im großen und ganzen dieselben Arten wie im KK auf. Bemerkenswert war aber das zahlenmäßig stärkere Vorkommen einer Desmidiaceenart (*Closterium cornu*: 126800 Ind./l), die im KK nicht angetroffen wurde. Die Desmidiaceen waren mit 5 Arten vertreten. Es wurden bei weitem nicht die Individuenzahlen der KKS erreicht. Interessant war auch, daß hier *Cryptomonas*, die im Mai noch dominierte, im Juni keine Rolle mehr spielte. *Dinobryon* fehlte völlig.

Das tierische Leben in den Hochmoorschlenken ist vor allem durch Protozoen gekennzeichnet. Der spirotriche Ciliat *Uroleptus dispar* STOKES war sehr häufig. Er wurde in allen Schlenken zwischen Sphagnen gefunden. Besonders auffallend war neben einigen weiteren Ciliaten *Stentor coeruleus*, der ebenfalls im *Sphagnum*-Preßsaft vorkam. Von den Rhizopoden waren erwartungsgemäß die Thekamöben am stärksten vertreten. Sehr häufig wurden *Centropyxis*-Arten und *Arcella discoides* var. *scutelliformis* PLAY. beobachtet. Seit HARNISCHS (1929) eingehenden Untersuchungen ist bekannt, daß auch eine reiche Testaceenfauna die Hochmoorgewässer charakterisiert. Besonders kennzeichnend für echte Hochmoore sollen *Amphitrema*-Arten sein, die jedoch in den Kranichseen nicht angetroffen wurden.

Die wichtigsten Rotatorien waren *Lecane acus* HARR. und *Rotatoria* sp. *Chydorus sphaericus* O. F. M. dominierte unter den Krustazeen eindeutig. Andere Cladoceren und Copepoden waren selten. In den beiden tieferen Schlenken KK 1 und KK 2 wurde *Chaoborus pallidus* gefunden, im Winter unter Eis in KK 1 auch *Ch. crystallinus*.

Die biologischen Ergebnisse bestätigen also die bekannte Tatsache, daß Hochmoore als extreme Biotope sehr artenarm sind. Dafür kommen einige wenige bestangepaßte Formen massenhaft vor. Der Mannigfaltigkeits-Index *d* einer Phytoplanktonprobe nach MARGALEF (1960, zit. nach UHLMANN 1964) ist in diesem Zusammenhang recht aufschlußreich. Die diversity-indices liegen in den beiden Hochmooren zwischen 0,42 und 1,15, d. h., die Mannigfaltigkeit ist außerordentlich gering.

Fragt man nach den die Zusammensetzung der Biozönose der Hochmoorwässer bedingenden Faktoren, so muß man mit HARNISCH (1929) dem niedrigen pH-Wert im Verein mit dem geringen Elektrolyt- und Nährstoffgehalt die Hauptbedeutung zuerkennen. Die Rolle der HS ist noch unklar, doch dürfte sicher sein, daß sie negative Einflüsse nicht ausüben (s. SHAPIRO 1957, WETZEL 1968, PRAKASH & RASHID 1968 und SCHNEE 1968). Nach HARNISCH (1929) und RUTTNER (1962) ist die Hochmoorlebensgemeinschaft durch das Fehlen ganzer Organismengruppen besonders gekennzeichnet, noch besser als durch das Vorkommen anderer für sie charakteristischer Lebewesen. So fehlen z. B. die Tricladiden, die meisten Rotatorien und Oligochaeten, *Daphnia*, *Moina*, die Ostracoden und Malacostraken, die Ephemeriden und Culiciden (außer *Chaoborus*), die Mollusken, Amphibien und Fische.

Die eigenen Untersuchungen konnten das mit einer Ausnahme bestätigen: Anfang Mai kamen in den Schlenken der beiden Hochmoore massenhaft *Culex*-Larven vor.

Da die Phytoplanktonzahlen nur ein grobes Maß für das „standing crop" sind, wurde versucht, mittels der Hell—Dunkel-Flaschenmethode Aussagen über die Primärproduktion zu gewinnen. Die Flaschen wurden in 55 cm Tiefe exponiert.

Am 3. 5. 66 wurden in KK 1 folgende Werte erhalten:

Bruttoproduktion: 1,4 mg O_2/l · 24 h = 0,53 mg C/l · 24 h
Nettoproduktion: 0,8 mg O_2/l · 24 h = 0,32 mg C/l · 24 h

Die Zahlen sind Mittelwerte zweier Parallelen. Trotz hoher Individuendichte (etwa 44 Mill. Ind./l) und strahlendem Sonnenschein war die Produktion gering. Zum selben Zeitpunkt durchgeführte Messungen in GKS ergaben noch niedrigere Resultate. Hier waren die Planktonzahlen bedeutend kleiner als in KK 1. Trotz hohem „standing crop" sind die Hochmoorschlenken also zumindest zum Zeitpunkt der Untersuchung in der Zeiteinheit wenig produktiv gewesen.

6.2. Wildenhainer Bruch

Die biologischen Untersuchungen beschränkten sich auf den Flügeldammteich. In diesem oligohumosen, nicht durch extreme Lebensbedingungen gekennzeichneten Biotyp wurde ein großer Reichtum an Arten festgestellt. Dafür waren die Individuenzahlen relativ niedrig. Der Mannigfaltigkeitsindex beträgt am 23. 6. 1966 17,6! Die weitaus vorherrschende Algengruppe waren die Desmidiaceen, von denen am selben Tag 48 Arten festgestellt werden konnten. Sicherlich ist das noch längst nicht die Höchstzahl. Das schwach saure Gewässer scheint für die Desmidiaceen ein optimaler Lebensraum zu sein. Sie gehörten z. T. dem Benthos an. Von einem echten Plankton kann bei der Flachheit des Wassers ohnehin nicht die Rede sein.

Der compound-index beträgt 0,23 und zeigt Oligotrophie an, ein Urteil, das man auch auf Grund der Nährstoffverhältnisse und der Planktonzahlen fällen muß (*Scenedesmus quadricauda* (TURP.) BRÉB. 48000 Ind./l).

Unter den Krustazeen waren verständlicherweise räuberische Copepoden recht artenreich. *Chydorus sphaericus* fehlte aber auch hier nicht.

6.3. Zadlitzbruch

In den durch extreme chemische Eigenschaften charakterisierten Schlenken war die Artenzahl äußerst gering. Dafür herrschte eine einzige Fadenalgenart vor: *Tribonema vulgaris* PASCHER. Ihr Wachstum wurde nur im Winter unter Eis unterbrochen, sonst erfüllten die dunkelgrünen Watten fast die gesamte Schlenkenausdehnung.

Am 23. 6. 66 wurde ferner eine Massenentwicklung einer planktischen, leider unbekannten einzelligen Alge (20 μm) beobachtet (19200000 Ind./l). Es handelte sich wahrscheinlich um eine Heterokonte oder Chrysophycee, da Stärke als Assimilationsprodukt nicht in Frage kam.

Die Diatomeen waren durch die Gattungen *Eunotia* und *Pinnularia* in vier Arten vertreten, wobei die acidobionte *E. exigua* (BRÉB.) GRUN. am häufigsten auftrat. Auffallend war das völlige Fehlen der Cyanophyceen und Desmidiaceen.

Das tierische Leben war ebenfalls nur gering entfaltet. Neben einigen Ciliaten, Thekamöben und Rotatorien kamen nur das Allerweltstier *Chydorus sphaericus* und *Chaoborus crystallinus* recht häufig vor.

Im Frühjahr wurden die Schlenken von großen Mengen von *Culex*-Larven bevölkert.

6.4. Serrahner Moorgewässer

Die Braunwässer KB und BB zeigten vor allem im Benthos ein reiches Leben. Am häufigsten waren die Diatomeen *Pinnularia subcapitata* GREY, *P. viridis* (NITZSCH) EHR. und *Eunotia lunaris* (EHR.) GRUN. sowie die Heterokonte *Tribonema* sp. Zahlreiche Vertreter der Thekamöben, der Rotatorien und der Krustazeen beherrschten das tierische Benthos.

Die Mannigfaltigkeit des Planktons war viel geringer (BB: $d = 0{,}45$). Es dominierten im Frühjahr drei Flagellatenarten: *Glenodinium* sp. und zwei *Chlamydomonas* sp. (z. B. BB 34075000 bzw. 26450000 Ind./l). Anhand des Vorkommens von Euglenen und Bakterien ist zunehmende Saprobie in der Reihenfolge KB-BB-CB angedeutet. Daß die Desmidiaceen völlig fehlen, verwundert nicht. Das saprobe Milieu der Fallaubgewässer sagt ihnen offenbar nicht zu, obwohl der Elektrolytgehalt niedrig war und auch der pH-Wert günstig lag.

Ein Produktionsversuch im KB am 23. 3. 1966 ergab infolge trüben und regnerischen Wetters auch in den Hellflaschen nur Atmungsverluste und keinen Assimilationsgewinn.

Die dominierenden Formen der planktischen Fauna stellten im Frühjahr Copepoden (Hauptvertreter *Diacyclops bicuspidatus* CLAUS). Im Sommer kam dagegen *Daphnia pulex* in Massen vor (am 12. 8. 66 680 Ind./l). Diese Cladocere verhinderte zusammen mit den suboptimalen Lichtbedingungen in der warmen Jahreszeit ein Aufkommen von Phytoplankton.

6.5. Herthasee

Die überragende Algengruppe des Phytoplanktons waren hier mit 44 Arten die Diatomeen. Der diversity-index, $d = 3{,}56$, zeigt eine recht beachtliche Mannigfaltigkeit an. Zum Zeitpunkt der Untersuchung wurde eine außerordentlich starke Massenentfaltung einer sehr kleinen Form von *Stephanodiscus hantzschii* festgestellt (82000000 Ind./l). Ein Teil der größeren Diatomeen fand sich im Schlamm, in dem durch Sedimentation der toten Zellen eine allmähliche Anreicherung vonstatten ging. Trotz der geringen Sichttiefe wurden die meisten *Stephanodiscus*-Zellen in 4 m Tiefe gezählt (110 Mill. Ind./l). Die Beobachtung lebenden Materials ließ erkennen, daß ein Teil auch in dieser Tiefe noch lebte. Bei 8 m wurden nur noch abgestorbene Algen angetroffen. Alle anderen Arten spielten quantitativ keine Rolle.

Da Desmidiaceen völlig fehlten und andererseits Centrale und Cyanophyceen auftraten, resultiert ein compound-index von ∞. Durch das massenhafte Vorkommen von *St. hantzschii* ist jedoch der beste Beweis für eine starke Eutrophie gegeben.

Das Zooplankton war weder qualitativ noch quantitativ besonders entwickelt. Am meisten fielen noch die Rotatorien *Filinia longiseta* (Ehr.) und *Kellicottia longispina* (Kell.) auf. Die erste Art wurde von der Oberfläche bis zum Grund festgestellt. Im Schlamm fanden sich einige Thekamöben und holotriche Ciliaten. Krustazeen hatten untergeordnete Bedeutung. Bakterien waren insbesondere in 8 m Tiefe in großer Zahl anzutreffen.

6.6. Straminke

Alle im Zähler des compound-index stehenden Algengruppen waren in der Straminke vertreten. Es dominierte eine *Microcystis*-Art (1431000000 Ind./l), dann folgten je zwei Diatomeen- und Euglenophyceen-Arten. Das Gewässer ist also eutroph. Bakterien kamen massenhaft vor. Erwartungsgemäß wurden unter den 27 Diatomeen-Arten viele halophile Formen angetroffen, z. B. *Coscinodiscus lacustris* Grun., *Melosira nummuloides* (Dillw.) Ag., *Diploneis interrupta* u. a. *Keratella cochlearis* Gosse war die häufigste Zooplanktonform. Es folgten mit *K. quadrata* O. F. Müller und *Filinia longiseta* Ehr. weitere Rotatorien. Der Mannigfaltigkeitsindex betrug 2,7.

7. *Der Haematococcus-Test*

Haematococcus pluvialis Flot. eignet sich sehr gut für die Produktionsversuche, weil die Alge leicht kultivierbar ist und auf suboptimale Nährstoffverhältnisse mit der Bildung von Haematochrom (Pringsheim 1914) reagiert, worauf die Zellen sich sehr schnell festsetzen. Der Zeitpunkt, bei dem das Wachstumsoptimum überschritten wird, die Nährstoffe also ins Minimum geraten, ist so leicht feststellbar.

Die Versuche wurden im allgemeinen bald nach dem Beginn der Haematochrombildung abgebrochen. Die Kulturen wurden in vier Teilen zu je 3 Parallelen angesetzt (s. Abb. 12a). War von vornherein sehr viel Phosphat vorhanden (BB, S), erübrigte sich eine künstliche P-Zugabe.

Die stark sauren Braunwässer mußten mit $NaHCO_3$ neutralisiert werden, da die Flagellaten sonst nicht wuchsen und zerfielen.

Die Tests mit Hochmoorwasser aus den beiden Kranichseen hatten weitgehend ähnliche Ergebnisse (Abb. 12a). Sowohl in den Schlenkenproben als auch im Wasser

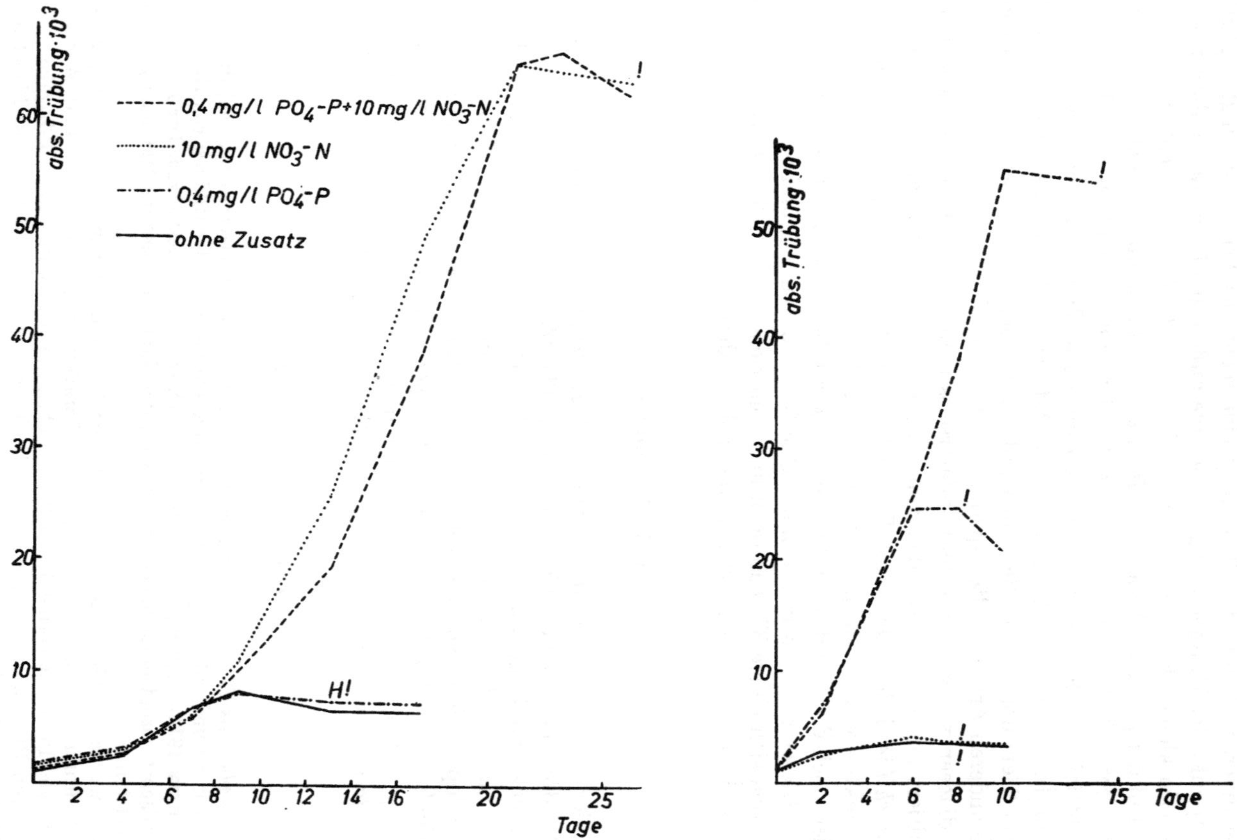

a b

Abb. 12a—e. Ergebnisse der *Haematococcus*-Tests. a — Kleines Kranichseehochmoor (3. 3. 66); b — Zadlitzbruch (31. 3. 66); c — Beilbruch (29. 3. 66); d — Herthassee (26. 4. 66); e — Straminke (27. 4. 66); ! = Beginn der Haematochrombildung

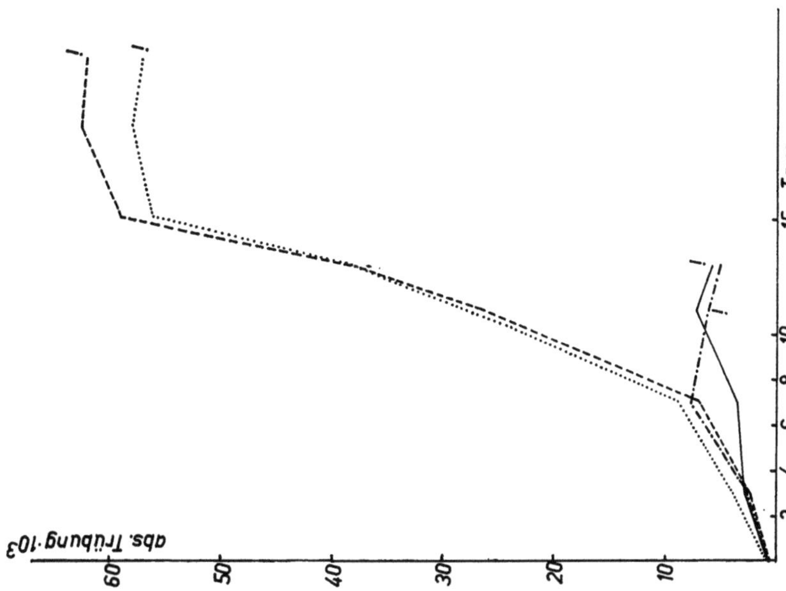
abs. Trübung·10³
10
20
30
40
50
60
2
4
6
8
10
15 Tage

d

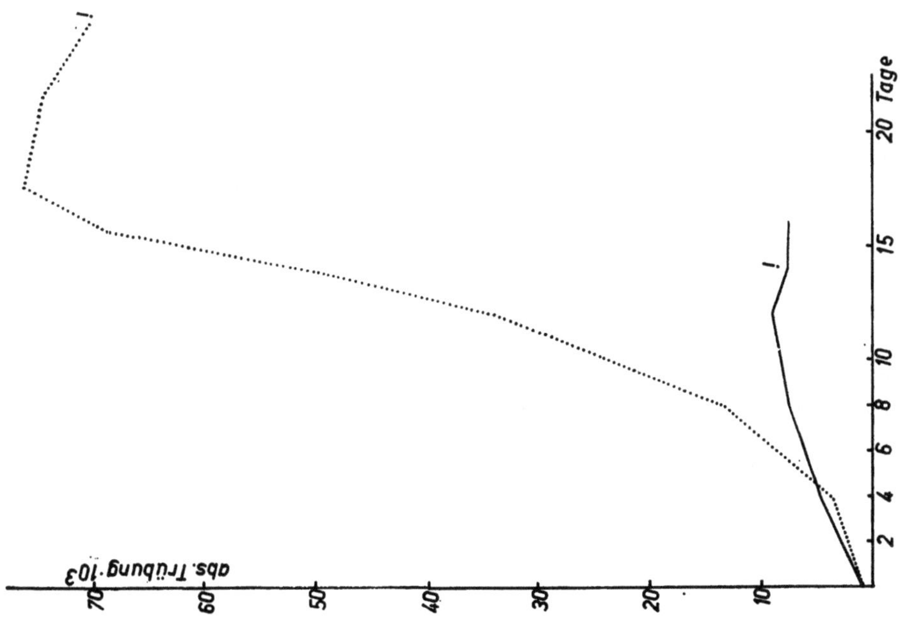
abs. Trübung·10³
10
20
30
40
50
60
70
2
4
6
8
10
15
20 Tage

c

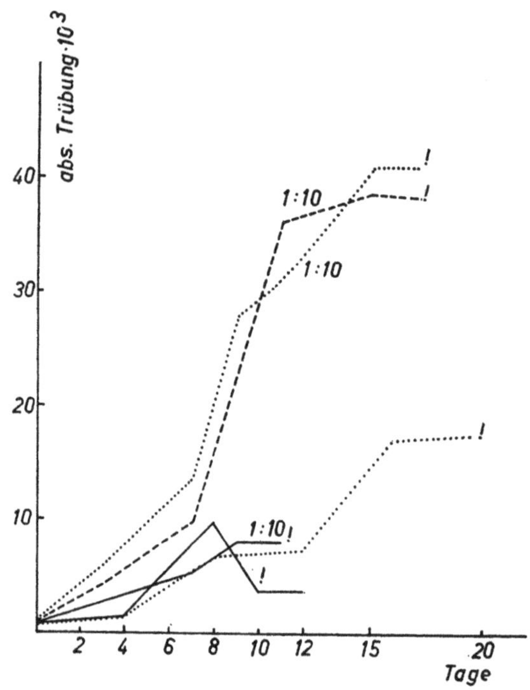

Abb. 12e

aus den Abflußbächen war das Wachstum ohne Nährstoffzusatz und bei alleiniger Phosphatzugabe minimal. Nach Nitratapplikation war in jedem Fall eine sprunghafte Produktionssteigerung zu verzeichnen, ebenso bei Verwendung von Nitrat und Phosphat.

Interessant ist die Betrachtung der natürlichen Nährstoffmengen in den Testproben: So enthielt das Wasser von KK1 (Abb. 12a) 10 µg PO_4–P/l, 0,06 mg NO_3–N/l und 0,56 mg NH_4–N/l. Auf Grund des geringen Nitratgehalts verwundert es nicht, daß der NO_3–N der das Wachstum limitierende Faktor ist, obwohl NH_4–N in größeren Mengen vorhanden war und er nach PRINGSHEIM (1914) ebensogut verwendet werden kann. Die ebenfalls niedrige PO_4-P-Konzentration reichte offenbar aus.

Noch anders konnte der Beweis für die Wirkung des NO_3–N als Minimumstoff erbracht werden. Nachdem das Wachstum in den Kulturröhrchen abgeschlossen war, wurden wieder Nährstoffe zugegeben. Es zeigte sich, daß nur bei Nitratzusatz die Zellen erneut grün und beweglich wurden und ein stürmisches Wachstum begann, auch in den im Hauptversuch unversetzten Proben.

Wie aus Abb. 12b hervorgeht, war im Z das Phosphat der primäre, die Produktion begrenzende Stoff. Die zum Test verwendete Probe hatte folgende Nährstoffkonzentrationen:

6 µg PO_4–P/l, 0,11 mg NO_3–N/l, 1,6 mg NH_4–N/l.

Nur bei PO_4-Zugabe fand eine Produktion statt. Auch der NH_4–N dürfte hier als Stickstoffquelle bedeutsam sein. Die Verabreichung sowohl von Phosphat als auch von Nitrat hatte eine noch weitergehendere Produktionssteigerung zur Folge. Beeinträchtigt wird dieses Ergebnis jedoch dadurch, daß bei der Neutralisation das Eisen ausfiel, so daß ein Teil der HS und evtl. auch des Phosphats aus der Lösung eliminiert wurde.

Abb. 12c veranschaulicht die beim Test von BB-Wasser erhaltenen Kurven. Auch hier ist ganz eindeutig der NO_3–N Minimumstoff. Sein Gehalt betrug 0,09 mg/l. Der etwas höhere NH_4–N-Gehalt ändert daran nichts. — PO_4–P ist genügend vorhanden: 560 µg/l.

Im Herthasee war nach den Analysen der assimilierbare Stickstoff als Mangelfaktor zu erwarten: NO_3–N 0,04 mg/l, NH_4–N 0,12 mg/l, PO_4–P 32 µg/l. Abb. 12d bestätigt diese Vermutung. Nur Nitrat sowie Nitrat und Phosphat vermochten eine Produktion in Gang zu bringen.

Das Wasser aus der halinen Straminke wurde einmal unverdünnt verwendet (280 µg PO_4–P/l, 0,02 mg NO_3–N/l, 0,1 mg NH_4–N/l, zum anderen mit destilliertem Wasser 1:10 verdünnt (Abb. 12e). Wie man aus den Kurven ersieht, fehlten auch hier wieder die anorganischen Stickstoffverbindungen, vor allem Nitrat. Phosphat wurde unverdünntem Wasser natürlich nicht zugegeben, und eine Zugabe wäre eigentlich auch im 1:10 verdünnten nicht nötig gewesen. In beiden Wässern vermochte nur Nitrat die Produktion zu forcieren. Doch kann *Haematococcus* im unverdünnten Brackwasser offenbar auch dann nicht gut gedeihen.

Die Tatsache, daß in allen untersuchten humosen Gewässern einer der Hauptnährstoffe, meist der assimilierbare Stickstoff, für *Haematococcus pluvialis* Mangelfaktor war, darf nun aber nicht zu der Annahme verleiten, die Gewässer seien unproduktiv. Ganz im Gegenteil wurden ja erhebliche Planktonentwicklungen beobachtet. Abgesehen von den stark sauren Gewässern der Hochmoore und des Zadlitzbruches, die ohnehin nur wenig Nitrat enthalten können, sind wahrscheinliche diese Massenentfaltungen erst die Ursache der niedrigen Konzentrationen. Die Testergebnisse lassen aber den Schluß zu, daß der betreffende Nährstoff bei andauernder Produktion auch in der Natur zuerst zum wachstumsbegrenzenden Element werden kann, in den meisten humosen Gewässern also der Nitrat-N.

8. *Versuche zur Erfassung der heterotrophen Bioaktivität humoser Gewässer*

Über den biochemischen Abbau der HS im Wasser ist leider nur wenig bekannt. Nach allgemeiner Ansicht sind sie schwer abbaubar (Krogh & Lange 1932, Shapiro 1957, Chalupa 1966 u. a.). Zur Klärung dieser Frage wurden BSB_5-Versuche durchgeführt (vgl. Tab. 8).

Die BSB_5-Werte liegen also überraschend nahe beieinander auf einem niedrigen Niveau und das trotz der verschiedensten Gewässertypen. Nur eine leicht steigende Tendenz von den sauren zu den alkalischen Braunwässern ist zu verzeichnen. Durch Neutralisation der sauren Gewässer konnte die geringe bakterielle Aktivität derselben nicht erhöht werden. Die BSB-Werte nahmen sogar etwas ab.

Das Verhältnis der O_2-Äquivalente des PV und des BSB_5 als ein grobes Maß der Abbaubarkeit organischer Substanzen ist in diesem Zusammenhang recht aufschlußreich (Tab. 8). Die insgesamt sehr hohen Zahlen weisen darauf hin, daß die HS in der Tat schwer abbaubar zu sein scheinen. Der BSB_5 zeigt eine gewisse Abhängigkeit

Tab. 8. BSB_5-Werte, O_2-Äquivalente PV/BSB_5- und N:P-Verhältnisse der untersuchten Gewässer (die Ergebnisse sind Mittelwerte mehrerer Versuche)

	BSB_5 [mg O_2/l]	$\frac{PV\ [mg\ O_2/l]}{BSB_5\ [mg\ O_2/l]}$	N:P
Z	1,3	38,5	244
KK	1,4	27,3	74
H	1,85	11,2	46
BB	2,1	23,2	4
S	2,75	10,8	8

vom N:P-Verhältnis, indem die niedrigen Werte den ungünstigen N:P-Relationen zukommen und die höheren den günstigeren (Tab. 8). Obwohl das N:P-Verhältnis für die mikrobielle Aktivität große Bedeutung hat, ist das noch keine ausreichende Erklärung für die ähnlich niedrigen BSB_5-Werte in den so unterschiedlichen humosen Gewässern. Man könnte eine gewisse Toxizität des Braunwassers in Betracht ziehen. In *Sphagnum*-Torfen sollen nach KUNZ (1957) Antibiotika vorkommen. Das würde aber nur die Erzgebirgshochmoore betreffen, in deren Wasser Bakterien unter dem Mikroskop nie zu sehen waren, während sie sich in S, BB und H in mehr oder weniger großen Mengen fanden. Selbstverständlich spielt der pH-Wert eine Rolle, da ja das saure Mileu aktivitätshemmend wirkt. Doch trifft das eben auf die alkalischen Braunwässer nicht zu, die nach HENRICI & MC COY (1938, zit. n. ÅBERG & RODHE 1942) sehr hohe Bakterienzahlen aufweisen können. Es ist auch auf Grund dieser Tatsachen erstaunlich, daß die Aktivitätsunterschiede so gering sind.

Es wurde mittels ZZ-Tests versucht, durch Zugabe eines genau definierten, leicht zu veratmenden Substrats (Glukose) Aufschluß über eine eventuell potentielle Bioaktivität der heterotrophen Mikroorganismen in den verschiedenen Braunwässern zu erhalten. Die durch die Glukose verursachte ZZ wurde in Prozent des theoretischen O_2-Bedarfs der verwendeten Zuckerkonzentration berechnet (1 mg Glukose $\triangleq$ 1,06 mg O_2). In den Tab. 9a—9e sind die Ergebnisse dargestellt. Neutralisation der stark sauren KK- und Z-Proben vermochte wie beim einfachen BSB keine erhöhte Abbauintensität zu erzeugen. Der vorhandene Bakterienbestand war also in der Lage, im natürlichen Aziditätsbereich optimal zu wirken.

Selbst bei 20 und 100 mg Glukose/l war in den stark sauren KK- und Z-Wässern am Versuchsende stets noch genügend Restsauerstoff vorhanden. Dagegen wurde der Sauerstoff in den schwach sauren bis alkalischen Braunwässern BB, H und S bei Verwendung der beiden höchsten Glukosemengen völlig aufgebraucht. Es mußte die Bebrütungszeit deshalb auf zwei Tage verkürzt werden. In der S war auch dann bei diesen hohen Konzentrationen kein O_2 mehr nachweisbar, im BB nur noch bei 100 mg Glukose/l nicht, während im Herthasee stets Restsauerstoff übrig blieb. Auch bei den niedrigen Konzentrationen 5 und 10 mg/l wurde in BB, S und H mehr Glukose veratmet als in KK und Z. Der im BSB_5 gerade angedeutete Unterschied in der Abbauintensität zwischen den stark sauren bakterienarmen und den schwach sauren bis alkalischen Gewässern tritt somit deutlich hervor.

Tab. 9. Ergebnisse der ZZ-Tests (Die Indizes bei ZZ bedeuten mg Glukose/l).

a) Kleiner Kranichsee | b) Zadlitzbruch

BSB_5	1,25 mg O_2/l pH 3,4	BSB_5	1,55 mg O_2/l pH 3,5
ZZ_5	2,6 = 49,1%	ZZ_5	2,25 = 42,5%
ZZ_{10}	3,7 = 34,9%	ZZ_{10}	3,35 = 31,6%
ZZ_{20}	4,0 = 18,9%	ZZ_{20}	3,95 = 18,6%
ZZ_{100}	4,6 = 4,3%	ZZ_{100}	3,9 = 3,7%

c) Beilbruch

BSB_5	2,35 mg O_2/l pH 5,9	BSB_2	1,8 mg O_2/l
ZZ_5	3,35 = 63,2%	ZZ_5	2,15 = 40,6
ZZ_{10}	5,6 = 52,8%	ZZ_{10}	4,15 = 39,1%
ZZ_{20}	—	ZZ_{20}	6,4 = 30,2%
ZZ_{100}	—	ZZ_{100}	—

d) Herthasee

BSB_5	1,85 mg O_2/l pH 9,2	BSB_2	1,0 mg O_2/l
ZZ_5	2,65 = 50,0%	ZZ_5	1,85 = 34,9%
ZZ_{10}	5,5 = 51,9%	ZZ_{10}	3,2 = 30,2%
ZZ_{20}	—	ZZ_{20}	3,4 = 16,0%
ZZ_{100}	—	ZZ_{100}	3,7 = 3,5%

e) Straminke

BSB_5	2,75 mg O_2/l pH 8,1	BSB_2	1,2 mg O_2/l
ZZ_5	2,95 = 55,7%	ZZ_5	1,9 = 35,8%
ZZ_{10}	4,15 = 39,2%	ZZ_{10}	3,8 = 35,8%
ZZ_{20}	—	ZZ_{20}	—
ZZ_{100}	—	ZZ_{100}	—

Das leicht verwertbare Substrat Glukose hat also erkennen lassen, daß die potentielle Aktivität der Braunwässer größer ist als die aktuelle, was auf Grund von Bakterienreichtum und günstigerer Milieubedingungen insbesondere für die alkalischen Humusgewässer zutrifft. Diese Tatsache ist ein deutlicher Hinweis auf eine schwere Abbaubarkeit der HS. Ob die Widerstandsfähigkeit aber auf eine strukturell bedingte Stabilität oder auf die ungünstigen Nährstoffverhältnisse (Mangel an assimilierbarem N) zurückzuführen ist, läßt sich nach den vorliegenden Ergebnissen nicht entscheiden. Es ist beides wahrscheinlich. Für die letztere Deutungsmöglichkeit sprechen auch zwei neuere Literaturangaben, deren Aussagen sich entgegenstehen. So konnte Ryhänen (1968) in finnischen Humusgewässern durch N- und P-Zugabe eine Steigerung der Abbauintensität erreichen, weshalb er meint, daß die HS doch nicht so widerstandsfähig gegen mikrobiellen Abbau seien, wie allgemein angenommen werde. Demgegenüber waren in BSB-Versuchen von Geisler (1969) mit Schwarzwasser aus dem Rio Negro die sehr niedrigen natürlichen BSB_5-Werte (0,8—1,4 mg/l) durch Nährstoffzufuhr nicht zu erhöhen.

Daß die durch den Glukoseeinsatz ausgelöste zusätzliche Aktivität ohne künstliche Nährstoffzufuhr möglich war, liegt darin begründet, daß es von seiten der Bakterien zur Veratmung der Glukose ohne Zweifel weit geringerer enzymatischer „Anstrengungen" bedarf als zur Zerlegung der komplizierten HS-Moleküle. Zusammenfassend kann folgendes gesagt werden: Die schwache heterotrophe Bioaktivität in den untersuchten humosen Gewässern ist sehr wahrscheinlich sowohl auf suboptimale Nährstoffbedingungen als auch auf die relative Widerstandsfähigkeit der Huminstoffe gegen bakterielle Destruktion zurückzuführen.

Abschließend seien noch die Ergebnisse von Zelluloseabbauversuchen im KK, BB, KB und H erwähnt. Während einmonatiger Expositionszeit von Zellulosefäden (5 cm unter der Wasseroberfläche) war in keinem Fall ein signifikanter Abbau eingetreten. Da die hier in Frage kommenden aeroben Zellulosezersetzer saure Reaktion meiden, ist das im Falle der Hochmoorgewässer nicht weiter verwunderlich. Im Herthasee sollte dagegen ein Abbau durchaus möglich sein, da die Zellulose in eutrophen Seen nach v. BRANDT (1954) sehr schnell starker Zersetzung anheimfällt. Doch sind die zellulosespaltenden Bakterien auf Nitrat als N-Quelle angewiesen, das aber in nur minimaler Konzentration zur Verfügung steht, so das der fehlende Abbau erklärlich wird.

Diese Ergebnisse weisen ebenfalls darauf hin, daß die heterotrophe Aktivität im freien Wasser humoser Gewässer vor allem auf Grund von Nährstoffmangel gering ist.

9. *Zusammenfassung*

1. In dieser Arbeit wurden folgende Gewässer untersucht: die Erzgebirgshochmoore Großer und Kleiner Kranichsee, die in der Dübener Heide gelegenen Übergangsmoore Wildenhainer und Zadlitzbruch, drei kleinere Hochmoore in Mecklenburg, der Herthasee auf Rügen und die haline Straminke bei Zingst.

2. Der Sauerstoffhaushalt dieser Braunwässer ist durch extreme Verhältnisse gekennzeichnet. Von Anaerobie bis zu hohen O_2-Übersättigungen waren alle Möglichkeiten gegeben. Die freie Kohlensäure zeigte ein dem Sauerstoff entgegengesetztes Verhalten. Gebundene Kohlensäure fehlte teilweise völlig.

3. Es wurden pH-Werte zwischen 3,2 und 9,2 registriert.

4. In den Gewässern der Kranichseehochmoore kommen fast ausschließlich huminoide organische Stoffe vor. In den Übergangsmooren sind dagegen auch ungefärbte organische Substanzen gelöst. In diesen Braunwässern konnten signifikante lineare Korrelationen zwischen dem Huminstoffgehalt, der Oxydabilität und dem Farbgrad festgestellt werden. Der Konzentrationsbereich, innerhalb dessen Huminstoffe gefunden wurden, ist mit 2,5—3300 mg/l ungeheuer groß. Da die Huminstoffe das Eisen in Lösung halten, fanden sich auch bei hohen O_2-Spannungen oft erhebliche Fe-Konzentrationen.

5. Die Hochmoorgewässer des Erzgebirges sind relativ phosphorarm, doch war das Orthophosphat nur während starker Planktonentfaltungen nicht nachweisbar. Der Ges.-P-Gehalt weist Beziehungen zur Huminstoff-Konzentration auf. In den Übergangsmooren wurden ähnliche P-Mengen festgestellt. Die mecklenburgischen Gewässer sind demgegenüber durch einen hohen P-Standard gekennzeichnet.

Bezüglich des Stickstoffhaushaltes herrschten im Gegensatz zum Phosphat weit ungünstigere Bedingungen, wobei besonders starker Mangel an NO_3-N auftrat. Den größten Anteil am Ges.-N hatte stets der organisch gebundene Stickstoff.

6. Sehr große Unterschiede traten in bezug auf Elektrolytgehalt und Ionenrelationen in Erscheinung. Extrem elektrolytarm waren die Hochmoorgewässer. Die mit Abstand höchste Leitfähigkeit kam der brackigen Straminke zu. Der Herthasee ist ein $Ca(HCO_3)_2$-Gewässer. Die Kranichseen und der Zadlitz sind als $CaSO_4$-Typ anzusprechen, während der Wildenhainer und der Beilbruch gemischte $CaSO_4-Ca(HCO_3)_2$-Gewässer darstellen. In der Straminke dominieren NaCl und $MgCl_2$, in einem Braunwassertümpel Na_2SO_4.

7. Mit Ausnahme des Wildenhainer Bruchs wurden in allen untersuchten Braunwässern Phytoplanktonmassenentwicklungen in den Größenordnungen 10^3-10^6 Ind./ml festgestellt. In jedem Gewässer waren andere Arten dominant. Die durch extreme Lebensbedingungen gekennzeichneten Hochmoore und der Zadlitzbruch wiesen eine sehr geringe Mannigfaltigkeit des Phytoplanktons auf. In den alkalischen Braunwässern Herthasee und Straminke sowie vor allem im „harmonischen" Wildenhainer Bruch herrschte dagegen eine große Artenvielfalt.

8. Wie *Haematococcus*-Tests zeigten, erwies sich im Zadlitzbruch das Orthophosphat als Minimumstoff. In allen anderen Fällen begrenzte das Nitrat das Wachstum der Testflagellaten. Dieser Befund steht im Einklang mit den chemischen Analysen.

9. Die heterotrophe Bioaktivität der untersuchten humosen Gewässer war trotz sehr differenter chemischer und mikrobieller Verhältnisse nur wenig unterschiedlich und auf niedrigem Niveau. Als Ursachen kommen die Widerstandsfähigkeit der Huminstoffe gegen bakterielle Destruktion und suboptimale Nährstoffbedingungen in Betracht.

10. Literatur

ÅBERG, B., & W. RODHE, 1942: Über die Milieufaktoren in einigen südschwedischen Seen. — Symb. Bot. Upsal. **5** (3), 256 S.

ASCHAN, O., 1908: Die wasserlöslichen Humusstoffe (Humussole) der nordischen Süßgewässer. — J. prakt. Chemie N. F. **77**, 172—188.

Atlas der Bezirke Rostock, Schwerin, Neubrandenburg, 1962. — VEB Topographischer Dienst, Schwerin.

Ausgewählte Methoden der Wasseruntersuchung, 1965: 207 S., Amt für Wasserwirtschaft, Berlin.

BAYLY, J. A. E., 1964: Chemical and Biological Studies on some Acidic Lakes of East Australian sandy coastal Lowlands. — Aust. J. Mar. Freshw. Res. **15**, 56—72.

BENADE, W., 1954: Mikrobiologie der Moore. — Linz, Wien, Frankfurt (Main).

BERG, A., 1962: Exposé des méthodes d'analyse chimique et physicó-chimique des eauc humique. — Mem. Ist. Ital. Idrobiol. **15**, 183—206.

BLACK, A. P., & R. F. CHRISTMAN, 1963a: Characteristics of colored surface water. — J. AWWA **55**, 753—770.

— 1963b: Chemical characteristics of fulvic acids. — J. AWWA **55**, 897—912.

BRANDT, A. v., 1954: Unterschiedliche Netzverrottung in Fischgewässern. — Archiv Fischereiwiss. **5**, 58—73.

CHALUPA, J., 1963: Humic acids in water. 1. Methods of preparation and determination. — Sci. Pap. Inst. Chem. Technol., Prague, Technol. Water. **7**: 57—131.

— 1966: A contribution to the knowledge of humic substances in peat waters. — Verh. internat. Ver. Limnol. **16**, 485—490.

— 1968: Chemische Probleme der gefärbten organischen Stoffe in Oberflächengewässern. — Mitt. internat. Ver. Limnol. **14**, 5—13.

CZENSZNY, R., 1960: Wasser-, Abwasser- und Fischereichemie. — Verl. Technik, Berlin.

Deutsche Einheitsverfahren zur Wasseruntersuchung, 1954, 1960: Verl. Chemie, Weinheim.

ELSTER, H. J., 1955: Beiträge zur Limnochemie der Hochschwarzwaldseen, I. — Arch. Hydrobiol. Suppl. **22**, 1—64.

FRENZEL, H., 1930: Entwicklungsgeschichte der sächsischen Moore und Wälder seit der letzten Eiszeit. — Abh. Sächs. Geol. Landesamtes **9**.

FOTT, B., 1959: Algenkunde. — VEB Gustav Fischer Verl., Jena.

FUESS, W., 1932: Die gegenwärtige Flora des Zadlitzbruches, eines Moores in der Dübener Heide — Verh. bot. Vereins Prov. Brandenburg **1**, 100—111.

GEISLER, R., 1969: Untersuchungen über den O_2-Gehalt in einem tropischen Schwarzwasser. — Arch. Hydrobiol. **66**, 307—325.

GESSNER, F., 1933: Nährstoffgehalt und Planktonproduktion in Hochmoorblänken. — Arch. Hydrobiol. **25**, 394—406.

— 1934: Nitrat und Phosphat im dystrophen See. — Arch. Hydrobiol. **27**, 131—161.

— 1955, 1959: Hydrobotanik Bd. 1, 2. — VEB Dt. Verl. Wiss., Berlin.

GHASSEMI, M., & R. F. CHRISTMAN, 1968: Properties of the yellow organic acids of natural water. — Limnol. & Oceanogr. **13** (4), 583—597.

GJESSING, E. T., 1964: Ferrous iron in water. — Limnol. & Oceanogr. **9** (2), 272—274.

GOLDSCHMIDT, J., 1950: Das Klima von Sachsen. — Abh. meteorol. Dienst. DDR, Nr. 3.

GORHAM, E., 1956a: The ionic composition of some bog and fen waters in the English Lake District. — J. Ecology **44**, 142—152.

— 1956b: On the chemical composition of some waters from the Moor House Nature Reserve. — J. Ecology **44**, 375—382.

GORHAM, E., & J. B. CRAGG, 1960: The chemical composition of some bog waters from the Falkland Islands. — J. Ecology **48**, 175—181.

HARKEWITSCH, N. S., 1964: Die Wechselbeziehungen der Formen des Stickstoffes und des Phosphors in den Oberflächengewässern Südkareliens. — In: Die biologischen Grundlagen der Fischwirtschaft in den Binnengewässern des Baltikums, 160—165, Minsk [russ.].

HARNISCH, O., 1929: Die Biologie der Moore (Die Binnengewässer **7**). — Schweizerbart'sche Verlagsbuchhandlung, Stuttgart.

HEDLICH, R., 1965: Vortrag, gehalten anläßlich der Tagung der Biologischen Gesellschaft der DDR in Erfurt.

HÖLL, K., 1960: Untersuchung, Beurteilung, Aufbereitung von Wasser. — De Gruyter, Berlin.

HUSTEDT, F., 1956: Kieselalgen. — Kosmos-Verlag, Stuttgart.

HUTCHINSON, G. E., 1957: A Treatise on Limnology Vol. 1. Geography, Physics and Chemistry. — John Wiley & Sons., New York.

IMSCHENEZKI, A. A., 1959: Mikrobiologie der Zellulose. — Akademie-Verlag, Berlin.

KARCHER, F.-H., 1939: Untersuchungen über den Stickstoffhaushalt in ostpreußischen Waldseen. — Arch. Hydrobiol. **35**, 177—266.

KÄSTNER, M., & W. FLÖSSNER, 1933: Die Pflanzengesellschaften des westsächsischen Berg- und Hügellandes. II. Teil. Die Pflanzengesellschaften der erzgebirgischen Moore. — Veröff. Landesver. Sächs. Heimatschutz Erforsch. Pflanzengesellsch. Freistaates Sachsen.

KIVINEN, E., 1935: Über Elektrolytgehalt und Reaktion der Moorgewässer. — Bull. Soil Division Agricult. Experiment Stat. Finland Nr. 38.

KLOTTER, H.-E., 1953: Die Algen in den Seen des südlichen Schwarzwaldes I. — Arch. Hydrobiol. Suppl. **20**, 442—485.

— 1955: Die Algen in den Seen des südlichen Schwarzwaldes II. — Ebenda **22**, 106—252.

KLUST, G., & H. MANN, 1955: Untersuchungen über den Einfluß chemischer Faktoren auf den Zelluloseabbau im Wasser. — Arch. Fischereiwiss. **6**, 249—275.

KJENSMO, J., 1968: Iron as the primary factor rendering lakes meromictic, and related problems. — Mitt. internat. Ver. Limnol. **14**, 83—93.

KNÖPP, H., 1964: Die „zusätzliche Zehrung" — eine neue biochemische Kennzahl zur Bewertung von Selbstreinigungskraft und Verschmutzung. — Gas- u. Wasserfach **105**, 92—98.

KROGH, A., & E. LANGE, 1932: Quantitative Untersuchungen über Plankton, Kolloide und gelöste organische und anorganische Substanz in dem Furesee. — Internat. Rev. ges. Hydrobiol. **26**, 20—53.

KUNZ, H., 1957: Antibiose-Erscheinungen im Moorboden. — Ber. V. Internat. Kongreß universelle Moorforsch. Lindau, 44—48.

LINSTOW, O. v., 1919: Erläuterungen zur Geologischen Karte von Preußen etc., Blatt Mockrehna.

MARTISONE, M. N., 1958: Die Veränderungen der hydrochemischen Verhältnisse moorbeeinflußter Teiche der Ekabpilsker Teichwirtschaft in der Lettischen SSR unter dem Einfluß komplexer Düngung. — Rybn. choz. vnutr. vod. Latv. SSR II, Trudy **7**, 197—220 [russ.].

MÜNCH, H.-D., & H.-G., PETZOLD, 1955/1956: Zur Fauna des Küstengrundwassers der Insel Hiddensee I. Chem.-physikalische Verhaltnisse und ihr Einfluß auf die Grundwasserfauna. — Wiss. Z. Ernst-Moritz-Arndt-Univ. Greifswald **5**, 413—436.

MÜCKE, D., & R. OBENAUS, 1963: Die Adsorption freier und komplex gebundener Huminsäuren an organischen Trägern. — Wiss. Z. Univ. Rostock **12**, 103—106.

NÜMANN, W., 1941: Der Stickstoffhaushalt eines mäßig eutrophen Sees (Schleinsee). — Z. Fischerei **39**, 387—405.

OBENAUS, R., 1962: Huminsäuren natürlicher Gewässer. Quantitative Bestimmung und Charakterisierung durch ihre oxydimetrischen Eigenschaften. — Diss., Math.-Nat. Fak. Halle—Wittenberg.

— 1963: Huminsäuren natürlicher Gewässer. Quantitative Bestimmung und Charakterisierung durch ihre oxydimetrischen Eigenschaften. — Schweiz. Z. Hydrol. **25**, 9—29.

— 1964: Eine neue Methode zur Bestimmung des spez. PV von Huminsäuren. — Z. anal. Chemie **201**, 428.

ODÉN, S., 1919: Die Huminsäuren. Chemische, physikalische und bodenkundliche Forschungen. — Kolloidchem. Beih. **11**, 75—260.

OHLE, W., 1933: Über organische Stoffe in Binnensseen. — Verh. internat. Ver. Limnol. **6**, 249 bis 262.

— 1934: Chemische und physikalische Untersuchungen norddeutscher Seen. — Arch. Hydrobiol. **26**, 386—464, 584—658.

— 1935: Organische Kolloidgele in ihrer Wirkung auf den Stoffhaushalt der Gewässer. — Naturwiss. **23**, 480—484.

— 1937: Kolloidgele als Nährstoffregulatoren der Gewässer. — Naturwiss. **25**, 471—474.

— 1940: Chemische Eigenart der småländischen Seen. — Verh. internat. Ver. Limnol. **9**, 145—159.

— 1952: Die hypolimnische Kohlendioxydakkumulation als produktionsbiologischer Indikator.— Arch. Hydrobiol. **46**, 153—285.

— 1953: Die chemische und die elektrochemische Bestimmung des molekular gelösten Sauerstoffs der Binnengewässer. — Mitt. intern. Ver. Limnol. **3**, 44 S.

— 1963: Kolloidkomplexe als Kationen- und Anionenaustauscher in Binnengewässern. — Vom Wasser **30**, 50—64.

PEUKERT, V., 1964: Biologische, chemische und physikalische Untersuchungen an den Zuflüssen geplanter Trinkwassertalsperren. — Dipl.-Arb., Math.-Nat. Fak., Univ. Leipzig.

— 1966: Die natürliche Eutrophierung durch Abflüsse aus Hochmooren. — Vortrag, Fortbildungskurs Wasserwirtschaftler in Leipzig.

PIETSCH, K., 1962: Geologie von Sachsen. — Dt. Verl. Wiss., Berlin.

POVOLEDO, D., 1964: Some comparative physical and chemical studies on soil and lacustrine organic matter. — Mem. Ist. Ital. Idrobiol. **17**, 21—32.

PRAKASH, A., & M. A. RASHID, 1968: Influence of humic substances on the growth of marine phytoplankton Dinoflagellates. — Limnol. & Oceanogr. **13** (4), 598—606.

PRINGSHEIM, E., 1914: Kulturversuche mit chlorophyllführenden Mikroorganismen. IV. Mitteilung: Die Ernährung von *Haematococcus pluvialis* FLOT. — Beitr. Biol. Pflanzen **12**, 413—434.

RUTTNER, F., 1962: Grundriß der Limnologie. — 3. Aufl., de Gruyter, Berlin.

RYHÄNEN, R., 1968: Die Bedeutung der Humussubstanzen im Stoffhaushalt der Gewässer Finnlands. — Mitt. internat. Ver. Limnol. **14**, 168—178.

SCAMONI, A., 1959: Geologie, Boden und Klima. — In: Das Naturschutz- und Forschungsgebiet Serrahn.

SCHNEE, B., 1968: Modellversuche über den Einfluß der Oberflächenspeicherung auf die Zusammensetzung von Braunwässern. — Dipl.-Arb., TU Dresden.

SCHREIBER, H., 1927: Moorkunde. — Parey, Berlin.

SHAPIRO, J., 1957: Chemical and Biological Studies on the Yellow Organic Acids in Lake Water. — Limnol. & Oceanogr. **2** (3), 161—179.

— 1964: Effect of Yellow Organic Acids on Iron and other Metals in Water. — J. AWWA **56**, 1062—1082.

— 1966a: The relation of humic color to iron in natural waters. — Verh. internat. Ver. Limnol. **16**, 477—484.

— 1966b: On the measurement of ferrous iron on natural waters. — Limnol. & Oceanogr. **11** (2), 293—298.

SIMM, H., 1969: Zur Humusforschung der Seen Estlands. — Arch. Hydrobiol. **66**, 273—282.

Standard Methods for the Examination of Water and Wastwater, 1960. — 11. Aufl., New York.

TOLPA, S., & E. GORHAM, 1961: The ionic Composition of Waters from three Polish Bogs. — J. Ecology **49**, 127—133.

UHLMANN, D., 1964: Beziehungen zwischen Stoffhaushalt und Planktonmassenwechsel extrem nährstoffreicher Flachgewässer. — Hab.-Schr., Math.-Nat.-Fak., Univ. Leipzig.

VERBEEK, F., 1961: Dosage spectrophotométrique du fer ferreux en présence d'un excès de fer ferrique, et du fer ferrique en présence d'un excès de fer ferreux. — Bull. Soc. Chim. Belg. **70**, 423—434.

VOGLER, P., 1965: Beiträge zur Phosphatanalytik in der Limnologie I. — Fortschr. Wasserchemie u. Grenzgebiete **2**, 100—108.

WARNKE, R., 1964: Biologische und chemische Untersuchungen an kleinen stehenden Fallaub-Gewässern. — Dipl.-Arb., Math.-Nat. Fak., Univ. Leipzig.

WETZEL, R. C., 1968: Dissolved organic matter and phytoplanktonic productivity in marl lakes. — Mitt. internat. Ver. Limnol. **14**, 261—270.

Eingegangen am 30. 11. 1970.

Anschrift des Verfassers:

Dipl.-Biol. HEINZ SCHNEE
DDR — 7113 Markkleeberg
Birkenweg 18

Limnologica (Berlin)	11	1	1976	101–123

KLAUS-DIETER THIEL & HANS-JOACHIM SUBKLEW

Der Bierbach im Naturschutzgebiet „Eldena" bei Greifswald

Faunistische Teilaufnahme eines Fließgewässers

Mit 11 Abbildungen und 3 Tabellen

Inhalt

1. Einleitung

Seit STEINMANNS fundamentaler Arbeit „Die Tierwelt der Gebirgsbäche, eine faunistisch-biologische Studie" (1907) wurde die Lebensgemeinschaft „Bach" mehrmals Gegenstand hydrobiologisch-limnologisch-ökologischer Untersuchungen. Meist jedoch wurden Forschungen an Hoch- und Mittelgebirgsbächen betrieben, während die Fließgewässer der Tiefebene erst in letzter Zeit bearbeitet wurden. Bisher liegen Aufsätze von NIETZKE (Kossau, 1937), BERG (Susaa, 1943/48) und ALBRECHT (Plane, 1952) vor.

In der vorliegenden Arbeit, angefertigt im Zoologischen Institut der Ernst-Moritz-Arndt-Universität Greifswald, wurde nun ein weiteres Fließgewässer im südlichen Ostseeraum untersucht. Dem Charakter der Arbeit entsprechend war der Untersuchungszeitraum von September 1956 bis Mai 1957 relativ kurz befristet. Obwohl nur die Makrofauna und ihre biotopmäßige Gliederung bearbeitet wurden, kann die Arbeit dennoch als Beitrag zur Fließgewässerforschung gewertet werden. Die grundlegenden Arbeiten von ILLIES (1961 und 1963) über Klassifizierung von Fließgewässern wurden eingehend berücksichtigt.

Der Elisenhain steht seit 1961 als „Revier Eldena" des Staatlichen Forstwirtschaftsbetriebes unter Naturschutz.

Das 407 ha große Gebiet „naturnaher Waldgesellschaften" ist damit in seinem Bestand gesichert und wird in bodenkundlicher, bioklimatischer und biozönotischer Hinsicht untersucht (vgl. JESCHKE in BENTHIN 1968).

Pflanzengeographisch nimmt der „Hain" eine Übergangsstellung zwischen den westlichen subatlantisch beeinflußten Wuchsgebieten und den Laubwaldgebieten der östlichen Küstenlandschaft ein. Hier erreicht der Eschen-Buchenwald (Fraxino-Fagetum balticum) die östliche Grenze seiner Verbreitung im Küstengebiet Mecklenburgs. Es ist ein großer Anteil alter Stieleichen (*Quercus robur*) und Hainbuchen (*Carpinus betulus*) vorhanden. An der Baumschicht sind in großem Maße beteiligt Bergahorn (*Acer pseudoplatanus*), Esche (*Fraxinus excelsior*), Buche (*Fagus sylvatica*), Spitzahorn (*Acer platanoides*) und Bergulme (*Ulmus glabra*).

Abhängig von der Bodenfeuchte und den Nährstoffverhältnissen lassen sich etwa drei Untergesellschaften des Eschen-Buchenwaldes unterscheiden. Die feuchteste Ausbildung wird durch Sumpfpippau (*Crepis paludosa*) und Winkelsegge (*Carex remota*) gekennzeichnet. Sie steht am Übergang zum Giersch-Eschenwald (Aegopodio-Fraxinetum). Auch die Lerchenspornuntergesellschaft mit zwei Vertretern dieser Gattung und mit zwei Goldsternarten besiedelt hier sehr feuchte, oft leicht hängige Standorte. Auf etwas trockneren Standorten tritt kleinflächig eine Waldgersten-Untergesellschaft auf. Sehr verbreitet auf Standorten mit mittleren Feuchteverhältnissen ist u. a. das Bingelkraut (*Mercurialis perennis*). Ein echter Erlenbruch (Carici-elongatae-Alnetum) entwickelt sich, Erlen-Eschenbestände (Aegopodio-Fraxinetum) nehmen dagegen einen größeren Raum ein.

Die Vegetation der Bäche setzt sich in der Hauptsache aus Kryptogamen-Gesellschaften (Cratoneurum-filicinum-Gesellschaft, Conocephalum-conicum-Gesellschaft und Rhynchostegietum usciformis) zusammen. Ein Bach-Eschenwald, in der Regel mit vorherrschender Schwarzerle, tritt in inniger Verzahnung mit dem Giersch-Eschenwald auf.

2. *Physiographischer Teil*

2.1. Bachverlauf

Der Bierbach beginnt als Wiesengraben in dem sumpfig-nassen Wiesengelände südöstlich des Naturschutzgebiets Eldena (Abb. 1, 2). Gespeist durch zahlreiche Entwässerungsgräben zieht er sich, teilweise stark zugewachsen, in vielen Windungen zum Waldrand (Abb. 3). Von hier verläuft er in südlicher Richtung, wendet sich scharf nach W und mündet nach kurzer Fließstrecke durch Laubwald in die Fischteiche. Nach deren Durchfluß läuft er mit stärkerer Strömung durch das hier sandig-kiesige Bachbett in die Rehwiese, wo die Fließgeschwindigkeit durch den Pflanzenwuchs im Bachbett außerordentlich verlangsamt wird. Hier nimmt der Bach wieder den typischen Wiesengrabencharakter an. Anschließend beginnt das Laubwaldgebiet des Elisenhains, in dem sich der Bach in nordwestliche Richtung wendet. Der Pflanzenwuchs des an dieser Stelle 2 m breiten und schlammigen Bachbetts geht wegen häufiger Kultur- und Säuberungsarbeiten zurück. Die Wasserführung ist hier sehr unterschiedlich. Nach Verlassen des Elisenhains und Durchfließen der Eldenaer Obstplantage nimmt der Bewuchs im Bachbett wieder stark zu. Hier trocknet in längeren Trockenperioden das Gewässer völlig aus. Der letzte Teil des Baches wurde nur noch ungenau beobachtet, denn bei seinem Eintritt in die Ortschaft Eldena ist die Zuführung von Abwasser aus der ehemaligen Brennerei so stark, daß die natürlichen Verhältnisse grob gestört sind. Nach 300 m mündet er als „mesosaprobes" Gewässer in die Dänische Wiek, eine Buch des Greifswalder Boddens.

Abb. 1. Quellgebiet des Bierbachs

Abb. 2. Pflanzenwuchs im Bierbach im Quellgebiet

Zur Bearbeitung wurde der Bierbach in Regionen eingeteilt. Diese klassifizieren keine Zonen, da sich die physiographischen Faktoren im Untersuchungsgebiet nicht kontinuierlich wandeln.

Quellgebiet: Beginn des Baches bis zum Eintritt in das Naturschutzgebiet,
Oberlauf: Eintritt in das Naturschutzgebiet bis zur Untersuchungsstation 4,
Mittellauf: zwischen den Untersuchungsstationen 4 und 6,
Unterlauf: zwischen den Untersuchungsstationen 6 und 9.

Abb. 3 gibt ein Bild des Bierbachverlaufs mit den eingezeichneten Untersuchungsstationen.

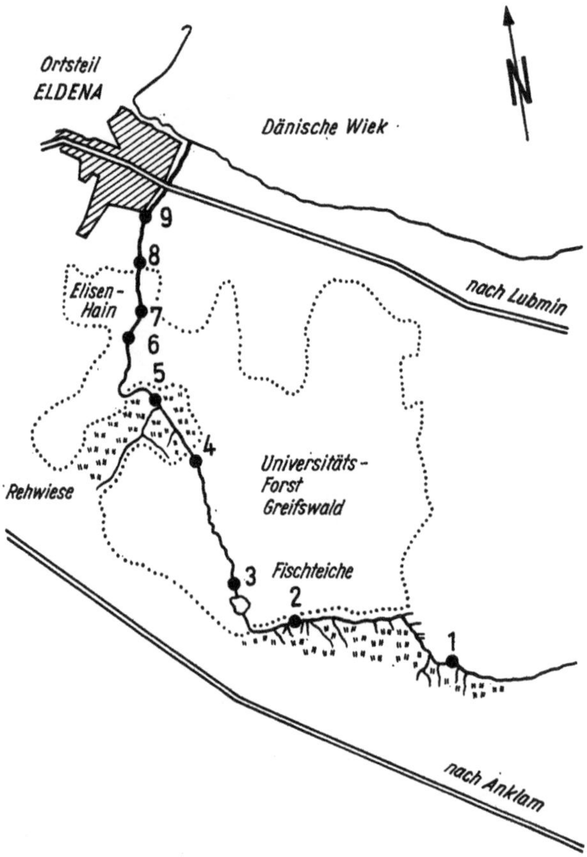

Abb. 3. Bierbachverlauf mit eingezeichneten Untersuchungsstationen (Auszug des Meßtischblattes 1946, Maßstab 1:25000)

2.2. Untersuchungsstationen

Um ein annähernd genaues Bild über die zu verschiedenen Jahreszeiten herrschenden Verhältnisse zu bekommen, wurden an charakteristischen Stellen Stationen festgelegt, an denen von September 1956 bis Mai 1957 regelmäßig Untersuchungen durchgeführt wurden.

Charakterisierung der einzelnen Untersuchungsstationen:

Station 1: Quellgebiet. Wasser fast stagnierend, Wasserführung schwankend, Bachbett stark verkrautet, Untergrund Mutterboden, wenig Schlamm, Detritus, Erlen am Bachrand (Abb. 4).

Station 2: Linkes Ufer Wiese, rechtes Ufer Wald, geringe Strömung, nie völlig austrocknend, relativ tiefes Bachbett, Untergrund Schlamm und Detritus (Blätter), kein Pflanzenwuchs im Bachbett.

Station 3: Waldstation. Mäßige Strömung bei klarem Wasser, nie austrocknend, hoher bemooster Prallhang und flacher Gleithang, Untergrund Sand, Schlamm und Blätter, kein Pflanzenwuchs im Bachbett.

Abb. 4. Bierbach an der Untersuchungsstation 1

Abb. 5. Bierbach an der Untersuchungsstation 6

Abb. 6. Bierbach an der Untersuchungsstation 7

Station 4: Waldstation. Ort der stärksten Strömung im Bierbach, Wasserführung schwankend, nie völlig austrocknend, Untergrund grober Kies und Sand, kein Pflanzenwuchs im Bachbett, vom physiographischen Aspekt her besitzt der Bach hier Rhithroncharakter.

Station 5: In der Rehwiese. Geringe Strömung bis stagnierend, Wasserstand niederschlagsbedingt, austrocknend, Untergrund Schlamm und Detritus, starker Pflanzenwuchs im Bachbett, Erlen am Bachufer.

Station 6: Waldstation, hinter Straßenüberführung. Wasser aus Drainagerohr in Sammelbecken des Baches, Strömung und Wasserführung variierend, breites Bachbett, Untergrund Steine, Schlamm, Sand und Kies, geringer Pflanzenwuchs im Bachbett (Abb. 5).

Station 7: Waldstation. Wasser trübe, geringe Strömung, meist stagnierend, Wasserstand und Wasserführung stark schwankend, jedoch nie völlig austrocknend, Untergrund Schlamm und Detritus, Äste und Holzreste verursachen oft Stauungen, kein Pflanzenwuchs im Bachbett (Abb. 6).

Station 8: Ende des Elisenhains. Linkes Ufer Feld, rechtes Ufer Wald, Strömung und Wasserführung stark niederschlagsbedingt, zeitweise austrocknend, Untergrund Schlamm, lehmiger Sand und Blätter, geringer Pflanzenwuchs im Bachbett.

Station 9: Obstplantage Eldena. Bachuferränder mit Gras bewachsen, geringe Strömung, Wasserführung schwankend, austrocknend, Untergrund Lehm, Schlamm und Detritus, Pflanzenwuchs im Bachbett.

Die Größe des Bachbetts an den einzelnen Untersuchungsstationen ist den Querprofilen der Abb. 7 zu entnehmen.

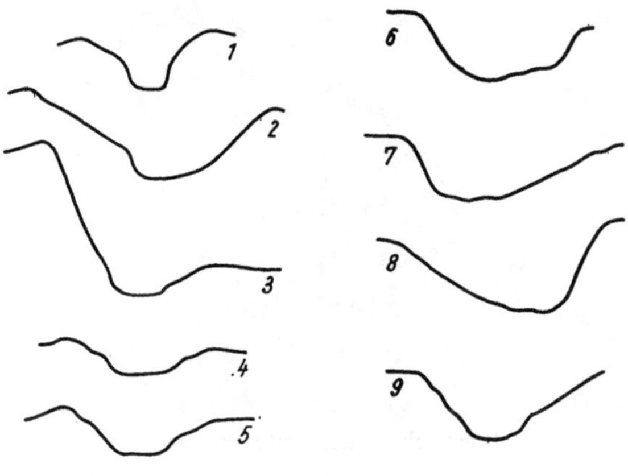

Abb. 7. Querprofile des Bierbachs an den Untersuchungsstationen 1 bis 9 (Maßstab 1:40)

2.3. Untersuchungsmethoden

2.3.1. Feststellung der physikalisch-chemischen Verhältnisse

An den neun Untersuchungsstationen wurden in der Zeit von September 1956 bis Mai 1957 Wasserproben zur chemischen Untersuchung entnommen sowie Temperatur- und Pegelmessungen durchgeführt.

Zu den Temperaturbestimmungen wurde ein Thermometer mit Zehntel-Grad-Einteilung verwendet. Pegelmessungen wurden in der Regel mit einem Lineal von 30 cm Länge, das für diese Zwecke vollauf genügte, durchgeführt.

Die Wasserproben — hauptsächlich zur Sauerstoffbestimmung — wurden in Winkler-Fläschchen entnommen. Jeweils zwei Proben wurden sofort an den Untersuchungsstationen fixiert. In zwei weiteren Flaschen blieb das Wasser unbeeinflußt, zur Bestimmung der Sauerstoffzehrung. Letztere wurde aus zeitlichen Gründen immer erst nach achtzehn Stunden vorgenommen, u. zw. nach der WINKLER-Methode. Die Sauerstoffsättigung des Wassers wurde nach den Tabellen von WERESTSCHAGIN errechnet (Arch. Hydrobiol. **23**, 1932). Die Wasserstoffionenkonzentrationsuntersuchungen erfolgten an Ort und Stelle mit Hilfe des CZENSNY-Indikators.

2.3.2. Feststellung der faunistisch-biologischen Verhältnisse

Es lag nicht im Charakter der Arbeit, eine genaue quantitative Erfassung der Wasserfauna zu erreichen, das Hauptaugenmerk richtete sich auf die qualitative Feststellung der Bachbesiedlung. Die einzelnen Proben wurden zum größten Teil mit dem Pfahlkratzer entnommen, in dichtem Pflanzenbestand die einzelnen Pflanzen abgelesen. In Emailleschalen wurden die Tiere ausgelesen und nach den für die einzelnen Tiergruppen üblichen Methoden konserviert.

2.4. Hydrographisch-chemische Verhältnisse

2.4.1. Laufentwicklung

Der Bierbach mißt von der Quelle bis zur Mündung in die Dänische Wiek 6,52 km, die Entfernung beider Endpunkte ergibt in der Luftlinie nur 4,35 km. Die Laufentwicklung beträgt folglich 66,71% der wirklichen Bachlänge. Die übrigen 33,29% entfallen auf den großen Bogen, den der Bierbach beschreibt, und auf zahlreiche Mikromäanderbildungen. „Dieser im Verhältnis zu Gebirgsbächen hohe Prozentsatz ist charakteristisch für Fließgewässer des Hügel- und Flachlandes“ (ALBRECHT 1952).

2.4.2. Gefälle

Entsprechend den Höhenlinien auf dem Meßtischblatt liegt das Quellgebiet 22 m ü. NN. Das würde einem Gefälle von 3,38 m pro Kilometer Bachverlauf entsprechen. Der Höhenunterschied ist jedoch ungleich auf den Gewässerlauf verteilt. Auf der Strecke zwischen den Stationen 2 und 5 (rund ein Drittel der Gesamtlänge des Baches) finden sich etwa zwei Drittel des Gefälles. Auf einer Länge von etwa zwei Kilometern beträgt der Höhenunterschied ca. 14 m; das entspricht einem Gefälle von 7 m pro Kilometer Bachlänge, während die übrigen 4,5 Kilometer 8 m Höhenunterschied aufweisen, das entspricht einem Gefälle von nur 1,77 Metern pro Kilometer Bachlänge. Die Darstellung des Bierbachgefälles gibt darüber ein genaues Bild (Abb. 8).

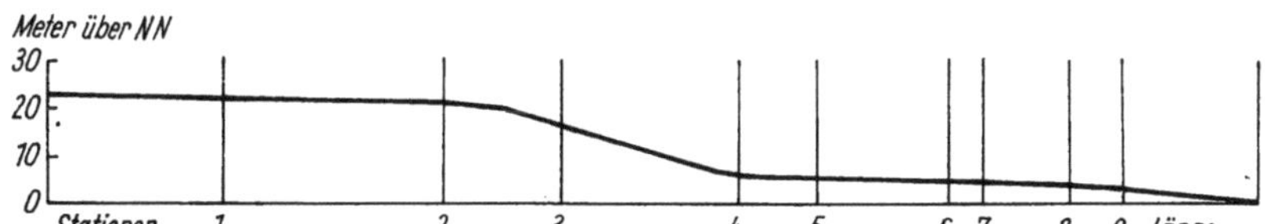

Abb. 8. Gefälle des Bierbachs (Längenmaßstab 1:50000, Höhenmaßstab 1:2000)

2.4.3. Wasserstand

Nach der Wasserführung kann für den Gesamtlauf die Bezeichnung Bach nicht aufrecht erhalten werden, da der Bierbach ein periodisch partiell austrocknendes Gewässer von Wald- und Wiesengrabencharakter ist. Die Wasserführung ist größtenteils von den Niederschlägen direkt abhängig, nur ein geringer Teil des Bachwassers stammt aus den Zubringergräben des sumpfignassen Wiesengeländes der Quellregion. In den Wintermonaten war überall im Bachbett genügend Wasser vorhanden (durchschnittliche Tiefe = 25 cm). Die maximale Wassertiefe von 60 cm wurde am 31. Oktober 1956 an der Untersuchungsstation 8 gemessen, hervorgerufen durch reichliche Niederschläge der vorhergegangenen Tage. Die monatlichen Niederschläge sind in Abb. 9 graphisch dargestellt. Die Werte wurden uns freundlicherweise von der meteorologischen Station in Wieck überlassen.

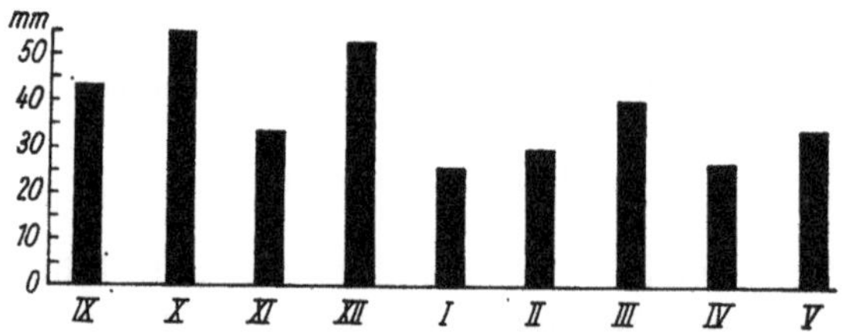

Abb. 9. Monatliche Niederschläge von September 1956 bis zum Mai 1957

2.4.4. Chemismus

Die Wasserstoffionenkonzentration hat im Verlauf des Baches einen Schwankungsbereich von 6,5—7,5, der Jahresmittelwert beträgt 7,25. Der erwähnte Schwankungsbereich bezieht sich auf die gesamte Untersuchungsperiode. An vielen Tagen wurden die beiden Grenzen nicht erreicht. Es ist weder ein systematisches Ansteigen noch ein Abfallen des pH-Wertes vom Quellgebiet bis zur Mündung, noch eine Konstanz bei einer Untersuchungsstation festzustellen. Das Wasser ist im allgemeinen als neutral zu bezeichnen. Die unter dem Neutralpunkt liegenden Werte sind durch Huminsäureeinflüsse zu erklären, da hier der schlammige Untergrund und die geringe Strömungsgeschwindigkeit den Abbau organischer Substanzen, insbesondere der Zellulose und des Lignins abgestorbener Pflanzenteile, sehr begünstigt. Schlagartig aber ändert sich das Bild bei Eintritt des Baches in die Ortschaft Eldena. Hier wurde von der damaligen Brennerei 40 °C warmes Wasser mit einem pH-Wert von 4,5 ins Bachbett geleitet. Deshalb konnte der letzte Teil nicht berücksichtigt werden. Das biologische Gleichgewicht war durch diese Maßnahme erheblich gestört.

Der Sauerstoffgehalt des Bierbachwassers schwankte zwischen den Mittelwerten 7,0 und 10,0 mg/l. Der niedrigste Wert wurde an der Station 8 mit 3,5 mg je l, der höchste an der Station 4 mit 13,4 mg/l gemessen. Während bis zur Station 6 — ohne jahreszeitliche Schwankungen zu berücksichtigen — regelmäßige Werte um 9,0 mg/l gemessen wurden, kam es bei den beiden folgenden Stationen zu sehr starken Abweichungen, die sich kaum 400 m weiter an der Untersuchungsstation 9 nicht mehr auswirkten.

Das kann folgende Gründe haben: Bei den Untersuchungsstationen 1 und 5 sowie mit Einschränkungen an der Station 2 wird die fehlende Strömung und die dabei entstehende Schlammsedimentation, die sauerstoffzehrend wirkt, durch eine üppige, Sauerstoff an das Wasser abgebende, submerse Vegetation ausgeglichen, und der Sauerstoffgehalt bleibt in den bereits angeführten Grenzen relativ konstant. Das sandig-kiesige Bachbett und die durch das stärkere Gefälle entstehende gleichmäßige Strömung, die einen innigeren Kontakt mit der Luft und damit eine Sauerstoffanreicherung fördert, scheinen die an den Untersuchungsstationen 3 und 4 gefundenen Höchstwerte zu erklären. Ähnlich gelagert sind die Verhältnisse an der Station 6. Bei der nun beginnenden Strecke mit geringerer Wasserführung und fast völliger Stagnation erhält der Bierbach tümpelähnliche Eigenschaften, z. B. schnelles Reagieren auf die äußeren Temperaturverhältnisse. Auch hier muß bezweifelt werden, ob Fließgewässer mit dem Charakter des Bierbachs die Bezeichnung „Bach" verdienen, da der eine „Generalfaktor jeden Fließgewässers" (Illies 1961), nämlich die Strömung und der dazu proportional verlaufende Faktor des Sauerstoffgehaltes, kaum erkennbar sind. Zwei Beispiele der Station 7 mögen dies dokumentieren:

24. 8. 1956	Wassertemperatur:	17,2 °C
	Sauerstoffgehalt	3,6 mg/l
4. 5. 1957	Wassertemperatur:	7,3 °C
	Sauerstoffgehalt:	10,7 mg/l

Außerdem fehlt dieser Untersuchungsstation jeglicher Pflanzenwuchs, während durch Blattfall und Humus die sauerstoffzehrende Schlammschicht bis auf 25 cm anwächst, der in den heißen Tagen bei Probeentnahmen im Bachbett übelriechende Gase entweichen. An der letzten Untersuchungsstation (9) ist nur eine geringe Schlammsedimentation vorhanden, Pflanzenwuchs und langsame Strömung bewirken beim Bierbach wieder Durchschnittsverhältnisse.

Die mittlere Sauerstoffsättigung betrug 85%, der Hauptschwankungsbereich lag zwischen 80% und 95%, vollständige Sättigung oder Übersättigung des Wassers wurden selten erreicht, doch machte sich ein Ansteigen in den Frühjahrsmonaten bemerkbar. Der niedrigste Wert von 34,31% entstammt einer Messung bei 8, als sich ein Minimum an Wasser in einer Vertiefung des Bachbettes angesammelt hatte, worin sich eine große Anzahl *Gasterosteus aculeatus* L. und viele Trichopterenlarven angesammelt hatten. Der Höchstwert wurde hier mit 121,71% festgestellt.

Die Sauerstoffzehrung wurde nach achtzehn Stunden bestimmt. Sie war am niedrigsten bei 5 mit 0,03 mg/l, am höchsten bei 2 mit 1,6 mg/l. Die durchschnittlichen Zehrungswerte lagen zwischen 0,6 mg/l und 1,1 mg/l. Auch hier sind aus den vielen vorliegenden Werten weder ein jahreszeitlicher Rhythmus noch sonstige Gesetzmäßigkeiten herauszulesen, nur scheint bei geringer Wasserführung und damit verbundener Stagnation die Zehrung anzusteigen.

Abb. 10 zeigt eine typische Tagesaufnahme, in der pH-Wert, Wasserstand (Pegel), Luft- und Wassertemperatur sowie die Sauerstoffverhältnisse (Sauerstoffgehalt, Sättigungswert, Zehrungswert und das Sauerstoffdefizit) in einer zusammenhängenden Darstellung aufgeführt werden.

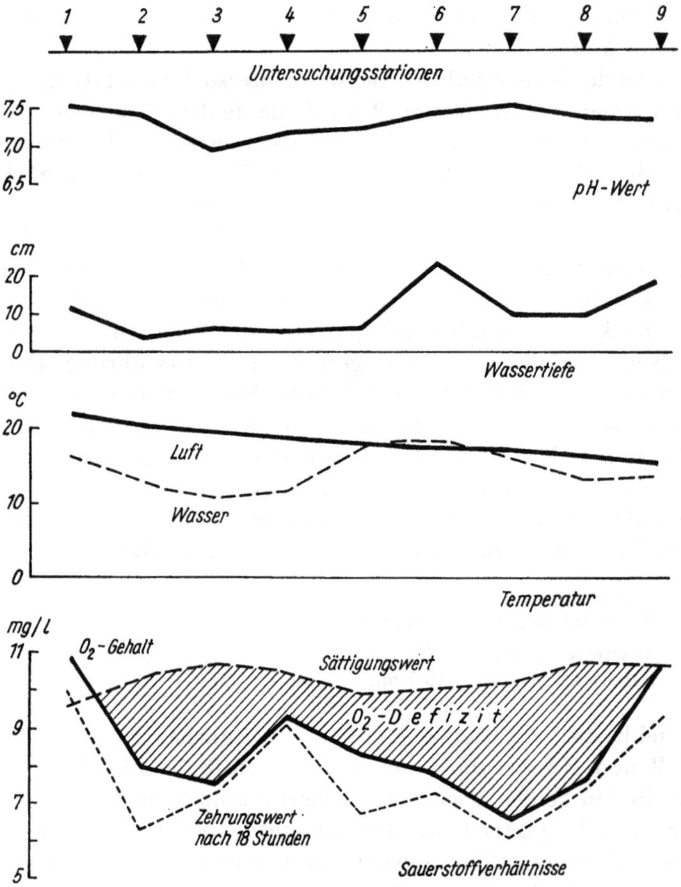

Abb. 10. Typische Tagesaufnahme vom 15. 5. 1957

3. *Biologischer Teil*

3.1. Die Flora

Da die Wasservegetation vielen Tieren erst eine Besiedlung des Bachbiotops ermöglicht, soll in diesem Abschnitt kurz darauf eingegangen werden. Jedoch werden nur die Pflanzen genannt, welche direkt im Bachwasser wachsen.

Bedingt durch den Verlauf des Baches sind zwei große Biotope 1. Ordnung zu unterscheiden: Einmal der durch reichhaltigen Bewuchs des Bachbetts ausgezeichnete Biotop Wiesengraben und zum anderen der Biotop Waldgraben, mit spärlichem bzw. nicht bewachsenem Bachbett, die in wiederholter Folge hintereinander auftreten und charakterbestimmend für den Bierbach sind.

Insgesamt wurden folgende Arten festgestellt:

Alisma plantago-aquatica L.
Calla palustris L.

Callitriche palustris L.
Carex acutiformis Ehrh.
Equisetum fluviatile L. em. Ehrh.
Iris pseudacorus L.
Lemna minor L.
Mentha aquatica L.
Myosotis palustris (L.) Nathh.
Phragmites communis Trin.
Ranunculus aquatilis L.
Ranunculus sceleratus L.
Ranunculus repens L.
Sium erectum Huds.
Veronica anagallis-aquatica L.
Veronica beccabunga L.

Die Verteilung der Arten in den einzelnen Abschnitten des Bierbachs ist aus Abb. 11 ersichtlich, die auch halbquantitat:ve Angaben bringt.

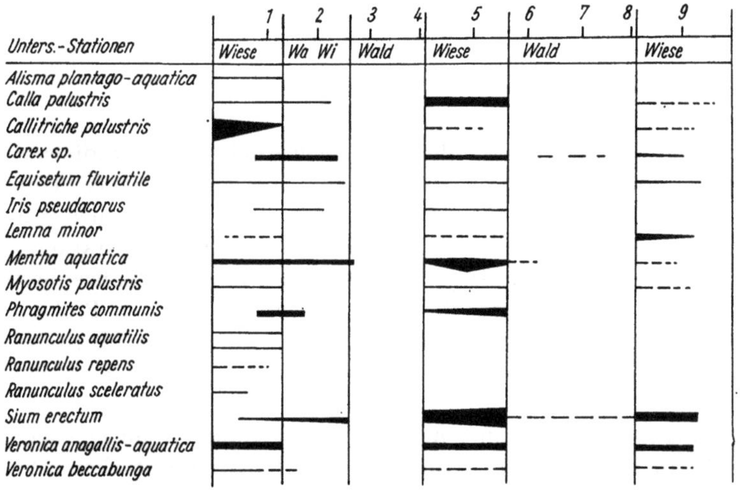

Abb. 11. Verteilung der Pflanzenarten im Bierbach

3.2. Die Fauna

Es wurde versucht, eine möglichst große Zahl von Tiergruppen zu erfassen, doch wurden Protozoa, Hydrozoa, Nematoda, Oligochaeta, Polychaeta, Entemostraca und Diptera nicht berücksichtigt. Die Liste enthält die Ergebnisse einer faunistischen Teilaufnahme. Die beim systematischen Überblick angeführten Bezeichnungen der Fundorte entsprechen den im Abschnitt 2.2. genannten Stationen 1—9.

3.2.1. Systematischer Überblick

Spongiaria

Das Fehlen von *Spongilla lacustris* L. und *Ephydatia fluviatilis* L. scheint am Mangel von Anheftungsmöglichkeiten der Kolonien, also an Steinen und Ästen, in der Zone des konstant strömenden Wassers zu liegen. Dies trifft für den Abschnitt zwischen 2 und 4 zu.

Turbellaria

Typische Vertreter im Bierbach sind die folgenden Tricladenarten: *Planaria torva* M. SCH., *Polycelis nigra* EHRBG. und *Dendrocoelum lacteum* (MÜLLER).

Planaria torva wurde in einigen Proben im Unterlauf zwischen Pflanzenrasen und Blättern gefunden. Der Prozentsatz auf die Gesamtbesiedlung bezogen ist minimal. Etwas häufiger ist *Polycelis nigra* im Unterlauf anzutreffen, doch auch im Quellgebiet waren manchmal einige Exemplare zu beobachten. Die Hauptaufenthaltsorte waren die dichten Bestände von *Sium*, *Callitriche* und *Ranunculus aquatilis*. *Dendrocoelum lacteum* hielt sich meistens auf oder zwischen den den Bachgrund bedeckenden Blättern auf. Stellen stärkerer Strömung wurden gemieden. Nur bei 6 kam *Dendrocoelum lacteum* auch an den wenigen hier vorhandenen Steinen vor, der Hauptfundort lag jedoch nahe 8.

Hirudinea

Wesentlichen Anteil an der Besiedlung des Baches haben die Hirudinea, denn das an vielen Stellen stagnierende Wasser und die zahlreichen Orte geringer Strömung bieten ihnen außerordentlich gute Lebensmöglichkeiten. Die vorhandenen Saugnäpfe gestatten ihnen auch, aber nur vereinzelt, Aufenthalt an den Orten stärkerer Strömung.

Glossiphonia complanata (L.) lebte hauptsächlich in den Pflanzenbeständen von *Sium* und *Callitriche*, wurde jedoch einmal — bei 7 — in einer Probe an pflanzenfreiem Orte gefunden. Die größten Exemplare kamen im unteren Abschnitt des Baches vor, die hier massenweise vorhandenen Limnaeen boten dem Egel eine reichliche Nahrungsgrundlage.

Haemopis sanguisuga L. kommt in wenigen Exemplaren im Quellgebiet, häufig dagegen bei 6, an Steinen vor, was bei der hier doch etwas stärkeren Strömung erstaunlich anmutet.

Herpobdella octoculata L. lebt vereinzelt am Boden des Baches und auf *Phragmites*-Stengeln.

Malacostraca

Der Zahl nach und hinsichtlich seines Volumens steht *Gammarus* (*Rivulogammarus*) *pulex pulex* (L.) an erster Stelle des Untersuchungsgebiets. Seine eurytopen und eurythermen Eigenschaften kennzeichnen ihn als Ubiquisten, der den gesamten Bachlauf besiedelt. Submerse lebende und abgestorbene Teile höherer Pflanzen, die nach WILLER die wesentlichste Nahrungsquelle darstellen, sind reichlich vorhanden, die Pflanzen sind zugleich die Hauptaufenthaltsorte der zahlreichen Jungtiere. Dazu kommen noch die günstigen pH-Werte des Wassers, denn der durchschnittliche Wert von 7,25 des Bierbachs liegt in dem von 7,2—7,8 reichenden Optimum.

Eine große Rolle für die Färbung spielen die Bodenverhältnisse des Bachbetts. Der schlammige Untergrund bei 2 wurde in der überwiegenden Mehrheit von dunkelgefärbten Individuen besiedelt, während auf dem sandig-kiesigen Bachbett bei 4 gelblich gefärbte Gammariden überwogen. Auffallend war die bei den meisten Gammariden vorhandene Protozoenepifauna (Ciliata).

Gammarus (*Rivulogammarus*) *roeselii* GERVAIS, nach SCHELLENBERG (1942) „in den langsam fließenden Bächen des Tieflandes und des Hügellandes überall gemein“, der auch an den Sauerstoffgehalt nicht so große Ansprüche stellen soll (EIDEL) und somit im Bierbach zu vermuten wäre, wurde trotz sorgfältiger Nachprüfungen besonders in den Pflanzenbeständen ruhiger Buchten nie gefunden.

Asellus aquaticus (L.) ist überall im Bachgebiet anzutreffen, doch längst nicht in dem Maße wie *Gammarus pulex pulex*. Nur dort, wo verwesende pflanzliche Bestandteile angehäuft waren, wurde die Besiedlung stärker.

Hydracarina

Alle Arten sind für Flachlandgewässer typisch.

Hygrobates (*Hygrobates*) *longipalpis* (HERM.) ist nach VIETS (1936) „bei uns die häufigste der Gattung und in den stehenden Gewässern ebenso wie in fließenden zu Hause“. Die dichten Bestände von *Callitriche, Sium* und *Ranunculus aquatilis* sind der Hauptaufenthaltsort. Proben stagnierender Stellen waren weitaus ergiebiger als solche an Orten strömenden Wassers.

Arrhenurus (*Megaluracarus*) *cylindratus* PIERS. wurde nur ganz vereinzelt bei 8 im Mai 1957 gefunden. Auch hier ist eine Übereinstimmung mit den Fundortsangaben von VIETS (1936) vorhanden. Er schreibt u. a. „. . . die Art ist verbreitet in stehenden und langsam fließenden Gewässern, sie tritt aber nicht häufig auf, sie scheint kühles Wasser zu bevorzugen. . .“.

Eine ebenfalls selten auftretende Art ist *Piona nodata*. Sie ist durchscheinend rot gefärbt. Die Fundstellen lagen ausnahmslos nahe 1.

Rhynchota

Die Wasserwanzen sind über den ganzen Bachlauf verteilt anzutreffen.

Velia currens F. besiedelt als positiv-rheotaktisches Tier die Wasseroberfläche. In den Buchten des Baches kam es manchmal zu Massenentwicklungen. Nach einer Reihe trockener Tage sammelten sich viele Veliiden auf den restlichen Wasserflächen des austrocknenden Bachunterlaufs.

Gerris sp. lebt ebenfalls gesellig als Oberflächenbewohner, doch in weit weniger Exemplaren als *Velia currens*.

Nepa rubra L. wurde vereinzelt an seichten Stellen des Bierbachs im Pflanzenbewuchs beobachtet. Über 5 gingen die Fundstellen nicht hinaus.

Trichoptera

Gegenüber anderen Insektenordnungen wurden die Larven von Trichopteren am häufigsten gefangen. Viele Tiere sind typische Vertreter stehender Gewässer, die

Hauptanzahl der Arten wurden in dem Biotop Waldgraben erbeutet. Bemerkenswert ist die Erscheinung, daß nur Larven aus der Familie der Limnephilidae gefunden wurden. Im einzelnen handelt es sich um folgende Arten: *Allophylax dubius* STEPH., *Glyphotaelius pellucidus* RETZ., *Halesus* sp., *Limnophilus bipunctatus* CURT., *L. griseus* L., *L. lunatus* CURT. und *L. rhombicus* L.

Limnophilus rhombicus wurde während der Untersuchungsperiode im gesamten Bach (1—9) fast in jeder Probe angetroffen. Die Köcher sind nach den unterschiedlichen Fundorten verschieden aufgebaut: In der Hauptsache wurden Vegetabilien, daneben aber auch Sandkörnchen und in einigen Fällen Conchylien zum Bau benutzt. Die Larven lebten sowohl zwischen Pflanzenbeständen wie auch auf sandigem Boden und verwesendem Laub des Bachbetts, nur Stellen schlammigen Untergrundes wurden gemieden.

Weniger häufig kam *Limnophilus bipunctatus* vor. Mit ihrem konischen Sandkörnchenköcher findet sie sich nur an pflanzenreichen Stellen mit geringer Strömung (1 und 5). Der häufigste Fundort war die Wiesenquellregion südöstlich des Naturschutzgebiets Eldena, dort wo der Grabenrand mit Erlen bestanden ist.

Larven von *Limnophilus lunatus* wurden nur ganz vereinzelt in Pflanzenbeständen bei 9 im Januar 1957 gefunden.

Als Vertreter stehender Gewässer ist die Larve von *Limnophilus griseus* ebenfalls selten. Unter *Ranunculus aquatilis* wurden einige Exemplare am Grunde des Baches bei 1 gefunden.

Halesus sp. dagegen ist sehr häufig anzutreffen. Wahrscheinlich handelt es sich hierbei um die zwei Arten *Halesus tesselatus* RAMB. und *Halesus digitatus* SCHRK., doch war eine genaue Bestimmung nicht möglich. Die Larven wurden sowohl an Stellen mit starker Strömung (4) auf kiesigem Bachgrund wie auch an stagnierenden Orten in Pflanzenbeständen und in dem im Bach vorhandenen Laub gefunden.

Die Köcher waren dem Untergrund entsprechend sehr verschieden: fast völlig aus Sandkörnchen hergestellte mit selten eingebauten Rindenstückchen, z. T. aus Steinen und Vegetabilien gebaute und beschwerte sowie Gehäuse aus Stengel- und Blattstückchen mit kurzen Belastungsteilen. Eigenartig war die Beobachtung, daß nur die zur Gattung *Halesus* gehörenden Individuen ausschließlich am Kopf-Thorax-Abschnitt von zahlreichen Protozoen (Ciliaten) besiedelt waren, der Epifauna von *Gammarus pulex* gleichend.

Larven von *Glyphotaelius pellucidus* lebten ausschließlich an Orten stärkerer Strömung. Die zwei Fundorte nahe 3 und 6 ähnelten sich auffallend: Nach Weg- bzw. Straßenüberführung strömt das Wasser aus Drainageröhren in ein Sammelbecken. Das hier lagernde Laub eignet sich hervorragend zum Bau des breiten Köchers. Nach ULMER (1909) ist auch diese Art Bewohner von Tümpeln und flachen Teichen, an deren Boden Laub liegt.

Die Larve von *Allophylax dubius* ist dagegen ein Vertreter langsam fließenden pflanzenreichen Wassers. Es wurde nur ein Exemplar nahe 9 zwischen Pflanzen unter Erlen gefunden. Weder NIETZKE (1938), ILLIES (1952) noch ALBRECHT (1952) erwähnen *Allophylax dubius* in ihren Untersuchungen. Aus dem einmaligen Fund im Bierbach könnte man schlußfolgern, daß es sich dabei um eine im Norden der DDR wenig verbreitete Art handelt.

Ephemeroptera

Arm war die Ephemeridenfauna des Bierbachs. Die einzigen Vertreter waren Larven von *Ephemera vulgata* L., von denen viele noch nicht das letzte Larvenstadium erreicht hatten. Sie lebten an sandig-lehmigen Orten im Boden eingegraben, nahe 5 und 6 in Ufernähe.

Einen Einzelfund stellte die zwischen Pflanzen bei 1 gefundene Larve von *Cloëon* sp. dar.

Odonata

Odonatenlarven wurden in den Fischteichen wiederholt beobachtet, da sie aber als typische Vertreter stehender Gewässer im Bierbach fehlten, wurden die Funde nicht ausgewertet.

Neuroptera

Sialis sp. kam nur nahe 9 vor, wurde hier allerdings häufiger gefunden. Die Larve scheint Stellen stärkerer Strömung zu meiden und findet sich in Blatt- und Detritusablagerungen oder im Schlamm unter Pflanzenbeständen.

Plecoptera

Arten- und individuenarm waren auch die Steinfliegen im Bierbach vertreten.

Nemurella picteti Klap. lebt zwischen Pflanzenbeständen von *Sium*, *Ranunculus aquatilis* und *Callitriche* in den Wiesenabschnitten des Baches. Die Fänge beschränkten sich auf die Monate Januar bis Mai 1957.

Coleoptera

Neben einer nicht bestimmbaren Larve wurden nur Imagines gefangen. Die geringe Strömung und der reiche Pflanzenwuchs im Bachbett gewährten den stillwasserliebenden Käfern einen guten Aufenthalt in den ruhigen Buchten. Die Regionen des Wiesengrabens bildeten ebenfalls das Hauptareal für sie.

Gyrinus natator L. ist der bei weitem häufigste Käfer. Er findet sich bei gutem Wetter in Scharen auf der Oberfläche der Buchten, kreisförmige Boge beschreibend, im gesamten Untersuchungsgebiet oftmals gesellig mit *Velia currens* und *Gerris* sp.

Dytiscus marginalis L., nach Lampert (1925) „fast stets in Teichen und sonstigen stehenden Gewässern zu finden“, wurde viel seltener und nur in Regionen langsam fließenden Wassers zwischen Pflanzen beobachtet, kam jedoch auch dort vor, wo das saure Abwasser der Brennerei dem Bierbach zugeleitet worden war (August 1956). An dem begradigten Teil des Baches, der durch die Obstplantage fließt, war der Käfer verschwunden.

Ilybius obscurus Mrsh. lebte in einem begrenzten Bachbezirk. Hauptfundorte waren der mit Laub bedeckte Boden und das freie Wasser der Regionen Wald- und Wiesengraben an Orten schwächerer Strömung oder stagnierenden Wassers.

Hydroporus palustris L. kam vereinzelt im Quellgebiet zwischen *Ranunculus aquatilis* vor. *Haliplus* sp. sowie ein nicht näher bestimmbarer kleiner Vertreter aus der Familie der Hydrophilidae wurden nur in den Monaten April und Mai selten bei 1 und 9 gefangen. Auch hier waren es die Pflanzenbestände, die den Tieren als Siedlungsraum dienten.

Mollusca

Nach den Malacostraca stellten die Mollusken individuenmäßig den Hauptteil der Besiedlung des Baches. Die Pflanzen der Stillwasserzonen waren überaus reich mit Jungtieren besetzt. Die Überbevölkerung einzelner Bachbezirke führte zu typischen Kümmerformen.

Planorbis corneus L. war im gesamten unteren Bereich des Baches zu allen Jahreszeiten häufig anzutreffen, besonders ihre an der feinen Gitterzeichnung deutlich erkennbaren Jugendformen. Die Posthornschnecke besiedelte vorwiegend Steine und ins Wasser gefallene Äste, kroch an Pflanzenstengeln, und selbst im Schlamm bei 7 war sie anzutreffen. Man kann sie als Wanderform (im Sinne von Nietzke 1938) zwischen Pflanzen und Schlamm bezeichnen.

Limnaea (*Radix*) *ovata* Drap. war ebenfalls dort weit verbreitet, wo dichter Pflanzenwuchs reichlich Nahrung bot, doch waren voll ausgewachsene große Schneckengehäuse selten. In einer Pfahlkratzerprobe vom 19. 11. 1956 wurden bei 9 elf größere und 104 kleinere Exemplare gezählt. Starke Temperatur- und Sauerstoffschwankungen hatten auf ihre Entwicklung keinen sichtbaren Einfluß.

Als anspruchsloser Detritus- und Pflanzenfresser wurde *Pisidium* (*Eupisidium*) *subtruncatum* Malm. überall im Schlamm gefunden. Einige Exemplare zeigten Merkmale, die auf *Pisidium casertanum* Poli. deuteten, doch kann es sich um Umweltanpassungen (Kümmerformen) gehandelt haben. An manchen Orten kam es oft zu Massenentwicklungen, so wurden z. B. in der schon erwähnten Probe vom 19. 11. 1956 619 Pisidien gezählt, die ca. einen halben Quadratmeter des schlammigen Untergrunds besiedelten. Einige wenige Tiere wurden auch im sandigen Bachbett bei 4 gefunden.

Pisces

Gasterosteus aculeatus L. ist häufig im Bach. Er steigt bis zu den Fischteichen bachaufwärts, von der Quelle bis zu den Fischteichen fehlt der Stichling. In derselben Bachregion lebt der seltenere *Gasterosteus pungitius* L. Nach Angaben einiger Dorfbewohner sollen auch *Salmotrutta* f. *fario* und *Anguilla vulgaris* im Unterlauf vorkommen.

Im Frühjahr 1956 sind in den in Eldena gelegenen und vom Bach gespeisten Sammelbecken — nach Berichten — mehrere Aale und einige Bachforellen gefangen worden. Die genaue Nachprüfung verlief leider ergebnislos.

Anhang zur Fauna

Einige Beobachtungen über die nicht bestimmten Tiergruppen sollen hier kurz erwähnt werden.

Nematoden traten an schlammigen und sandigen Stellen im Detritus auf. Oligochaeten und besonders Tubificiden kamen häufig im schlammigen Untergrund vor, hauptsächlich bei 9, mit Einschränkung bei 5. Darunter befand sich *Tubifex barbatus* GRUBAY. Larven aus der Familie der Tipulidae fanden sich im Oktober des Jahres 1956 im Bodenbelag des Unterlaufs. Im Schlamm, Sand und den dichten Pflanzengefilzen zählten Larven der Tendipediden zu den häufigsten Vertretern der Bachfauna. Zahlenmäßig scheinen sie alle anderen Gruppen zu übertreffen. Die Hauptentwicklung dieser Larven erfolgte an Orten geringer Strömung.

3.3. Ökologische Betrachtungen

Die genauere Betrachtung der Artenliste ergibt, daß der Bierbach hauptsächlich von eurythermen, stillwasserliebenden Organismen besiedelt wird. Verglichen mit anderen Fließgewässern ähnlicher Beschaffenheit ist die Artenarmut auffallend. Nun ist der Bierbach aber ein periodisch austrocknender Bach und nur solche Organismen, die eine kurze oder längere Trockenperiode überstehen können, haben die Aussicht, im Unterlauf entsprechende Lebensmöglichkeiten vorzufinden. So muß von der bisher üblichen, besonders auf den physiographischen Besonderheiten eines Fließgewässers beruhenden Einteilung in Quellgebiet, Ober-, Mittel- und Unterlauf abgesehen werden, da weder die Strömungsgeschwindigkeit in Bodennähe, Temperaturamplituden der Monatsmittel, Sauerstoffsättigung des Wassers noch Wasserführung und Untergrundbeschaffenheit eine solche Einteilung zulassen.

Sowohl bei flüchtiger Betrachtung als auch bei genauerer Untersuchung sind zwei Biotope zu charakterisieren, der relativ arten- und individuenreiche Biotop „Wiesengraben“ und der artenarme Biotop „Waldgraben“. Tab. 1 zeigt die Verteilung der Fauna auf die im Bachverlauf auftretenden Biotope erster Ordnung. Von den 40 bestimmten Arten des Bierbachs wurden 39 im Biotop Wiesengraben gefunden, davon 10 selten. Im Biotop Waldgraben ergaben die Untersuchungen 23 Arten, davon kamen 5 nur selten vor. *Glyphotaelius pellucidus* ist typisch für den letzten Biotop, alle anderen sind auch in den Wiesengegenden im Bach zu finden.

3.3.1. Verteilung der Fauna auf die Biotope erster Ordnung des Bierbachs

(s. Tab. 1, Seite 118)

3.3.2. Besiedlung der Biotope zweiter Ordnung (Besiedlung der Substrate bzw. Habitate)

(s. Tab. 2, Seite 119)

Als weiterer Versuch der limnologischen Klassifizierung eines Fließgewässers genügt die Aufstellung einer Faunenliste nicht, da sie dafür nicht genügend aussagekräftig ist, sondern es empfiehlt sich eine Aufteilung der Bachfauna auf Substrat-Kategorien (Habitata), wofür besonders die Benthos-Organismen bedeutsam sind. „Die Substrate unterscheiden sich von den Zonen vor allem dadurch, daß sie mosaikartig und in wiederholter Folge hintereinander auftreten können“ (ILLIES & BOTOSANEAU 1963), was besonders im Potamon bisher untersuchter größerer Fließgewässer zu beobachten ist.

Tab. 1. Die Faunenverteilung in den Biotopen erster Ordnung

Biotop 1. Ordnung	Wiese	Wald, Wiese	Wald	Wiese	Wald	Feld, Wiese
Art	Untersuchungsstation					
	1	2	3	4–5	6–7	8–9
Planaria torva					×	×
Polycelis nigra	××			×		××
Dendrocoelum lacteum					××	×
Glossiphonia complanata	×	×		××	××	×××
Haemopis sanguisuga	××				×××	××
Herpobdella octoculata	×	×	×	×	××	××
Nematoda (sp.)	××		××	×		××
Gammarus pulex pulex	×××	××××	×××	×××	××××	××××
Asellus aquaticus	××	××	××	×××	×××	×××
Hygrobates longipalpis	××			××	×	×××
Arrhenurus cylindratus					×	×
Piona nodata	××					
Velia currens	××	××	×	××	×××	××
Gerris sp.	××			×	××	×
Nepa rubra	××	×		×		
Allophylax dubius						×
Glyphotaelius pellucidus		××			××	
Halesus sp.	××	×	×	×××	×××	××
Limnophilus bipunctatus	×			××		
Limnophilus griseus	×					
Limnophilus lunatus						×
Limnophilus rhombicus	××	××	××	××	×××	×××
Ephemera vulgata				××	×	
Cloeon sp.	×					
Tendipediae (sp.)	×××	××	×	×××	××	××××
Tipulidae (sp.)						××
Sialis sp.						×××
Nemurella picteti	×			×××		××
Gyrinus natater	××	×		××	×××	×××
Dytiscus marginalis						××
Ilybius obscurus	××	×		×	××	××
Hydroporus palustris	×					
Haliplus sp.	×					×
Hydrophilidae (sp.)	×					×
Planorbis corneus	×	××	×	××	×××	×××
Limnaea ovata	××	×	×	×××	×××	××××
Bathyomphalus contortus						×
Pisidium subtruncatum	×××	×	×	××	××	××××
Gasterosteus aculeatus				×	××	××
Gasterosteus pungitius				×	×	×

Erklärung der Zeichen (Versuch einer quantitativen Auswertung):

× × × × – **sehr zahlreich;**
× × × – **häufig;**
× × – **vorkommend;**
× – **selten**

Tab. 2. Besiedlung der Biotope 2. Ordnung

Biotop 2. Ordnung Art	 Ökolog. Bezeichn.	Sand	Schlamm	Steine	Detri-tus	freies zen	Pflan-Wasser	Wasser-ober-fläche
Nematoda		×	×		×			
Planaria torva	W. Pd.				⊗	⊗		
Polycelis nigra						○		
Dendrocoelum lacteum				×	⊗			
Glossiphonia complanata	W. Ps.					⊗		
Haemopis sanguisuga	W. Pst.			⊗		×		
Herpobdella octoculata			×	×		×		
Gammarus p. pulex	U.	○				○		
Asellus aquaticus	U.				○			
Hygrobatis longipalpis						○		
Arrhenurus cylindratus					×			
Piona nodata						×		
Velia currens								○
Gerris sp.								○
Nepa rubra	W. Pd.				×	⊗		
Allophylax dubius						×		
Glyphotaelius pellucidus					○			
Halesus sp.	W. Psd.			○	×	⊗		
Limnophilus bipunctatus						×		
Limnophilus griseus			×			×		
Limnophilus lunatus						×		
Limnophilus rhombicus	W. Psd.	⊗			×	○		
Ephemera vulgata		×						
Cloëon sp.						×		
Tipulidae			×		×			
Tendipedidae			○					
Sialis sp.	W. Sd.		⊗		×			
Nemurella picteti						○		
Gyrinus natater							×	○
Dytiscus marginalis						×		
Ilybius obscurus					⊗		×	
Hydroporus palustris						×		
Haliplus sp.						×		
Hydrophilidae						×		
Planorbis corneus	W. Ps.		⊗	×		⊗		
Limnaea ovata			×			○		
Bathyomphalus contorus						×		
Pisidium subtruncatum		×	○					
Gasterosteus aculeatus							○	
Gasterosteus pungitius							×	

W. Pst. – Wanderform zwischen Pflanzen und Steinen; W. Ps. – Wanderform zwischen Pflanzen und Schlamm; W. Psd. – Wanderform zwischen Pflanzen and Sand; P. Pd. – Wanderform zwischen Pflanzen und Detritus; W. Pd. – Wanderform zwischen Schlamm und Detritus; U. – Ubiquisten; ○ – für den betreffenden Biotop typisch; ⊕ – als Wanderform zwischen den betreffenden Biotopen typisch; × – weder typisch noch Wanderform

Im Bierbach wurden folgende Substrate — der Begriff ist in abgewandelter Form auch als Biotop 2. Ordnung zu verstehen — für bestimmte Besiedler charakteristisch festgelegt: Sand, Schlamm, Detritus, Pflanzen, freies Wasser und die Wasseroberfläche. Unter dem Substrat Detritus ist der Bachuntergrund zu verstehen, der vollkommen mit Blättern, pflanzlichen Resten und gröberen Holzstücken bedeckt war.

Zum Verständnis der folgenden Tabelle wird eine Erklärung von NIETZKE über die Kossau (1938) vorausgeschickt: „Als Wanderformen sind solche Arten bezeichnet, die innerhalb zweier Biotope anzutreffen sind. Die entsprechenden Biotope sind gewissermaßen die Pole, zwischen denen die Wanderformen pendeln. Unter Ubiquisten sind solche Arten zu verstehen, die in allen Strömungsgruppen und Biotopen anzutreffen sind."

3.3.3. Vorkommen der im Bierbach festgestellten Arten in anderen Fließgewässern mit teilweise ähnlichem Charakter

In Tab. 3 wird die Artenliste des Bierbachs mit der anderer Fließgewässer teilweise ähnlichen Charakters verglichen. Leider fehlen der Arbeit von NIETZKE über die Kossau quantitative Angaben.

Von den bestimmten 40 Arten des Bierbachs wurden je 23 ebenfalls in der Kossau (NIETZKE 1938) und in der Plane (ALBRECHT 1952), nur 9 jedoch in der Mölle (ILLIES 1952) gefunden.

Dendrocoelum lacteum, Gerris sp., *Halesus* sp., Tendipedidae, *Sialis* sp. und *Limnaea ovata* sind in allen genannten Fließgewässern vorhanden. *Planaria torva, Arrhenurus cylindratus, Piona nodata, Allophylax dubius, Hydroporus palustris* und *Pisidium subtruncatum* beschränken sich nur auf den Bierbach.

Die meisten bestimmten Arten zeigen eurythermen Charakter, was auch für die Arten der Plane und der Kossau zutrifft.

3.4. Bemerkungen zu Klassifizierungsversuchen

Inwieweit ein Fließgewässer vom Typ des Bierbachs als periodisch-partiell austrocknendes Gewässer von Wald- und Wiesengrabencharakter im Sinne von ILLIES (1961, 1963) klassifiziert werden kann, ist weiteren Arbeiten zu überlassen. Hierbei müßten mehrere Bäche und Gräben ähnlicher Art über längere Zeiträume erfaßt werden. Limnologische Zonen mit „isozönotischen Gesetzmäßigkeiten" (ILLIES 1961) können nicht aufgestellt werden, da weder Crenon-, Rhithron- noch ausgesprochene Potamonregionen den Bachverlauf in gesetzmäßiger Folge charakterisieren. Die kontinuierliche Veränderung der physiographischen Faktoren ist nicht eindeutig abzulesen.

Die Berücksichtigung der biozönotischen Verhältnisse ergibt kein klares Bild. Möglicherweise kann eine Klassizierung nach Substraten als Biotope 2. Ordnung ein erster Ansatz sein. Ein „dauerhaft fixiertes Mosaik der Substrate und Synusien" (ILLIES & BOTOSANEANU 1963) deutet sich in diesen Gewässern nur im Potamon an.

Die Arbeit zeigt, wie wertvoll Klassifizierungsversuche für Fließgewässer vom Bierbachcharakter sein könnten. Die Probleme wurden deutlich aufgezeigt.

Tab. 3. Vergleich der Artenlisten einiger Fließgewässer

Art	Kossau (NIETZKE)	Mölle (ILLIES)	Plane (ALBRECHT)	Bierbach
Nematoden			×')	×')
Planaria torva				×
Polycelis nigra			×	×
Dendrocoelum lacteum	×	×	×	×
Glossiphonia complanata	×		×	××
Haemopis sanguisuga			×	××
Herpobdella octoculata	×		××	×
Gammarus pulex pulex	×		×××	×××
Asellus aquaticus	×		×	××
Hygrobatis longipalpis	×			××
Arrhenurus cylindratus				×
Piona nodata				×
Velia currens	×		×	×
Gerris sp.	×	×	×	×
Nepa rubra	×		×	×
Allophylax dubius				×
Glyphotaelius pellucidus			×	×
Halesus sp.	×	×	×	×
Limnophilus bipunctatus	nur		juvenile Larven	×
Limnophilus griseus	*Limnophilus*		von Limnophiliden	×
Limnophilus lunatus	sp.		im Frühjahr	×
Limnophilus rhombicus	genannt		häufig	×××
Ephemera vulgata	×		×××	×
Cloëon sp.	×			×
Tipulidae		×		×
Tendipedidae	×	××	×××	×××
Sialis sp.	×	×	×	××
Nemurella pecteti		×	×	×
Gyrinus natator	×		×	××
Dytiscus marginalis	×			×
Ilybius obscurus	×'')			×
Haliplus sp.	×		×	×
Hydrophilidae		×		×
Planorbis corneus	×			××
Limnaea ovata	×	×	×	×××
Bathyomphalus contortus	×			×
Pisidium subtruncatum				×××
Gasterosteus aculeatus			×	×
Gasterosteus pungitius	×		×	×

××× – sehr zahlreich; ') – quantitative Angaben fehlen;
×× – häufig; '') – *Ilybius* sp.
× – vorhanden;

4. *Zusammenfassung*

1. Der Bierbach ist ein periodisch-austrocknendes Fließgewässer von Wald- und Wiesengrabencharakter.

2. Die Länge beträgt 6,52 km mit einer Laufentwicklung von 66,71%. Der Höhenunterschied von der Quelle bis zur Mündung mißt 22 m.

3. Die Wasserführung ist niederschlagsbedingt.

4. Die Wasserstoffionenkonzentration variierte zwischen 6,5 und 7,5. Der Sauerstoffgehalt schwankte von 7,0—10,0 mg O_2/l bei einer Sättigung von durchschnittlich 85%, die Zehrungswerte lagen bei 0,85 mg O_2/l.

5. Die Untersuchungszeit belief sich von September 1956 bis einschließlich Mai 1957.

6. Im Wiesengebiet ist eine starke Wasservegetation vorhanden. Sie geht im Waldgebiet bis auf ein Minimum zurück.

7. Von den 40 bestimmten Arten wurden 39 im Biotop Wiesengraben gefunden. Im Biotop Waldgraben ergaben die Untersuchungen nur 23 verschiedenen Arten, davon kamen 5 selten vor.

8. Die Besiedlung der Biotope 2. Ordnung zeigt, daß die Wasserpflanzen der Aufenthaltsort der meisten Arten sind, es folgen Detritus und Schlamm.

9. In der Artenzahl überwiegen die Trichoptera und Coleoptera, in der Individuenzahl die Malacostraca und Mollusca.

10. Die Betrachtung der Faunenliste ergibt, daß viele Arten Vertreter stehender Gewässer sind. Die eurythermen Arten überwiegen.

11. Von den 40 bestimmten Arten des Bierbachs wurden je 23 ebenfalls in der Kossau und der Planc, 9 in der Mölle gefunden.

12. Eine eindeutige Klassifizierung konnte beim Bierbach nicht vorgenommen werden.

Summary

1. The "Bierbach" — that is a periodical drying up running water of woody and meadow-character.

2. The length is 6.52 km with a running development of 66,71%. The difference in elevation from spring to mouth measures 22 m.

3. The water-bearing depends on the quantity of precipitation.

4. The hydrogen ion concentration varies between 6.5 and 7.5. The percentage of oxygen ranged from 7.0 to 10.0 mg O_2/l at an average saturation of 85%, the consume value was of the order of 0.85 mg O_2/.l.

5. The test period extended from September 1956 till May 1957.

6. In the meadow-area there is a large water vegetation, which is diminishing to a minimum in the wooded area.

7.. From the 40 species, which had been determined, 39 where found in the biotope meadow-ditch. In the biotope wood-ditch only 23 different representatives were discovered by means of analysis, 5 of them appearing very scarcely.

8. The colonization of the biotopes of the second order shows that the aquatic plants are place of residence for the most species, detritus and mud following then.

9. Concerning the numbers of the species Trichoptera and Coleoptera predominate. Concerning the numbers of individuals Malacostraca and Mollusca are of numerical superiority.

10. The examination of the fauna-list proves that many species are representatives of stagnant water. The eurythermal species predominate.

11. From the 40 species of the „Bierbach", which were determined, 23 species had been each found in the "Kossau" and the "Plane", too, 9 in the "Mölle".

12. Unequivocal classification could not be made for the "Bierbach".

5. Literatur

Abderhalden, E., 1936: Handbuch der biologischen Arbeitsmethoden. Abt. IX: Methoden der Erforschung der Leistungen des tierischen Organismus. Teil 2 (2. Hälfte). Methoden der Süßwasserbiologie. — XXI + 1135 S., Berlin, Wien.

Albrecht, M.-L., 1952: Die Plane und andere Flämingbäche. — Z. Fisch. N. F. **1** (5—6), 390 bis 473.

Benthin, B. (Hrsg.), 1968: Greifswald und seine Umgebung. — Berlin.

BERG, K., 1948: Biological studies on the river Susaa. — Folia Limnol. Scand. **4**.

BROHMER, P., 1953: Fauna von Deutschland. 7. Aufl. — Heidelberg.

EIDEL, K., 1933: Beiträge zur Biologie einiger Bäche des Schwarzwaldes mit besonderer Berücksichtigung der Elz und Kinzig. — Arch. Hydrobiol. **25**, 543–615.

HEYMONS, R., & TH. KUHLGATZ, 1909: Collembola, Neuroptera, Hymenoptera, Rhynchota. (In: A. BRAUER: Die Süßwasserfauna Deutschlands, H. 7). — Jena.

ILLIES, J., 1952: Die Mölle. Faunistisch-ökologische Untersuchungen an einem Forellenbach im Lipper Bergland. — Arch. Hydrobiol. **46**, 424–612.

— 1955: Steinfliegen oder Plecoptera. (In: F. DAHL: Die Tierwelt Deutschlands und der angrenzenden Meeresteile, T. 43). — Jena.

— 1961: Versuch einer allgemeinen biozönotischen Gliederung der Fließgewässer. — Internat. Revue ges. Hydrobiol. **46** (2), 205–213.

ILLIES, J., & L. BOTOSANEANU, 1963: Problemes et méthodes de la classification et de la zonation écologique des eaux courantes, considerées surtout du point de vue faunistique. — Mitt. internat. Verein. Limnol. No. 12, 57 S., Stuttgart.

JOHANSSON, L., 1929: Hirudinea. (In: F. DAHL: Die Tierwelt Deutschlands und der angrenzenden Meeresteile, Teil 15). — Jena.

KUHN, H., 1952: Gewässerleben und Gewässerschutz. Eine allgemein verständliche Darstellung der Lebensgemeinschaften der reinen und verschmutzten Binnengewässer. — Zürich.

KUHNT, P., 1913: Illustrierte Bestimmungstabellen der Käfer Deutschlands. Ein Handbuch zum genauen und leichten Bestimmen aller in Deutschland vorkommenden Käfer. — Stuttgart.

LAMPERT, K., 1925: Das Leben der Binnengewässer. 3. Aufl. — Leipzig.

NIETZKE, G., 1938: Hydrobiologisch-faunistische Untersuchungen an schleswig-holsteinischen Fließgewässern. — Arch. Hydrobiol. **32** (1), 1–74.

REITTER, E., 1916: Fauna Germanica. Die Käfer des Deutschen Reiches. Bd. I. II. — Stuttgart.

ROTHMALER, W., 1953: Exkursionsflora (Ein Pflanzenbestimmungsbuch für Schulen und Hochschulen). — Berlin.

SCHELLENBERG, A., 1942: Krebstiere oder Crustacea. IV. Flohkrebse oder Amphipoda. (In: F. DAHL: Die Tierwelt Deutschlands und der angrenzenden Meeresteile, Teil 40). — Jena.

SCHOENEMUND, E., 1930: Eintagsfliegen oder Ephemeroptera. (In: F. DAHL: Die Tierwelt Deutschlands und der angrenzenden Meeresteile, Teil 19). — Jena.

STEINMANN, P., 1907: Die Tierwelt der Gebirgsbäche, eine faunistisch-biologische Studie. — Ann. Biol. lacustre **2**.

STEINER, G., 1919: Untersuchungsverfahren und Hilfsmittel zur Erforschung der Lebewelt der Gewässer. — Handb. mikrosk. Technik, Teil 7/8. — Stuttgart.

STICHEL, W., 1925–1938: Illustrierte Bestimmungstabellen der Deutschen Wanzen. — Berlin–Hermsdorf.

ULMER, G., 1909: Trichoptera. (In: A. BRAUER: Die Süßwasserfauna Deutschlands, H. 5, 6). — Jena.

VIETS, K., 1936: Spinnentiere oder Arachnoidea. VII. Wassermilben oder Hydracarina. (In: F. DAHL: Die Tierwelt Deutschland und der angrenzenden Meeresteile, Teil 31, 32). — Jena.

Eingegangen am 16. 5 1972.

Anschriften der Verfasser:

KLAUS-DIETER THIEL
DDR-50 Erfurt
Eobanstr. 3
Dr. HANS-JOACHIM SUBKLEW
Ernst-Moritz-Arndt Universität
Sektion Biologie, FG Allgemeine
Zoologie und Tierphysiologie
DDR-22 Greifswald
Bachstraße 11/12

Limnologica (Berlin)	11	1	1976	125—160

GUNTHER KLEMM

Zur Beziehung zwischen Wasserfaktor und Pflanzengesellschaften eines grundwassernahen Talsandgebiets (Unterspreewald-Randgebiet)

Mit 11 Abbildungen und einer Tabelle

Inhalt

1. *Einleitung*

Die erhöhte Ausnutzung aller natürlichen Wasserreserven durch Industrie, Bevölkerung und Landwirtschaft macht es notwendig, in verstärktem Maße Wasserspeicherwirtschaft zu betreiben. Neben der Anlage von Talsperren ist in neuerer Zeit auch eine Grundwasseranreicherung bereits verschiedentlich angewandt worden (vgl. HERRMANN 1963).

In der durch verschiedene Vorarbeiten (ANDREAE 1956, ROSENTRÄGER 1962) als Grundwasserabsenkungs- und -mangelgebiet charakterisierten Randzone des Unterspreewalds bot es sich somit an, die Möglichkeiten für eine oberflächennahe Grundwasseranreicherung zu erkunden, zum Zwecke der Erhöhung des Wasserdargebots im durch neu entstandene Industriezentren stark beanspruchten Spreegebiet und vor allem zu einer Verbesserung der land- und forstwirtschaftlichen Nutzflächen in diesem Raum. Die Untersuchungen wurden vom Hydrologischen Institut der Humboldt-Universität Berlin in den Jahren 1964—1966 mit einer komplexen Erforschung der hydrogeologischen, hydromechanischen, hydrochemischen und pflanzensoziologischen Verhältnisse dieses Gebiets durchgeführt (HERRMANN 1968, PESCHKE 1968, KLEMM 1967, HERRMANN et al. 1967). Die Untersuchung der Beziehungen zwischen den Bodenwasserverhältnissen und den Pflanzengesellschaften verfolgte den Zweck, durch eine Erfassung des jetzigen Zustandes der Vegetation in Abhängigkeit insbesondere vom Wasserfaktor Hinweise auf die möglichen land- und forstwirtschaftlichen Veränderungen als Folge einer Grundwasseranreicherung zu erhalten.

Für wertvolle fachliche Hinweise bin ich Herrn Prof. Dr. R. SCHUBERT zu besonderem Dank verpflichtet. Ebenso gilt mein Dank allen Mitarbeitern des ehemaligen Hydrologischen Instituts der Humboldt-Universität, die mich bei meiner Arbeit vielseitig unterstützten.

2. *Das Untersuchungsgebiet (UG)*

2.1. Geographie und Geologie

Das UG mit einer Gesamtfläche von 29 km² liegt etwa 15 km nördlich Lübben im nördlichen Bereich des Lübben-Pretschener Talsandgebiets, das den zentralen Unterspreewald nach E hin begrenzt und von einem Spreearm, der Pretschener Spree, durchflossen wird (Abb. 1). Die formgebenden geologischen Elemente sind Moränensande des Brandenburger Stadiums der Weichsel-Eiszeit, Talsandbildungen des Baruther Urstromtals mit aufgesetzten spät- und postglazialen Dünen und Flugsanddecken und die holozänen Alluvionen mooriger und anmooriger Senken und der Spreearmtäler. Moränensande und -kiese treten nur in einem flächenmäßig unbedeutenden Ausläufer des Pretschener Endmoränenzuges im Zentrum des nördlichen Teilgebiets zu Tage. Der überwiegende Teil dieser nördlichen Hälfte sowie zwei größere, von S in das UG hineinreichende, inselartige Flächen und der SE-Rand des Gebiets werden von ausgedehnten Talsandflächen eingenommen. Das Bodensubstrat ist hier Mittel- bis Feinsand mit nur unbedeutendem Grobsandanteil ohne lehmige oder tonige Schichten. Die Bodentypen sind nährstoffarme, saure Gleypodsole, Podsole und Podsolranker. Häufig werden die Talsande von Flugsanddecken unterschiedlicher Mächtigkeit, die stellenweise zu geschlossenen Dünenzügen formiert sind, überlagert. Hier ist die Bodenentwicklung neben typischen Podsolbildungen an exponierten Stellen oft noch kaum fortgeschritten. In den grundwasserbeeinflußten und zeitweise überstauten Senken, die im Talsandgebiet zerstreut und kleinflächig vorhanden sind, erfolgte eine Anmoor-, nur selten eine Seggen- oder Torfmoos-Moorbildung.

Im Süden des UG ist in der Talsandfläche von der Pretschener Spree eine Talniederung ausgeräumt worden und mit Flußsanden und stellenweise schlickreichen, tonig durchsetzten Humusablagerungen (den sogenannten Klockböden) unterschiedlicher Mächtigkeit angefüllt worden. Die Bodentypen sind hier neben selteneren Podsolen und Gleypodsolen vor allem Gleyböden. In tieferen Senken der Talniederung kam es zu Anmoorverlandung mit Ausbildung entsprechender Böden.

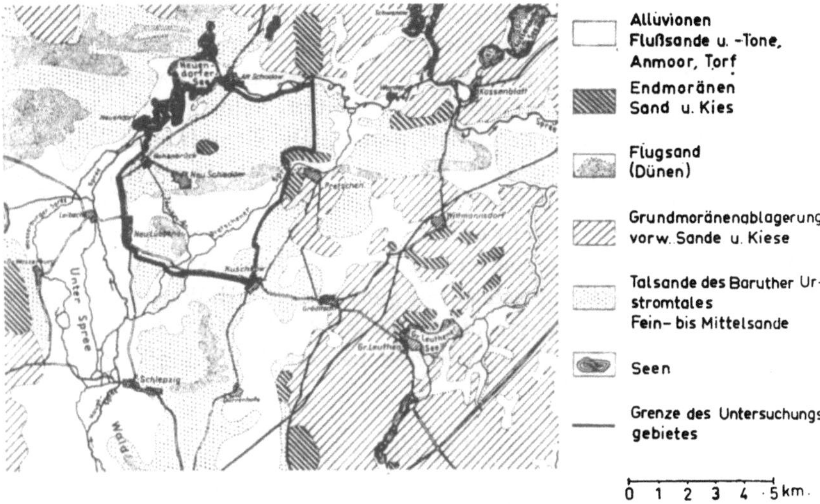

Abb. 1. Geologische Übersichtskarte des Untersuchungsgebiets (nach KEILHACK, ergänzt und verändert)

2.2. Hydrologie

Das UG hat trotz seiner Zuordnung zur naturräumlichen Großeinheit des Spreewalds nur noch recht wenig Ähnlichkeit mit dem sumpfigen, von zahlreichen Fließen durchzogenen Niederungsgebiet im landläufigen Sinne dieses Wortes. Der nördliche Teil wird fast ausschließlich von trockenen Talsandflächen mit nur unbedeutenden moorigen und anmoorigen Senken eingenommen. Den Südteil durchzieht ein Spreearm, die Pretschener Spree, mit einem Nebengraben (Jänicken-Graben), der heute seine Hauptwassermassen zum Neuendorfer See abführt. Die Hauptspree durchfließt das UG nicht, bildet jedoch mit ihrem Tal und dem Neuendorfer See im W und N die Begrenzung der untersuchten Flächen.

In den letzten 50—60 Jahren sind die hydrologischen Verhältnisse des UG entscheidenden Veränderungen unterworfen gewesen. Zur Verhütung von Hochwässern wurde in den Jahren 1906—1912 die Hauptspree weitgehend begradigt und vertieft, z. B. wurde auf der Strecke Leibsch—Schwielochsee eine Laufverkürzung um etwa 40% vorgenommen. Auch die Pretschener Spree wurde begradigt und 1934 weitgehend von der Hauptspree abgesperrt, so daß ihre Wasserführung wesentlich geringer ist als früher. Alle diese Maßnahmen haben sich in einer Grundwassersenkung um etwa 50 cm bemerkbar gemacht, was entscheidende hydrologische und landeskulturelle Folgen hatte (vgl. ANDREAE 1956, ROSENTRÄGER 1962). Die heute auftretenden Hochwässer sind vor allem auf tiefere, anmoorige Senken beschränkt und wirken sich meist nur bei zeitlich ungünstigem Eintritt schädigend aus, das Tal der Pretschener Spree wird nur noch an tiefgelegenen Stellen überflutet. Dagegen haben sich die Grundwasserverhältnisse vor allem für die höhergelegenen Acker- und Gründlandflächen bedeutend verschlechtert, so daß es sich als Ganzes gesehen um ein Grundwassermangelgebiet handelt.

In den Tal- und Dünensanden betragen die kapillaren Aufstiegswerte etwa 30 bis 50 cm, bei den mittleren gemessenen Durchlässigkeitsbeiwerten von $k_f = 5 \cdot 10^{-3}$ cm/s (Darcysches Gesetz) betragen die mittleren Einsickerungsgeschwindigkeiten 1—4 m je Tag, wobei die Werte durch die häufig beobachtete Schichtung unterschiedlicher Korngrößen-Bänder meist herabgemindert werden. Die Grundwasser-Filtergeschwindigkeiten sind infolge der geringen Gefällewerte von etwa 0,5—1 ‰ (im nördlichen Teilgebiet) gering und liegen im Mittel bei 0,5 cm/Tag (PESCHKE 1968, HERRMANN 1968). Die Grundwasserstandsschwankungen innerhalb eines Jahres betragen im Durchschnitt etwa 50 cm, die Schwankungen zwischen Extremwerten etwa 1 m. Das Grundwasser ist durchweg sauer, mit gemessenen pH-Werten (CZENSNY) von 5,2—6,5, wobei die höheren Werte oft in Nähe der Gehöfte oder unter gedüngten Flächen zu finden sind. Der Sauerstoffgehalt beträgt 7,5—9 mg/l (HERRMANN et al. 1967).

2.3. Klima

Das UG liegt nach dem Klimaatlas für das Gebiet der DDR (1953) im Klimagebiet des Ostdeutschen Binnenlandklimas. Die subkontinentale Klimatönung kommt sowohl in den Niederschlags- als auch in den Temperaturwerten zum Ausdruck. Die mittleren jährlichen Niederschläge (Station Lübben) liegen mit 537 mm relativ niedrig. Der niederschlagsärmste Monat ist der Februar (30 mm), die niederschlagsreichsten sind Juli und August (68 mm). Im Gegensatz zum zentralen Unterspreewald und zum Oberspreewald ist im UG die verdunstende Wasserfläche nicht besonders groß, so daß sich die erhöhte Luftfeuchtigkeit dieser Gebiete hier nicht in gleichem Maße in einem gewissen Ausgleich für die niedrigen Niederschlagsmengen bemerkbar macht. Die Hygrische Kontinentalität (nach REICHEL) liegt bei +1, und die Grenze zu 0 verläuft direkt westlich des UG. Die mittleren Jahrestemperaturen (Station Lübben) liegen bei 8,5 °C mit den höchsten Werten im Juli (18,2 °C) und den tiefsten im Januar (—0,7 °C), so daß eine relativ hohe mittlere Jahresschwankung zu verzeichnen ist (18,9 °C). Die in der Thermischen Kontinentalität (nach RINGLEB) zusammengefaßten Temperaturverhältnisse ergaben für die Periode 1901—1950 die höchsten Werte im Gebiet der DDR überhaupt (49%) (Werte der Station Lübben nach den Monatlichen Witterungsberichten).

2.4. Vegetation und Landeskultur

Der nördliche Teil des UG wird zum größten Teil von forstlich beeinflußten, armen Kiefernwäldern mit wenig Stieleichen-Birken-Wald-Resten und deren Kiefern-Forst-

Ersatzgesellschaften, zum geringeren Teil von Äckern und Ackerödländern sowie wenig ausgedehnten Grünländereien (v. a. Großseggen-Rieder, Feuchtwiesen und Weiden) eingenommen. Im südlichen Teil des Gebiets sind Wälder in drei Komplexen vor allem auf den Talsandstandorten verbreitet, es handelt sich fast ausschließlich um feuchte und trockene Stieleichen-Birken-Wälder, die erst seit jüngerer Zeit großflächiger in Kiefern- und Kiefern-Birken-Forste umgewandelt wurden, sowie unbedeutende Stieleichen-Hainbuchen- und Stieleichen-Buchen-Wald-Flächen. Den größten Raum nehmen Grünlandflächen (v. a. Groß- und Kleinseggen-Rieder, Feucht- und Frischwiesen sowie Weiderasen) mit inselartigen Erlen-Wald-Resten in den Flußtälern und moorigen und anmoorigen Niederungen ein. Die Äcker sind vorwiegend auf die höhergelegenen randlichen Talsandflächen begrenzt.

Für die Landeskultur hat die Grundwasserabsenkung tiefgreifende Veränderungen mit sich gebracht. Nur unbedeutend verringerten sich die für eine Bewirtschaftung zu nassen Flächen, dagegen verschob sich das Verhältnis von Großseggen-Riedern zugunsten der besseren Feuchtwiesen, andererseits mußten zahlreiche zu trocken gewordene Grünlandflächen in Äcker überführt und die Bewirtschaftung zu trocken gewordener Äcker in großem Umfange aufgegeben werden. Die so entstandenen Trockenödlandflächen werden seit einiger Zeit mit Kiefern aufgeforstet, so daß die Waldfläche vor allem im N-Teil zunahm. Die Laubwaldflächen des südlichen Teiles wurden in geringerem Maße gerodet.

Tab. 1. Vergleich des Bewirtschaftungszustandes zum Zeitpunkt der Erstaufnahmen der Meßtischblätter (1846) mit dem von 1966

		N-Teil		S-Teil		Gesamtes UG	
		1846	1966	1846	1966	1846	1966
Wälder	ha	973,5	1182,5	527,8	352,2	1501,3	1534,7
u. Forste	%	59,5	72,4	41,6	27,8	51,7	52,9
Grünland	ha	137,4	90,0	626,0	585,9	763,4	675,9
	%	8,4	5,5	49,4	46,2	26,3	23,3
Äcker	ha	507,8	257,4	107,8	315,6	615,6	573,0
	%	31,1	15,8	8,5	24,8	21,2	19,8
Ödland	ha	—	89,1	—	14,9	—	104,0
(zu trocken)	%	—	5,4	—	1,2	—	3,6
Ödland	ha	15,3	15,0	7,0	—	22,3	15,0
(zu feucht)	%	1,0	0,9	0,5	—	0,8	0,4
zusammen	ha	1634,0		1268,6		2902,6	

3. *Die Bedeutung des Wasserfaktors für die Vegetation des Untersuchungsgebiets*

3.1. Allgemeines

Der Wasserfaktor ist der entscheidende vegetationsdifferenzierende Faktor im UG. Da das Bodenausgangsmaterial im ganzen Gebiet vor allem armer Talsand, im Tal der Pretschener Spree nur stellenweise lehmiger bis toniger Flußsand ist, treten die chemischen Bodenfaktoren kaum differenzierend in Erscheinung. Der neben der

Wasserversorgung wichtigste Standortsfaktor ist der Humusgehalt des Oberbodens, wobei auch dieser als Folge des unterschiedlichen Grundwassereinflusses zu betrachten ist. Die große Durchlässigkeit der Sande und ihre geringe Wasserhaltekraft bedingen bei geringen Grundwasserabweichungen bereits eine deutliche Vegetationsänderung. So ist eine landwirtschaftliche Nutzung auf nicht grundwasserbeeinflußten Standorten im UG nicht möglich. Die Spanne zwischen vernäßten, für die Nutzung zu nassen Standorten bis zur Grenze der dafür bereits zu trockenen Flächen ist deshalb im Gebiet äußerst klein und beträgt nur etwa 150 cm Niveauunterschied, wobei allerdings sowohl nach der feuchten als auch nach der trockenen Seite hin $\pm$ unrentable Flächen mit hinzugezogen sind. Eine landwirtschaftliche Nutzung ist bei sommerlichen Tiefstständen des Grundwassers unter 150 cm bereits stark ertragsgefährdet. Innerhalb dieser engen Spanne ist jedoch eine Reihenfolge zahlreicher Acker- und Grünlandgesellschaften und -Ausbildungsformen entwickelt, die in den meisten Fällen gute Aussagen über den Wasserhaushalt erlauben. Letzterer ist — infolge Fehlens stärker wasserhaltenden Oberbodens oder stauender Schichten im oberflächennahen Untergrund — zugleich ein Anzeiger des Grundwasserstandes. In derartigen Gebieten läßt sich an Hand der Vegetation mit einiger Sicherheit auch ein Überblick über die Grundwasserverhältnisse geben. Für die Acker- und Grünlandgesellschaften spielt der Wasserfaktor insofern eine besonders bedeutsame Rolle, als eine befriedigende Massenproduktion nur bei ausreichender Wasserversogung möglich ist (vgl. WALTHER 1960), wobei die Substanz innerhalb der Vegetationsperiode erzeugt werden muß, während Waldgesellschaften nur in langjährigen Abständen genutzt werden. Außerdem ist das Wurzelsystem der landwirtschaftlichen Kulturpflanzen nicht in gleichem Maße befähigt, tiefere Wasserreserven auszunutzen wie das der Gehölze. Dennoch sind auch die Waldgesellschaften im UG als Zeiger der Grundwasserverhältnisse brauchbar, wenn auch nur unter Berücksichtigung der Feldschicht und im allgemeinen in großflächigerem Maßstab. Zum trockenen. grundwasserfernen Bereich hin wird der Wald nicht begrenzt. Doch auf den Flächen, die im UG von Kiefern- und trockenen Eichen-Birken-Wäldern eingenommen werden, versagt der genauere Zeigerwert der Vegetation für den Grundwasserstand verständlicherweise. Änderungen der Grundwasserstände müssen sich also insbesondere auf das Acker- und Grünland recht schnell und einschneidend auswirken, beim Wald kommt es zu einer langsameren Änderung, da namentlich die Gehölze weitgehend erhalten bleiben und auch in der Feldschicht sich einige Arten noch lange Zeit zu halten vermögen, während Arten, die den neuen Standortsverhältnissen entsprechen, sich bereits auszubreiten beginnen. Es kommt hier zu Mischbeständen. Das hat sich in der Vergangenheit als Folge der Grundwasserabsenkung gezeigt, deren Auswirkungen auf die Acker- und Grünlandvegetation sich in einer Verschiebung der Gesellschaften und verschiedenen Bewirtschaftungsformen zum Feuchten hin (s. o.) und im Wald in einer Veränderung und Durchmischung der Bodenvegetation sowie Verminderung der Ertragsleistungen bemerkbar gemacht hat. Auch eine Aufhöhung des Grundwasserspiegels, wie sie als Folge einer oberflächennahen Anreicherung zustande käme, würde sich in ähnlich entscheidender Weise auswirken. Zur Klärung der Abhängigkeit der Vegetation von den Bodenwasserfaktoren wurden in den Abflußjahren 1965—1966 Grundwassermessungen und in der Vegetationsperiode 1966 Bodenfeuchtemessungen unter verschiedenen Pflanzengesellschaften durchgeführt.

3.2. Wetterverlauf und Vegetationsentwicklung im Untersuchungszeitraum

Am Ende der Vegetationsperiode 1964 war das herbstliche Wasserdefizit, gefördert durch ungenügende Niederschläge, besonders ausgeprägt, während es bei Inbetriebnahme der Grundwasser-Meßstellen Ende November 1964 bereits wieder zu einem leichten Anstieg gekommen war. Der zu warme Januar mit meist als Regen gefallenen Niederschlägen brachte es mit sich, daß die Bodenwasserreserven relativ schnell aufgefüllt wurden. Verbunden mit Frühjahrshochwassern kam es so im Februar und März zu Grundwasserhöchstständen innerhalb des Jahres 1965. Die Vernässung der feuchteren Äcker und Überschwemmung der tiefergelegenen Grünländereien ist in dieser Zeit die Regel. In den folgenden Monaten bis Juli waren jedoch die Niederschläge sämtlich zu hoch und lagen zum Teil weit über den langjährigen Mittelwerten (im April 181%, im Juni 156%, im Juli 179%), so daß die von der Vegetation verbrauchten Wassermengen zum großen Teil wieder ersetzt werden konnten und es nur zu einer ungenügenden Abtrocknung der feuchteren Standorte kam. Eine Hochwasserwelle aus dem infolge zu hohen Staus abgelassenen Speicherbecken Spremberg tat das übrige. So blieb ein großer Teil der feuchten Grünlandgesellschaften noch bis Ende Mai, die Schlankseggen-Rieder blieben sogar bis in den Juni und Juli hinein überschwemmt und so naß, daß eine Mahd nicht möglich war und der erste Schnitt insgesamt sehr unbefriedigend ausfiel. Auch auf die feuchteren Äcker wirkte sich die lange Vernässung nachteilig aus. Dagegen wurden die Wuchsleistungen der trockeneren Äcker und Wiesen positiv beeinflußt. Allerdings waren sämtliche Monate mit Ausnahme des Junis zu kühl, was eine gewisse Verzögerung der Vegetationsentwicklung bewirkte. Der August brachte mit 43% des langjährigen Mittels nur geringe Niederschläge, so daß es nun zu einer stärkeren Abtrocknung kam, der Oktober mit nur 28% Niederschlag vom Mittel führte schließlich zu einem deutlichen Wasserdefizit, das jedoch infolge des feuchten Gesamt-Wetterverlaufs der Vegetationsperiode nicht so ausgeprägt wie im Vorjahre war. Während also für die feuchteren landwirtschaftlichen Nutzflächen größere Ausfälle zu verzeichnen waren, wirkte sich die Witterung des Jahres 1965 auf die trockeneren Flächen günstig aus.

Zu reiche Niederschläge in den Monaten Dezember bis März (bis auf Januar) 1966 führten in noch stärkerem Maße als im Vorjahr zu der Winter- und Frühjahrsvernässung und -überschwemmung. Doch die nur wenig zu feuchten Monate April und Mai ließen die Bodenwasservorräte und den Grundwasserstand schneller absinken, so daß die feuchten Acker- und Grünlandflächen abtrockneten und in ihrer Vegetationsentwicklung kaum gehemmt wurden. Bereits Mitte Mai waren die Schlankseggen-Rieder vom Überschwemmungswasser befreit, so daß diesmal auch der erste Schnitt genutzt werden konnte. In den feuchten Gründlandgesellschaften mit erschwerter Versickerung kam es Ende Juni nach wolkenbruchartigen Niederschlägen noch einmal zu einer stärkeren Überschwemmung und bald danach auch zu einem allgemeinen Grundwasseranstieg, der sich jedoch schnell wieder ausglich. Die folgenden Monate brachten (bis auf Juli) auch noch etwas zu hohe Niederschläge, was vor allem die Getreideernte erschwerte, andererseits jedoch zu einer relativ günstigen Grundwasseranreicherung führte und sich somit für die Hackfrüchte positiv auswirkte.

Insgesamt waren also beide Abflußjahre feuchter als normal, wobei das Jahr1966 jedoch nur wenig abnorm verlief.

3.3. Der Grundwassergang

In der angewandten Vegetationskunde ist den Beziehungen zwischen Pflanzengesellschaften und dem Grundwassergang bereits von zahlreichen Autoren nachgegangen worden (Tüxen 1954, v. Müller 1956, Freitag 1957a, Bochert 1958, Eskuche 1962, Niemann 1963, 1968, Schlüter 1969 u. a.). Insbesondere in Gebieten, in denen die Grundwasserabhängigkeit der Vegetation so stark ist wie im UG, ist eine derartige Untersuchung von guter Aussagekraft.

Um einen Überblick über den Grundwassergang in den wichtigsten Wald-, Acker- und Grünlandgesellschaften des UG zu gewinnen, wurden im November 1964 in ausgesuchten Beständen 27 Meßrohre (Eisenrohre) gesetzt und im allgemeinen einmal im Monat (mit Ausnahme der Wintermonate) der Grundwasserstand gemessen. Nur ein Grundwasserpegel der Wasserwirtschafsdirektion Cottbus (Meßstelle 1) mit seit 1955 zweimal monatlich gemessenen Werten konnte mit verwendet werden (Abb. 2).

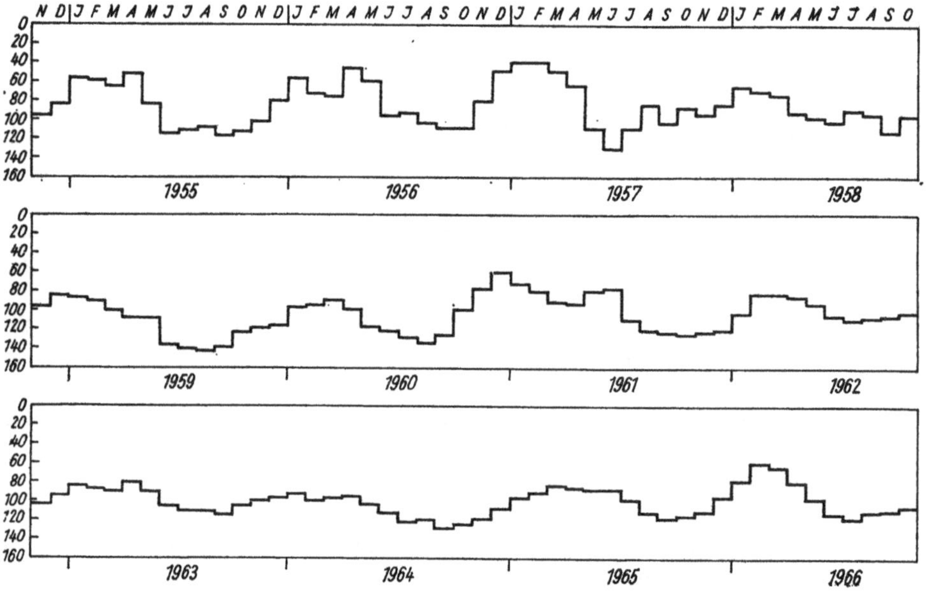

Abb. 2. Monatsmittel der Grundwasserstände (1955—1966) Meßstelle 1. Vegetation (1965/66): Papaveretum argemone, typ. Var., *Juncus*-Subvar.

Eine Reihe von Meßrohren konnte für andere, eng benachbarte Pflanzengesellschaften, deren Niveau einnivelliert wurde, mitbenutzt werden. Zahlreiche Einzelbohrungen und -messungen in weiteren Beständen ergänzten die Beobachtungen. Der allgemeine Grundwassergang in den beiden Untersuchungsjahren läßt sich an Hand der meisten Meßstellen folgendermaßen charakterisieren: Von absoluten Tiefstwerten im Herbst 1964 war ein regelmäßiger Anstieg bis zum Höchststand im März 1965 festzustellen. Im Vergleich zu den vorausgegangenen zehn Jahren lag er jedoch noch unter den mittleren Höchstständen. Charakteristisch für den weiteren Verlauf war das bereits erwähnte langsame Absinken des Grundwassers, das erst im Juli und August stärker zurückging, was in sämtlichen Jahren des vorangegangenen Jahrzehnts nicht zu beobachten war. Die Jahrestiefstwerte, die etwa mittleren Tiefstwerten entsprachen, wurden im September und Oktober erreicht, wonach ein allmählicher Anstieg bis zu Höchstwerten im Februar erfolgte. Diese Werte lagen wesentlich höher als im Vorjahr. Ende März, als die Messung in den neu gesetzten Rohren erfolgte, war jedoch schon wieder ein stärkerer Rückgang zu verzeichnen, der bis zum Juni anhielt. In diesem Monat lagen in den meisten Meßstellen (mit Ausnahme weniger Waldgebiete) bereits die Jahrestiefstwerte, die im allgemeinen unter die Tiefstwerte des Vorjahres absanken. Das wirkte sich zu diesem Zeitpunkt negativ auf die Vegetation aus. Dennoch ist dieser Verlauf innerhalb der zwölf Vergleichsjahre nicht ungewöhnlich. Die wolkenbruchartigen Regenfälle Ende Juni brachten nur einen vorübergehenden stärkeren Grundwasseranstieg, der sich dann langsam mit Unterbrechungen bis Oktober fortsetzte.

Im gesamten UG war der beschriebene Grundwassergang in entsprechender Form in den verschiedenen Pflanzengesellschaften mit relativ guter Übereinstimmung zu finden. Einige in der Nähe fließender oder zeitweise aufgestauter Gräben gelegene Meßstellen zeigten Abweichungen, die jedoch von der Grundtendenz nur unwesentlich differierten. Das war bei den recht einheitlichen Bodenverhältnissen zu erwarten. Innerhalb der verschiedenen Feuchtestufen kommt es jedoch zu unterschiedlicher Ausschöpfung des Grundwassers durch die Vegetation und damit zu verschieden hohen Schwankungen im Jahresgang. An den feuchten, während des ganzen Jahres bis in den Wurzelraum hinein mit Grund- und aufsteigendem Kapillarwasser versorgten Standorten, die zudem meist humusreiche und damit wasserhaltende Oberbodenschichten aufweisen, ist die sommerliche Absenkung relativ gering. Da es sich meist um Senken oder Standorte in Nähe von Gewässern handelt, ist auch der seitliche Zufluß oft groß genug, um den Grundwassergang ausgeglichener verlaufen zu lassen. Die höchsten Schwankungswerte weisen jene mittleren Standorte auf, in denen die sommerliche Wasserversorgung der Vegetation bei weitgehendem Fehlen von wasserhaltendem Oberboden aus den tieferen Schichten erfolgt, die Wurzeln aber noch längere Zeit den Saugsaum erreichen, so daß sie die Grundwasserreserven stark anzugreifen vermögen. Dagegen sind für die trockeneren Standorte, in denen während der Vegetationsperiode der Saugsaum vom Wurzelraum abreißt und somit die Wasserversorgung der Pflanzendecke unabhängig vom Grundwasser erfolgen muß, wieder ausgeglichenere Ganglinien charakteristisch. Auf die unterschiedlichen Werte der Schwankungen unter Wald-, Grünland- und Ackergesellschaften wird bei deren Besprechung hingewiesen.

3.4. Der Gang der Bodenfeuchte

Zur Untersuchung der Bodenfeuchteverhältnisse unter den Acker- und Grünlandgesellschaften wurden in der Vegetationsperiode 1966 (März—Oktober) monatlich je einmal Bodenfeuchteproben entnommen. Die Entnahme erfolgte mit einer Handbohrschappe. Die Proben wurden feucht und nach Trocknung bis zur Gewichtskonstanz erneut gewogen und die so gewonnenen Bodenfeuchte-Gewichtsprozente nach der Dichtebestimmung des Bodens auf Volumenprozente umgerechnet. Mit der Einschränkung, daß die so ermittelten Werte einer gewissen Zufälligkeit in Bezug auf ihre Niederschlagsabhängigkeit unterliegen und andererseits die Bedeutung der Bodenfeuchte für die Vegetation vom unterschiedlichen „nicht ausnutzbaren“ Wassergehalt (Eintritt des Welkepunkts) verschiedener Bodenarten abhängt, geben sie dennoch Anhaltspunkte über die unterschiedlichen Bodenfeuchteverhältnisse. Die Werte des „pflanzenverfügbaren Wassers“, also die Spanne zwischen Feldkapazität und Welkepunkt, liegen in den Böden des UG am höchsten im Torf, in den Sandböden je niedriger, je geringer ihr Feinbodenanteil und ihr Humusgehalt sind (vgl. v. MÜLLER 1956). In der gleichen Richtung sinkt auch ihr Wasserspeichervermögen, ausgedrückt in der Feldkapazität, so daß die humusärmsten Sande die ungünstigsten Pflanzenstandorte darstellen. Für den Vergleich der Bodenfeuchtewerte ist die Höhe der Niederschläge unmittelbar vor den Meßtagen von Bedeutung. In der Woche jeweils vor den Probeentnahmetagen waren folgende Niederschläge zu verzeichnen: Im März und Juli geringe bis mittlere Niederschläge an fast allen Tagen, im Juni einige Tage mit geringen Niederschlägen, im April der letzte Niederschlag (9,1 mm) am 5. Tag vor der Messung, im August nach kurzer Trockenperiode wolkenbruchartige Niederschläge mit unterschiedlicher Beregnung der einzelnen Meßstellen am Vortage, im Oktober stärkere Niederschläge nach niederschlagsarmen Tagen ebenfalls am Vortage und schließlich im Mai und September trockene Tage mit einzelnen unbedeutenden Regenfällen.

Die Bodenfeuchtemessungen bestätigen, daß an den Talsandstandorten des UG eine ausreichende Wasserversorgung der Acker- und Grünlandvegetation nur erfolgen kann, wenn im Sommer die Verbindung zwischen Saugsaum und Wurzelraum nicht auf längere Zeit vollständig abreißt. Die Wasserhaltekraft dieser Böden ist bei weitem zu gering, um eine landwirtschaftliche Nutzung auf grundwasserunbeeinflußten Standorten zu ermöglichen. Aber auch auf den feinerdereicheren Böden im Tal der Pretschener Spree ist die Acker- und Grünlandnutzung weitgehend an die Grundwassernähe gebunden, da bei zwar höherer Feldkapazität des Oberbodens der Anteil des pflanzenverfügbaren Wassers jedoch relativ geringer ist und zum anderen meist rein sandige Zwischenschichten bzw. schon in geringer Tiefe anstehender sandiger Untergrund das Wasserhaltevermögen des gesamten Profils nicht entscheidend von dem der reinen Sandböden abweichen lassen.

Geringe Unterschiede bestehen bei der Ausnutzung der Bodenwasservorräte durch die verschiedenen landwirtschaftlichen Kulturen, worauf bei Besprechung der Gesellschaften eingegangen wird.

4. *Der Wasserfaktor in den Pflanzengesellschaften des Untersuchungsgebiets*

4.1. Allgemeines

Im Vergleich zu anderen Gebieten sind die Pflanzengesellschaften des UG, bedingt durch die beinah ausschließlich vorkommenden armen, azidophilen Böden mit rein sandigem Untergrund als Ausgangsmaterial fast nur in Form ihrer ärmeren Gesellschaftsausbildungen bzw. Untergesellschaften entwickelt. Sowohl die Verbreitung der einzelnen Gesellschaften als auch ihre Untergliederung in Untereinheiten wird jedoch durch den Einfluß des Wasserfaktors und von ihm abhängender Faktoren

(Humusgehalt des Oberbodens!) bestimmt. So entsprechen die pflanzensoziologischen Einheiten fast stets zugleich verschiedenen Wasserstufen, und es sind bei Kenntnis jeweils einer Komponente Rückschlüsse auf die andere möglich. Im folgenden werden neben der Beschreibung der Bodenwasserverhältnisse bei den wichtigsten untersuchten Gesellschaften auch deren Untereinheiten genannt — wo der Wasserfaktor der bedeutendste differenzierende Standortsfaktor gegenüber den Nachbareinheiten ist, wird dies angegeben.

Es bedeuten:

F = feuchtedifferenziert gegenüber der Subassoziation, F = gegenüber der Variante und f = gegenüber der Subvariante bzw. Ausbildungsform. Die Reihenfolge der Untereinheiten folgt der Abstufung feucht — trocken.

Vegetationstabellen und eingehende Beschreibungen der genannten Pflanzengesellschaften, die hier aus Platzgründen nicht mit veröffentlicht werden können, sind bei KLEMM (1967, 1969/70) zu finden.

Abkürzungen: AF = Ausbildungsform(en)
MS = Meßstelle(n)

4.2. Wald- und Forstgesellschaften

In den Gehölzgesellschaften wurden die höchsten Grundwasserschwankungen innerhalb des Jahresgangs gemessen (bis zu 90 cm in den mittleren Beständen), da bei relativ hohem Wasserverbrauch der Vegetation die Baumwurzeln befähigt sind, die Grundwasserreserven langfristig und bis in größere Tiefen auszuschöpfen, was zu ausgeprägten herbstlichen Tiefstständen führt. Bodenfeuchtemessungen wurden in den Wald- und Forstgesellschaften wegen der hohen möglichen Abweichungen durch die Interzeption nicht durchgeführt.

4.2.1. Erlen-Wälder [Carici-elongatae-Alnetum (KOCH 26) TX. et BODEUX 55]

Hottonio-Alnetum **F**
Carici-Alnetum **F**
Macrophorbio-Alnetum **F**

Die Erlen-Wälder des UG stocken auf Bruchwaldtorf über Sand, in den trockeneren AF mit Übergang zu Anmoor-Gley. Der Einfluß des Grundwassers ist in den drei Assoziationen unterschiedlich, obwohl sämtliche Standorte im nassen bis stark feuchten Bereich liegen. Das Hottonio-Alnetum ist für längere Zeit des Jahres überschwemmt, nur im Sommer meist ohne Überstauung. Bereits im späten Frühjahr sind die Standorte des Carici-Alnetums ohne Überstauung. Am kürzesten dauert sie im Macrophorbio-Alnetum, in dem sie oft sogar ganz unterbleibt. Hier macht sich der Einfluß der Grundwasserabsenkung besonders deutlich bemerkbar. Die Tiefststände lagen aber auch hier kaum unter 50 cm (vgl. MS 11 u. 13). Innerhalb der Waldgesellschaften waren die jährlichen Grundwasserschwankungen in diesen feuchtesten Wäldern am geringsten, sie lagen bei etwa 40—50 cm. Eine gute Wassernachlieferung der in den Flußauen bzw. in abflußlosen Senken gelegenen Bestände und erhöhte Wasserhaltekraft des organischen Bodensubstrats ermöglichen den relativ ausgeglichenen Grundwassergang.

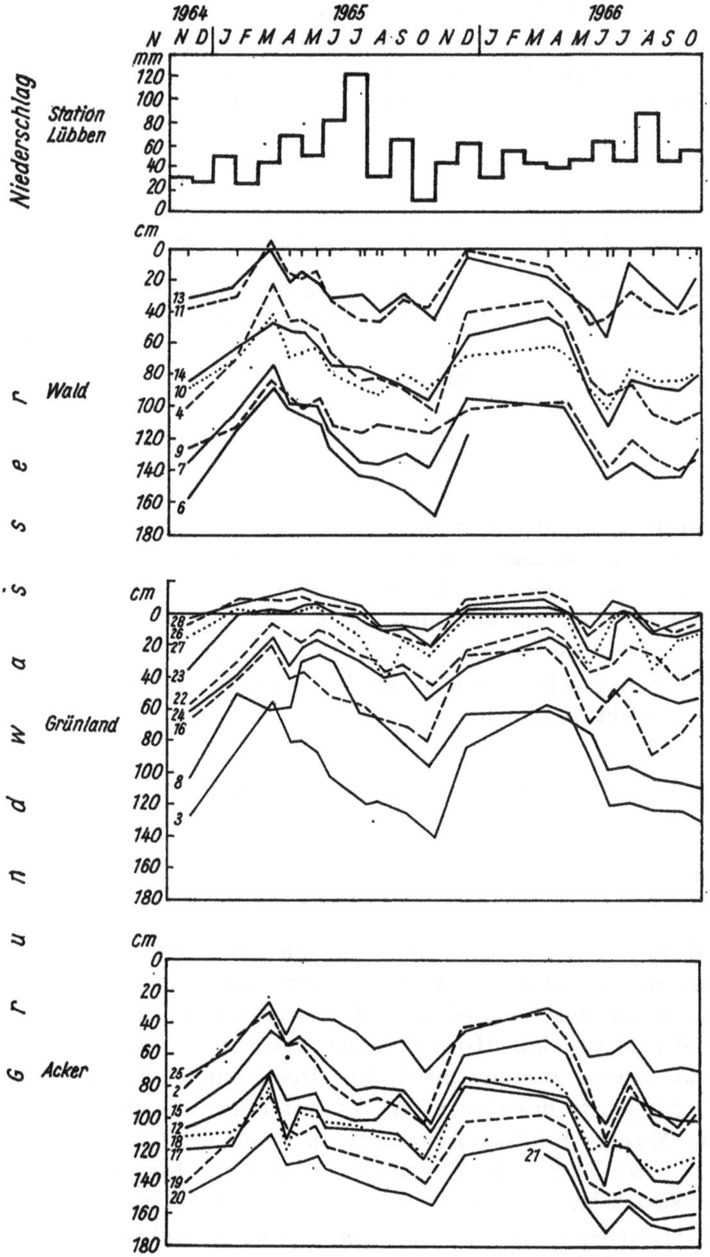
1964
1965
1966
N N D J F M A M J J A S O N D J F M A M J J A S O
mm
Niederschlag
Station Lübben
Grundwasser
Wald
Grünland
Acker
cm

Abb. 3

4.2.2. Grundwasserbedingter Stieleichen-Birken-Wald [Molinio-Quercetum(Tx. 37) SCAM. & PASS. 59]

Alnus-glutinosa-Subassoziation	F
Thelyteris-palustris-Variante	F
Typische Variante	F
Vaccinium-myrtillus-Variante	F
Typische Subassoziation	F
Hydrocotyle-vulgaris-Variante	F
Alnus-glutinosa-Subvariante	f
Typische Subvariante	f
Typische Variante	F
Dicranum-scoparium-Variante	F

Diese häufigste Laubwaldgesellschaft des UG nimmt fast alle mäßig bis schwach grundwasserbeeinflußten heutigen Waldstandorte ein und ist auf ihnen in einer Reihe überwiegend feuchtebedingter Untereinheiten entwickelt. Dabei sind sowohl Übergänge zu den feuchteren Erlen-Wäldern als auch zum trockenen Eichen-Birken-Wald vorhanden.

Die *Alnus*-Subass. nimmt die tiefstgelegenen Stellen innerhalb des Stieleichen-Birken-Wald-Gebiets ein, als Übergangsbildung zu den Erlen-Wäldern. Die Bestände stocken auf Sandböden mit ± mächtiger Humusauflage vom Typ eines Grundwasser-Gleys. Der G_{ox}-Horizont ist im Grundwasserschwankungsbereich meist sehr deutlich

Abb. 3. Grundwasserganglinien der Pflanzengesellschaften, benachbarte Gesellschaften mit Angabe der einnivellierten Flurhöhenunterschiede (a, b)

Waldgesellschaften:

MS 4: Molinio-Quercetum typicum (Forst); MS 6: Trockener Stieleichen-Birken-Wald, typ. AF (Forst), 1966 gerodet; MS 7: Molinio-Quercetum, typ. Subass., *Dicranum*-Var., a) Trockener Stieleichen-Birken-Wald, typ. AF (+80 cm); MS 9: Polytricho-Carpinetum, a) Molinio-Fagetum (+75 cm); MS 10: Molinio-Quercetum, *Alnus*-Subass., *Vaccinium*-Var.; MS 11: Macrophorbio-Alnetum; MS 13: Macrophorbio-Alnetum, a) Holcetum, *Carex-fusca*-Subass., typ. Var., *Deschampsia*-AF (±0 cm); MS 14: Molinio-Quercetum, typ. Subass., *Hydrocotyle*-Var., *Alnus*-Subvar., a) Panicetum, *Stachys*-Var. (±0 cm); b) Succisietum pratensis, *Mentha*-Subass. (−40 cm)

Grünlandgesellschaften:

MS 3: Festucetum rubrae, *Festuca-ovina*-Subass., typ. Var., a) Molinio-Quercetum, typ. Subass., *Dicranum*-Var. (±0 cm); MS 8: Festucetum rubrae typicum, a) Panicetum, typ. Var., typ. Subvar. (±0 cm); MS 16: Holcetum lanati typicum, a) Holcetum, *Veronica*-Subass., typ. Var. (+10 cm), b) Papaveretum, typ. Var., *Juncus*-Subvar. (+35 cm); MS 22: Holcetum, *Carex-fusca*-Subass., *Carex-gracilis*-Var., typ. Subvar.; MS23: Phalaridetum arundinaceae, a) Alopecuretum geniculati (±0 cm); M 24: Holcetum lanati (typicum); MS 26: Caricetum gracilis, typ. Var.; MS 27: Holcetum, *Carex-fusca*-Subass., *Carex-gracilis*-Var., *Caltha*–Subvar., *Deschampsia*-AF; MS 28: Caricetum gracilis, *Alopecurus*-Var.

Ackerunkrautgesellschaften:

MS 2: Papaveretum, *Stachys*-Var., a) Molinio-Quercetum, typ. Subass., *Dicranum*-Var. (Forst) (+15 cm); MS 12: Panicetum, typ. Var., *Juncus*-Subvar., a) Festucetum rubrae, *Festuca-ovina*-Subass., typ. Var. (±0 cm); MS 15: Panicetum, *Stachys*-Var., a) Panicetum, typ. Var., *Juncus*-Subvar. (+25 cm); MS 17: Panicetum, typ. Var., typ. Subvar., a) Festucetum rubrae, *Alopecurus*-Subass., *Arrhenatherum*-Var. (±0 cm); MS 18: Papaveretum, typ. Var., typ. Subvar., MS 19: Chenopodietum albi, *Echinochloa*-Var., typ. Subvar., a) Arnoseretum, *Corynephorus*-Subass. (+30 cm); MS 20: Chenopodietum albi, *Echinochloa*-Var., typ. Subvar. – 1965, Arnoseretum, *Myosotis*-Subass., typ. Var. – 1966; MS21: Arnoseretum minimae typicum, a) Chenopodietum albi, *Echinochloa*-Var., typ. Subvar. (−35 cm); MS 25: Chenopodietum polyspermi ausgebildet. Die Böden sind infolge der Grundwasserabsenkung heute stärker ausgetrocknet, so daß der Oberboden kaum noch längere Zeit durchnäßt wird, z. T. erreicht ihn das Grundwasser gar nicht mehr. So kommt es vereinzelt zu beginnender Podsolierung. Das Grundwasser steigt im Winterhalbjahr bis etwa 20—40 cm, und die Tiefstwerte liegen bei etwa 100 (—120) cm. Es sind mittlere, meist schon höher liegende Schwankungen als im Erlen-Wald bezeichnend. Die vor der Grundwasserabsenkung offensichtlich häufiger stattgefundene Überschwemmung der Standorte — wofür das Wachsen der Bäume auf ± deutlichen bultenartigen Erhebungen spricht — ist heute nicht mehr zu beobachten. MS 10 stellt einen solchen trocken gewordenen, noch mit dem Grundstock der charakteristischen Feuchtezeiger versehenen Bestand dar.

Die typische Subass. nimmt mittlere Standorte in bezug auf die Grundwasserverhältnisse innerhalb der Eichen-Birken-Wälder im UG ein. Die Bodenart ist in allen Fällen Mittel- bis Feinsand ohne bemerkenswerten Ton- oder Kiesanteil, die Bodentypen sind in den feuchteren AF ein Grundwasser-Gley mit beginnender Podsolierung und in den trockeneren AF ein typischer Gleypodsol mit ± gut ausgebildetem Bleichhorizont. Die Grundwasserstände in der *Hydrocotyle*-Variante sind etwa die gleichen wie in der *Alnus*-Subass., jedoch dürfte der begrenzende Faktor die größere Nährstoffarmut der Standorte sein. Die Grundwasserhöchststände liegen bei 30—50 cm, so daß die oberen Bodenschichten im Hauptwurzelraum der Feldschicht noch zeitweise vom aufsteigenden Kapillarwasser erreicht werden (vgl. MS 14). In der typischen Variante kann das Grundwasser zwar noch bis 50 cm (stellenweise auch höher) ansteigen, doch liegen die Tiefstwerte meist schon beträchtlich unter 1 m. Die Variante nimmt etwa mittlere Grundwasserstandorte innerhalb der Subassoziation ein. In der trockensten, der *Dicranum*-Variante steigt das Grundwasser höchstens bis etwa 70 cm und sinkt bis etwa 150 cm, an verschiedenen Stellen erreicht es jedoch nicht einmal diese Werte. Die oberste Bodenschicht wird nicht mehr vom Grundwasser oder Kapillarwasser erreicht, so daß eine Podsolierung des Bodens verstärkt einsetzt. Aber auch hier ist ein ehemals höherer Grundwasserstand u. a. an der Ausdehnung des G_{ox}-Horizontes auf einen heute nicht mehr vom Grundwasser berührten Bereich abzulesen. In diesen Beständen wurden die höchsten im Untersuchungszeitraum gemessenen Jahresschwankungen des Grundwassers verzeichnet. Sie lagen bei etwa 80 (70—90) cm, wie die MS 4, 5, 6 und ebenfalls die am Waldrand liegenden MS 2 und 3 zeigen. Das Wurzelsystem bleibt während der ganzen Vegetationsperiode in der Lage, die Grundwasserreserven zu erreichen und auszuschöpfen, so daß ein ausgeprägter herbstlicher Tiefststand entsteht, der 1966 teilweise den allgemeinen Juni-Tiefststand übertraf (MS 3,4). Innerhalb der Waldgesellschaften hat sich die Grundwasserabsenkung besonders hier in dem Eindringen trockenheitsliebender Arten und dem Zurückbleiben einiger Feuchtezeiger in der Feldschicht bemerkbar gemacht, so daß es bei weitgehend unveränderten Gehölzschichten und Erhaltenbleiben der Artenkombination der Feldschicht doch zu Mengenveränderungen und teilweise zu charakteristischen Mischbeständen kam.

4.2.3. Trockener Stieleichen-Birken-Wald

Typische AF	F
Trockene AF	F

Auf den sehr schwach grundwasserbeeinflußten bis grundwasserfernen Standorten löst ein trockener, armer, horstgrasreicher Stieleichen-Birken-Wald ohne Pfeifengras den vorher beschriebenen Waldtyp ab. Die Bodenart ist in allen Fällen Sand, der Bodentyp ein $\pm$ gut ausgebildeter Podsol mit nur dünner Humushorizontauflage. Nur in der typischen AF wird zeitweise ein Grundwassereinfluß wirksam, die Höchstwerte können noch bei etwa 90—100 cm liegen, die Tiefstwerte sind dann etwa 170—200 cm. Vor allem für die tieferwurzelnden Gehölze macht sich das in besserer Wüchsigkeit bemerkbar, und die Grundwasserschwankungen erreichen noch etwa Werte wie im Molinio-Quercetum typicum. Über Übergangsbestände kommen Abstufungen bis zu den Standorten vor, auf denen die trockene AF wächst und wo das Grundwasser erst in mehreren Metern Tiefe steht. Hier vermögen auch die Gehölze die Grundwasserreserven nur noch unvollkommen oder gar nicht mehr auszunutzen, wodurch ein ausgeglichenerer Grundwassergang unabhängig von der Vegetation zustande kommt.

4.2.4. Widerton-Stieleichen-Hainbuchen-Wald [Polytricho-Carpinetum (Prsg. 43) Scam. & Pass. 59]

Nur unbedeutende Flächen nimmt im UG dieser Laubwald ein, dessen Vorkommen vom Eichen-Birken-Wald standörtlich durch etwa nährstoffreicheren Sandboden vom Typ eines Gleypodsols differenziert ist. Der sonst in diesem Waldtyp zu findende Modergley (Scamoni 1960) ist durch die Grundwasserabsenkung wahrscheinlich so stark verändert worden. Das Grundwasser schwankte zwischen etwa 85 und 140 cm im Untersuchungszeitraum (MS 9). Auch hier ist der Unterschied zu den bei Scamoni (1960) angegebenen Verhältnissen mit Höchstbeständen von 20—30 cm deutlich. Trotzdem scheint die Vegetationszusammensetzung nur unwesentlich verändert. Mittlere Grundwasserschwankungen sind charakteristisch.

4.2.5 Pfeifengras-Stieleichen-Buchen-Wald [Molinio-Fagetum (Scam. 35) Scam. & Pass. 59]

Dieser Waldtyp, der bezeichnend für die Buchen-Vorkommen auf relativ armen, grundwasserbeeinflußten Sandstandorten ist, kommt kleinflächig im UG auf höhergelegenen, im Kontakt mit dem Eichen-Hainbuchen-Wald stehenden Kuppen vor. Der Boden ist ein (Gley-)Podsol. Der nur schwache Grundwassereinfluß mit etwa zwischen 160 und 220 cm Tiefe schwankenden Werten ist bezeichnend, trotzdem hat sich auch hier die Grundwasserabsenkung offensichtlich nicht auf die Vegetationszusammensetzung ausgewirkt.

4.2.6. Kiefernforste

Die Kiefernforste im UG sind fast ausschließlich Ersatzgesellschaften der Stieleichen-Birken-Wälder.

Nur sehr selten kommt ein *Sphagnum*-Kiefernforst als Ersatzgesellschaft eines nährstoffarmen Birken-Bruchs vor. Er stockt auf Anmoor mit Übergang zu Torfbildung in abflußlosen Senken, in denen durch die Torfschicht Staunässe auftreten kann, doch steht auch hier der Boden hauptsächlich unter dem Einfluß von Grundwasser, das jedoch nur noch im Winterhalbjahr den Oberboden durchnäßt und im Sommer bis etwa 80—100 cm absinkt.

Die Grundwasserverhältnisse in den übrigen Kiefernforsten entsprechen den Eichen-Birken-Wald-Ausbildungen, die sie ersetzen, doch ist die Aussagekraft der Vegetation über die Grundwasserverhältnisse gegenüber den Laubwaldausbildungen etwas geringer, da sich eine gewisse Nivellierung in der Zusammensetzung der Feldschicht durch den degradierenden Einfluß der Kiefern-Monokulturen auf die edaphischen Faktoren stärker bemerkbar macht und den Feuchtezeigerwert teils überdeckt. Es entsprechen ein *Molinia-Calluna*-Kiefernforst mit bereits fortgeschrittener Degradierung und ein erst seit kürzerem als Forst genutzter *Molinia-Deschampsia*-Kiefernforst etwa dem Molinio-Quercetum typicum sowie ein armer *Deschampsia-flexuosa*-Kiefernforst der typischen AF des trockenen Stieleichen-Birken-Waldes.

4.2.7. Kiefernwälder

Die Kiefern-Wälder des UG besiedeln die ärmsten Sandböden innerhalb der Waldkomplexe, die nur noch vereinzelt unter Grundwassereinfluß stehen. Neben der Armut der Standorte ist vor allem der äußerst geringe Humusgehalt des Oberbodens charakteristisch, so daß die unbedeutende Wasserhaltekraft nur eine sehr spärliche Feldschicht aus wenigen trockenheitsertragenden bzw. -bevorzugenden Blütenpflanzen gedeihen läßt. Das Bodenausgangsmaterial ist sehr selten kiesiger Moränensand, vor allem handelt es sich um mittlere Talsande mit meist höherem Flugsand (-Feinsand)-Anteil. Als bedingte Zeiger für noch zeitweisen Grundwassereinfluß können auf diesen durchlässigen Sanden im UG vor allem die beiden *Vaccinium*-Arten (*V. myrtillus*, *V. vitis-idaea*) gelten.

So besiedelt ein armer Beerkraut-Kiefern-Wald (Vaccinio-Pinetum KOBENDZA 30 em. PASS. 56) ohne Flechten die frischesten Standorte innerhalb des Kiefern-Wald-Komplexes. Der Grundwassereinfluß auf die Standorte ist im Sommer gering, die Tiefststände liegen bei 150—200 cm, während in den Winter- und Frühjahrsmonaten noch Höchststände bis etwa 1 m erreicht werden. Es ist also über den Saugsaum noch ein zeitweiser Einfluß des Grundwassers auf die Bodenvegetation vorhanden. Für die Baumschicht wird es meist während des ganzen Jahres wirksam, was sich insbesondere für die Verjüngung der Kiefer und deren noch relativ günstiges Wachstum positiv bemerkbar macht.

Von diesem Waldtyp leitet ein *Vaccinium*-Flechten-Kiefern-Wald mit den genannten Beersträuchern und bereits zahlreichen *Cladonia*-Arten in der Bodenschicht zu den trockenen Flechten-Kiefern-Wäldern über. Ein geringer zeitweiser und tiefer Grundwassereinfluß ist in diesem *Vaccinium*-Typ noch zu verzeichnen, die Höchststände liegen wenig über 150 cm, die Tiefststände oft schon beträchtlich unter 2 m. Zum Teil liegen die Vorkommen in Dünensenken, in denen eine oberflächliche Durchfeuchtung durch von den Dünenhängen abfließendes Niederschlagswasser hinzukommt.

Die Ausbildung der noch trockeneren Flechten-Kiefern-Wald-Typen, in denen neben Flechten und Moosen nur das Heidekraut (*Calluna vulgaris*) sowie einige Horstgräser (*Festuca ovina, Deschampsia flexuosa, Corynephorus canescens*) am Aufbau der Feldschicht stärker beteiligt sind, erfolgt unabhängig von bestimmten Grundwasserstufen, da in ihnen nur einige Baumwurzeln zeitweise bis in den Saugsaum hineinreichen. Hier sind vor allem kleinklimatische und edaphische Faktoren für die Vegetationsdifferenzierung verantwortlich.

4.3. Die Grünlandgesellschaften

Wie bereits betont, ist Grünlandnutzung auf den Böden des UG nur bei ganzjährigem Grundwassereinfluß möglich. Die Spanne im Grundwasserniveauunterschied beträgt dabei etwa 120 (—150) cm, wobei jedoch nur in einem eng begrenzten mittleren Bereich in Güte und Menge befriedigende Erträge erzielt werden können. Auch zu feuchte, gütemäßig nur wenig wertvolle und zu trockene, mengenmäßig unrentable Flächen werden noch wiesenwirtschaftlich genutzt. Die feuchtesten Bestände sind zudem hochwassergefährdet, insbesondere wenn dieses ungünstig eintritt und in die Erntezeit fällt (1965!). Gegenüber früher ist die düngende Wirkung der Frühjahrshochwässer heute gemindert. Auf den gedüngten Flächen können sie bei ungünstigem Eintritt schädigend wirken, indem sie den gestreuten Dünger mit fortführen. Die Grundwasserschwankungswerte liegen mit den höchsten gemessenen Werten von 60—70 cm (wiederum unter den mittleren, noch das Grundwasser erreichenden Beständen) etwas niedriger als im Wald.

4.3.1. Steifseggen-Ried (Caricetum elatae W. KOCH 26)

Comarum-palustre-Subassoziation

Typische Variante	F
Filipendula-ulmaria-Variante	F

Nur selten kommt dieses Großseggenried im UG vor. Es besiedelt Verlandungssenken ehemaliger stehender Gewässer mit Flachmoortorfböden. Während des ganzen Jahres bleibt der ohnehin durch eine hohe Wasserkapazität ausgezeichnete Boden vom aufsteigenden Kapillarwasser stark durchfeuchtet. Das Grundwasser sinkt kaum unter 20—30 cm ab und überstaut im Winterhalbjahr langfristig die Standorte, wobei in der typischen Variante infolge meist noch längerer, bis in den Frühsommer hinein andauernder Überschwemmung die Narbe zwischen den Seggenbullen nur unvollkommen geschlossen ist. Die trockenere *Filipendula*-Variante kann gelegentlich zur Streumahd genutzt werden.

4.3.2. Schlankseggen-Ried [Caricetum gracilis (GRAEBN. & HUECK 31) TX. 37]

Comarum-palustre-Subassoziation

Typische Variante	F
Alopecurus-pratensis-Variante	F

Das Ried besiedelt im UG Überschwemmungsstandorte in den tiefsten Schlenken der Flußtäler — wobei es sich häufig um verlandete Altwässer handelt — und in Senken der übrigen Wiesengebiete. Der Boden ist meist ein ± tiefgründiger Seggentorf, in den trockenen Beständen bereits mit Übergängen zu Anmoorböden. Die Standorte stehen ständig bis nahe unter die Oberfläche unter Grundwassereinfluß, das in den Winter- und Frühjahrsmonaten regelmäßig über Flur tritt (MS 26,28). Die Grundwasserstände sinken nicht unter 20—30 cm ab, so daß der Oberboden während des ganzen Jahres naß bis feucht ist. Die Dauer der Überschwemmung ist unterschiedlich und bei der *Alopecurus*-Variante, die durch das Eindringen bereits zahlreicher Feuchtwiesenarten gekennzeichnet ist, kürzer als bei der typischen Variante. Nur die flußnahen Schlankseggenrieder werden überwiegend durch Hochwässer überschwemmt, während in den Senken der flußfernen Bestände das Grundwasser über die Oberfläche drückt. Eine zusätzliche Nährstoffversorgung ist somit meist nicht vorhanden. Die Grundwasserschwankungen sind trotz hohen Wasserverbrauchs der massenmäßig produktivkräftigen, aber minderwertigen Bestände geringer als in den Feuchtwiesen, da eine genügende Wassernachlieferung und hohe Wasserhaltekraft des Torfbodens wirksam sind. Der typischen Variante entsprechende Standorte mit gleichem Wasserhaushalt besiedelt sehr selten auch ein Sumpfseggen-Ried (Caricetum ripario-acutiformis KOBENDZA 30).

4.3.3. Rohrglanzgras-Wiese (Phalaridetum arundinaceae LIBB. 31)

Die Gesellschaft besiedelt im UG nur an wenigen Stellen ufernahe Überschwemmungsstandorte, Die übrigen Vorkommen sind uferferne Senken in Wiesengebieten, in denen sie von fließendem Oberflächenwasser nicht erreicht werden und auch das Grundwasser ± stagnierend ist. Der Boden ist an diesen Stellen ein tiefgründiger Flachmoortorf. Die Ursachen für das Auftreten der Gesellschaft an diesen Standorten, das im Widerspruch zu den Verhältnissen in anderen Gebieten wie z. B. im Ober- und Unterspreewald steht (vgl. PASSARGE 1955, SCAMONI 1955, FREITAG 1957a), wo die Gesellschaft als Zeiger für fließendes, sauerstoffreiches Oberflächen- oder Grundwasser gilt, ist noch nicht geklärt. Während der Winter- und Frühjahrsmonate sind die Bestände ± überschwemmt, jedoch meist durch Druckwasser. Auch während der übrigen Zeit des Jahres sinkt das Grundwasser nicht unter 40—50 cm ab, so daß der Boden ständig naß bis gut durchfeuchtet bleibt (MS 23).

4.3.4. Knickfuchsschwanz-Rasen [Alopecuretum geniculati TX. (37) 50]

Die Gesellschaft besiedelt zeitweise überschwemmte Senken des Weidegrünlandes sowie Flutmulden innerhalb der Wiesenflächen, also vor allem Stellen mit oft stärkerer Störung (z. B. auch nach Grasneuansaaten auf zu nassen Flächen), so daß die Narbe nicht dicht geschlossen ist und sich sowohl Arten von Überschwemmungsrasen als auch Seggenried- und Feuchtwiesenarten in gleichem Maße ansiedeln können. Hierin bestehen im UG vor allem die begrenzenden Standortsfaktoren gegenüber anderen feuchten Grünlandgesellschaften. Eine verstärkte Nitratbeeinflussung dagegen, wie sie für die Gesellschaft verschiedentlich angegeben wird (z. B. PASSARGE 1964), ist nur in den Weidebeständen anzunehmen. Der Boden ist ein Flachmoortorf unterschiedlicher Mächtigkeit über Sand mit teilweisem Übergang zu Anmoor. Die Standorte werden regelmäßig in den Frühjahrs- und Wintermonaten durch Überschwemmungs- oder Druckwasser überstaut, und auch während der Sommermonate sinkt das Grundwasser nicht unter etwa 40 cm ab. Der Oberboden bleibt ständig gut durchfeuchtet (MS 23).

4.3.5. Hirseseggen-Gesellschaft

Carex-acutiformis-AF	f
Typische AF	f
Polygala-vulgaris-AF	f

Dieses sehr bedingt zum Junco-Caricetum fuscae Tx. (37) 52 gehörende Kleinseggen-Ried nimmt die am stärksten extensiv bewirtschafteten Flächen innerhalb der mäßig bis schwach feuchten Grünländereien ein. Drei AF entsprechen auch hier drei Feuchtestufen, wobei jedoch gegenüber den intensiver bewirtschafteten Feuchtwiesen eine etwas gröbere Differenzierung auffällt, vor allem einige Ödlandarten wie *Polygala vulgaris, Sieglingia decumbens* und *Viola canina* gehen dabei in feuchtere Bereiche hinunter. Die Standorte sind in der feuchtesten *Carex-acutiformis*-AF (Flachmoortorf über Sand) längere Zeit, vor allem im Winter und Frühjahr überschwemmt, während in der typischen AF (Bruchwaldtorf mit Übergang zu Anmoor) das Wasser nur noch selten über Flur tritt und die Grundwassertiefststände bei etwa 60 cm liegen. Die *Polygala*-AF stockt bereits auf stärker wechselfeuchten Standorten mit zeitweiser Austrocknung des Oberbodens (Typ eines nassen bzw. typischen Gleys), sie bildet den Übergang zur Rotschwingel-Wiese bzw. zu Borstgrasrasen.

4.3.6. Honiggras-Wiese (Holcetum lanati ISSLER 36 em. PASS. 64)

Carex-fusca-Subassoziation	F
Carex-gracilis-Variante	F
Caltha-palustris-Subvariante	f
Typische Subvariante	f
Typische Variante	F
Typische Subassoziation	F
Veronica chaemaedrys-Subassoziation	F
Arrhenatherum-elatius-Variante	
Typische Variante	

Feuchtwiesen, die im UG sämtlich zu diesem verarmten östlichen Typ, der Honiggras-Wiese, gehören, sind die wichtigsten und am weitesten verbreiteten Grünlandbestände im Gebiet. Da die Wirtschaftlichkeit des Grünlandes schon bei verhältnismäßig geringen Grundwasseränderungen (zu trocken oder zu feucht) sehr stark leidet, herrscht dieser mittelfeuchte Typ vor.

Die *Carex-fusca*-Subassoziation besiedelt die feuchtesten Standorte innerhalb der Gesellschaft. Die Böden sind Bruchwald- oder Seggentorf in den nassesten,

± mineralreiche Anmoorböden über Sand in den trockeneren Beständen. Das Grundwasser ist meist stagnierend bzw. nur gering fließend. Es sinkt auch in den Sommermonaten nie tief unter die Oberfläche, so daß der Oberboden ständig ± gut durchfeuchtet bleibt. Unterschiede ergeben sich bei den einzelnen AF. Die *Carex-gracilis*-Variante ist regelmäßig längere Zeit im Winter und Frühjahr und gelegentlich auch bei sommerlichen stärkeren Niederschlägen bzw. Hochwassern überstaut oder in der typischen Subvariante zumindest bis zur Oberfläche vom Grundwasser durchnäßt. Wie die Bodenfeuchte-Messungen (MS 22) zeigen, lagen die Feuchtewerte während des ganzen Jahres über 40%, der Höchstwert lag sogar bei 64% im Mai trotz vorausgegangener trockener Periode. Der Anmoor-Horizont über Sand hat hohe Feldkapazität und stand zudem fast ständig unter Grundwassereinfluß. Die darunterliegende Sandschicht lag ständig im Bereich des Grundwassers oder Saugsaums und war somit dauernd wassergetränkt. Das Grundwasser sinkt auch im Sommer und Herbst nicht unter etwa 30—45 cm ab. In der typischen Variante steigt das Grundwasser meist nur noch bis dicht unter Flur, Überflutungen sind selten. Im Sommer und Herbst kommt es hier zeitweise bereits zu einer stärkeren Abtrocknung des Oberbodens, wenn auch der Einfluß des Grundwassers in dieser Zeit noch stets vorhanden ist. Außerdem besitzt der humusreiche Oberboden eine genügend große Wasserhaltekraft, um eine ausreichende Wasserversorgung zu garantieren und die Erträge auch in Trockenzeiten nicht gefährden zu lassen.

Eine zu den Überschwemmungswiesen humoser, tonreicher Böden der Auen größerer Flüsse, dem Deschampsietum caespitosae HORVATIC 30, tendierende *Deschampsia-caespitosa*-AF innerhalb der Subass. mit vorherrschender Rasenschmiele besiedelt meist stärker tonig durchsetzte Anmoor- und Gleyböden, wie sie durch ehemalige Überschlickung im Tal der Spreearme entstanden sind. Aber im Unterschied zu den Standortsverhältnissen im Deschampsietum des Spreewalds und Oderbruchs (vgl. PASSARGE 1955, SCAMONI 1955, FREITAG 1957b) weichen die Grundwasserverhältnisse nicht von denen der reinen *Holcus*-Wiesen ab. Die Bestände der *Deschampsia*-AF werden im UG nur noch zum Teil und in der feuchtesten *Caltha*-Subvariante längere Zeit überschwemmt (MS 27), in den anderen Untereinheiten dagegen nur noch relativ selten und kurzfristiger. Bodenfeuchtemessungen in der typischen Variante ergaben auch hier eine gute ausreichende Durchfeuchtung, fast sämtliche gemessenen Werte lagen über 30%, meist über 40%. Lediglich in 20—25 cm Tiefe war im Mai nach trockenerer Periode eine stärkere Austrocknung bis auf 25% zu verzeichnen. Im sandigen G-Horizont, der fast ständig unter Grundwassereinfluß bleibt, änderten sich die Werte nur unbedeutend. Die nur kleinflächig entwickelten Übergangsbestände der typischen Subass. nehmen mittlere Standorte ein, sowohl in bezug auf die Wasser- als auch auf die Bodenverhältnisse (Humusgehalt). Die *Veronica*-Subass. mit zwei — keinen Wasser- sondern Trophiestufen entsprechenden — Varianten ist die trockenste AF der Feuchtwiesen. Sie besiedelt kleinere Erhebungen innerhalb des Feuchtwiesenkomplexes oder tiefere Senken in den Ackergebieten. Die Bestände unterliegen bereits einer stärkeren Wechselfeuchtigkeit. Der Oberboden trocknet während der Sommermonate ± stark aus, die Grundwassertiefststände liegen etwa bei 80—90 cm, die Höchststände etwa bei 15—20 cm in den Winter- und Frühjahrsmonaten, zum Teil aber auch tiefer. Eine Überschwemmung findet nicht mehr statt. In Trockenjahren sind also die Erträge bereits gefährdet, auch in anderen Jahren sind sie nur gering. Innerhalb der Wiesengesellschaften sind in dieser AF die größten jährlichen Grundwasserschwankungen zu verzeichnen (MS 16:60—70 cm). Da in Trockenzeiten bei ungenügender Wasserhaltekraft des relativ schwach humosen Oberbodens die tiefergehenden Wurzeln den Saugsaum noch erreichen, findet eine stärkere Ausnutzung und damit verbundene Absenkung des Grundwassers statt. Einen besonders interessanten Standort stellt die MS 24 mit einem zur typischen Subass. tendierenden Bestand dar. Der Boden ist eine künstlich aufgeschüttete 40—45 cm starke Sandschicht über Flachmoortorf (Moorkultur), deren oberste 5 cm bereits eine schwache Humusanreicherung zeigen. In der Sandauflage lagen die Feuch-

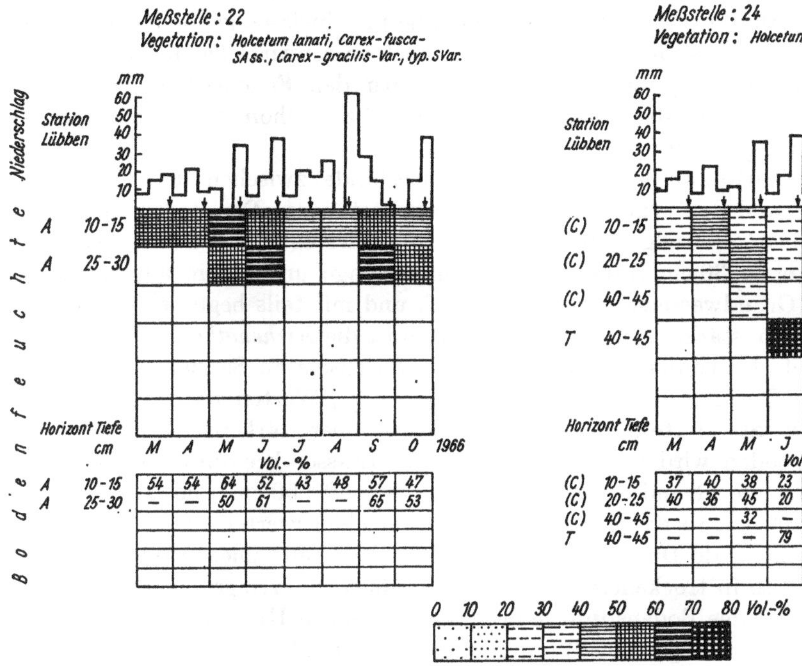

Horizont	Tiefe cm	M	A	M	J	J	A	S	O
A	10-15	54	54	64	52	43	48	57	47
A	25-30	—	—	50	61	—	—	65	53

Horizont	Tiefe cm	M	A	M	J	J	A	S	O
(C)	10-15	37	40	38	23	31	27	28	26
(C)	20-25	40	36	45	20	31	26	27	29
(C)	40-45	—	—	32	—	—	47	34	—
T	40-45	—	—	—	79	70	—	—	64

Abb. 4. Bodenfeuchtewerte der Meßstellen 22 und 24

tewerte in den Frühjahrsmonaten mit 36—45% recht hoch und günstig, auch in den übrigen Monaten waren noch günstige Feuchtewerte (26—31%) zu verzeichnen, lediglich im Juni führte das verstärkte Wachstum nach dem ersten Schnitt bei tiefen Grundwasserständen zu einer merklichen Austrocknung der oberen Sandschicht. Die direkt über dem Moorboden liegende Schicht blieb meist gut durchfeuchtet, und wo die Torfschicht bei der Probenentnahme selbst erfaßt wurde, ergaben sich die höchsten im UG überhaupt gemessenen Werte (64—78%). Wenngleich der moorige Untergrund einen gewissen Wasserstau erzeugt und tiefergehende Wurzeln ihn noch erreichen, so zeigt das Beispiel doch, daß auch auf sehr humusarmem Oberboden bei genügend hohen Grundwasserständen Grünlandkultur möglich ist (die Erträge lagen etwa gleichhoch wie in entsprechenden „natürlichen" Beständen), was bei Änderung der Grundwasserverhältnisse durch eine künftige Spiegelaufhöhung von Bedeutung sein kann.

4.3.7. Rotschwingel-Wiese [Festucetum rubrae (Scam. 55) Oberd. 57]

Alopecurus-pratensis-Subassoziation	F
Typische Variante	
Arrhenatherum-elatius-Variante	
Typische Subassoziation	F
Festuca-ovina-Subassoziation	F
Typische Variante	F
Hieracium-pilosella-Variante	F

Dieser in anderen Gebieten nur noch sehr selten vorkommende Wiesentyp armer, extensiv bewirtschafteter Standorte, zu dem hier alle Frischwiesen gehören, ist im UG noch relativ häufig. Er besiedelt die trockensten, noch wiesenwirtschaftlich genutzten Flächen innerhalb des UG, nimmt jedoch gegenüber den Feuchtwiesen geringere Fächen ein, da seine Ertragsleistungen an der Grenze bzw. schon jenseits einer möglichen Wiesennutzung liegen.

Die *Alopecurus*-Subass. besiedelt Übergangsbestände zwischen den trockenen Feuchtwiesen-AF und den typischen Frischwiesen-Beständen. Der Boden ist in der typischen Variante im allgemeinen ein nährstoffarmer Gley (auf Sand) mit einer durchschnittlich etwa 30 cm starken A-Horizont-Auflage und einem ausgeprägten G_{ox}-Horizont im Grundwasserschwankungsbereich und mit teils beginnender Podsolierung. Auf gleichen, stärker gedüngten Standorten ist die *Arrhenatherum*-Variante zu finden, die jedoch kleinflächig auch auf tonig durchsetzten Sanden (Bodentyp: Verbraunter Gley Kubiena 1953) mit relativ hohen p_H-Werten zwischen 6,4 und 7,0 vorkommt. Die Standorte der Subass. unterliegen einer stärkeren Wechselfeuchtigkeit. Der Oberboden wird nicht mehr vom Grundwasser beeinflußt, im Winterhalbjahr erreichen die Grundwasserstände etwa (30) 40—50 cm, die Tiefststände liegen bei etwa 100 cm, so daß bei niedrigen Niederschlagswerten in den Sommermonaten keine genügende Durchfeuchtung des Hauptwurzelraumes erfolgt (MS 17). Die Bestände sind so in trockeneren Sommern bereits stark ertragsgefährdet, zumal die Wasserhaltekraft des Bodens infolge des relativ geringen Humusanteils im Oberboden (etwa 8—10% im allgemeinen) nur gering ist, während in den tonig durchsetzten Böden der Anteil des nicht pflanzenverfügbaren Wassers relativ hoch ist.

Die im allgemeinen zweischürig genutzten Bestände der typischen Subass. kommen namentlich auf ortsfernen und schwerer zugänglichen Standorten vor. Für eine Wiesennutzung sind sie oft schon zu unrentabel, so daß sie an besser zugänglichen Stellen vielfach in Äcker umgewandelt wurden. Der Bodentyp ist ein nährstoffarmer Gley (auf Sand) mit beginnender Gleypodsol-Bildung. Die Subass. steht in ihren Grundwasser- und Bodenfeuchteverhältnissen als mittlerer Typ zwischen den beiden anderen Subass. Das Grundwasser steigt an diesen wechselfeuchten Standorten im allgemeinen nicht mehr über 50 cm an, die Tiefstwerte liegen bereits deutlich unter 1 m (MS 8). Nur die tiefsten Wurzeln reichen also im Sommer noch bis in den Saugsaum.

Die Bestände der *Festuca-ovina*-Subass. stellen die extremsten, noch wiesenwirtschaftlich genutzten Bestände dar, jedoch ist ihre Obergrasbildung praktisch unbedeutend, und ihre Erträge sind so niedrig, daß sie oft nur einmal im Jahre, z. T. überhaupt nicht oder nur nach günstiger Vegetationsentwicklung gemäht werden können. Der Boden ist ein $\pm$ ausgeprägter Gleypodsol, auf feinerdereicheren Standorten der Talaue ein schwach verbraunter Gley. Das Grundwasser kann in der typischen Variante zeitweise noch bis etwa 50—60 cm ansteigen, doch ist es während der längsten Zeit des Jahres und namentlich in der Vegetationsperiode dauernd unter 80 cm (bis 150 cm), so daß nur die tiefsten Wurzeln zeitweise in den Saugsaum dringen (vgl. MS 3,12). In der *Hieracium*-Variante sind noch tiefere Grundwasserstände zu verzeichnen. Nur in den Winter- und Frühjahrsmonaten erreicht der Saugsaum in der Regel den Wurzelbereich. Die Bestände bilden den direkten Übergang zum Grasnelken-Sandtrockenrasen (Diantho-Armerietum KRAUSCH 59), der überhaupt nicht mehr genutzt werden kann. Da das sommerliche Absinken des Grundwassers zu einem Abreißen zwischen Wurzelraum und Saugsaum führt, verläuft hier der Grundwassergang — soweit er von der Vegetation abhängt — wieder ausgeglichener. Namentlich vor dem zweiten Schnitt ist der Grundwasserstand in den Frischwiesen also recht ungünstig und führt oft zu größeren Ertragsausfällen in der ohnehin ertragsschwachen Wiesengesellschaft. Bodenfeuchtemessungen bestätigten, daß eine Wiesennutzung auf Standorten ohne sommerlichen Grundwassereinfluß bei den im UG vorliegenden armen Sanden nicht mehr rentabel sein kann. Es tritt hier eine zu starke sommerliche Austrocknung der oberen Bodenschichten ein, da der Hauptwurzelraum nicht mehr mit dem Saugsaum in Verbindung steht und die Wasserversorgung aus den (geringen) Bodenwasservorräten erfolgt. So zeigte MS 8 (Festucetum typicum) nach einer mittleren Durchfeuchtung im März—April eine deutliche und meist starke sommerliche Austrocknung des unteren A- und des B/G_{ox}-Horizontes bis in eine Tiefe von etwa 70 cm hinab. Während der obere A-Horizont die Niederschläge abfängt und nur im Mai nach einer trockeneren Periode eine stärkere Austrocknung zeigte (13%), werden die tieferen Schichten lediglich durch anhaltende Regenfälle durchfeuchtet und sind in den sommerlichen Trockenzeiten mit Feuchtewerten ständig unter 15% (bis 5,3%) zu trocken für eine ausreichende Versorgung der Vegetation. Bei etwa 85 cm Tiefe blieb der Feuchtegehalt unverändert, da diese Schicht sowohl von den Wurzeln kaum noch erreicht wird als auch zeitweise im Saugsaum lag. Demgegenüber lagen die Werte in einem Bestand mit höherem Feinbodenanteil (anlehmiger bis lehmiger Sand) in A- und B/G_{ox}-Horizont (schwach verbraunter Gley) stets verhältnismäßig hoch. Nur im A-Horizont sanken die August-Werte auf 24—25%, während sich sonst eine stärkere Durchfeuchtung zeigte. Namentlich inden tieferen Schichten ab 45 cm schwankten die Werte kaum und lagen ständig um 30%. Lediglich der von Niederschlägen beeinflußte Oberboden wies stärkere Schwankungen und oft relativ hohe Feuchtewerte auf, Ursache dafür dürften die höhere Wasserhaltekraft und die etwas stauende Wirkung des B/G_{ox}-Horizontes sein, so daß der Hauptwurzelraum besser mit Bodenwasser versorgt wurde und eine stärkere Ausschöpfung der tieferen Schichten unterbleiben konnte. Aber auch hier war die gleiche trockenheitszeigende Vegetation zu finden (*Festuca-ovina*-Subass.), da der höhere Totwasseranteil den Einfluß der größeren Wasserhaltekraft auf eine bessere Wasserversorgung der Vegetation weitgehend aufhebt.

4.3.8. Weidelgras-Weißklee-Weide (Lolio-Cynosuretum cristati Tx. 37)

Lotus-uliginosus-Subassoziation
  Agrostis-stolonifera-Variante F
    Ranunculus-flammula-Subvariante f

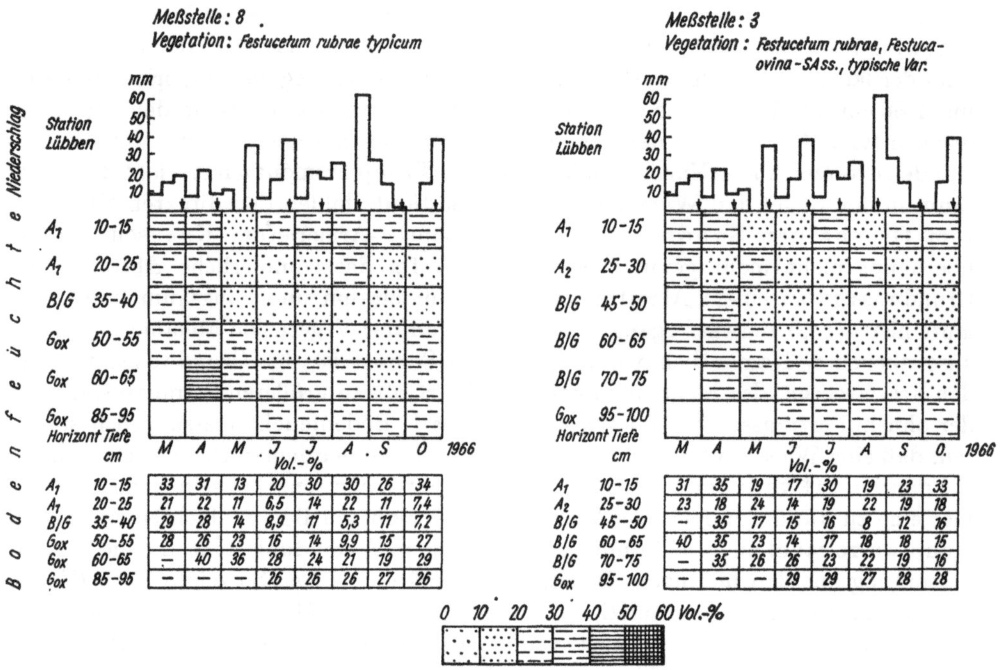

Abb. 5. Bodenfeuchtewerte der Meßstellen 8 und 3

Typische Subvariante	f
Typische Variante	F
Agrostis-tenuis-Variante	F

Dauerweiderasen befinden sich im UG nur auf wenigstens zeitweise grundwasserbeeinflußten Standorten, so entsprechen die drei Varianten in ihren Feuchtestufen den drei Subassoziationen des Holcetums. Auf trockeneren Standorten ist die Produktionskraft der Rasen infolge des sommerlichen Wassermangels nicht groß genug, um bei dem Weidebetritt und der ständigen Nutzung nicht zu stark geschädigt zu werden. Da jedoch auch die feuchtesten Bestände der *Ranunculus-flammula*-Subvar. für eine Beweidung nur schlecht geeignet sind (ungenügende Trittfestigkeit), genügen eigentlich allein die mittleren Ausbildungen den an Intensivweiderasen gestellten Anforderungen.

4.4. Ackerunkrautgesellschaften

Auch die Ackernutzung ist im UG nur auf Standorten möglich, die während der Vegetationsperiode wenigstens zeitweise noch unter Grundwassereinfluß stehen, auf Standorten mit länger anhaltenden sommerlichen Grundwasserständen unter 150 cm kann sie nicht mehr betrieben werden, weshalb als Folge der Grundwasserabsenkung zahlreiche Ackerflächen aufgelassen wurden und zu Silbergrasfluren wurden, die heute zu einem großen Teil mit Kiefern aufgeforstet sind. Zwischen den feuchtesten, vernässungsgefährdeten und den bereits zu trockenen Äckern liegen lediglich etwa 100 cm Niveauunterschied.

4.4.1. Vielsamen-Gänsefuß-Gesellschaft (Panico-Chenopodietum polyspermi Br.-Bl. 21)

Die Gesellschaft besiedelt die tiefstgelegenen Stellen innerhalb der Ackerkomplexe. Kulturarten sind meist Hackfrüchte, seltener Sommer- und nur an den trockensten Stellen Wintergetreide. Gleichgeartete Standorte werden oft noch von Grünland mittlerer Ertragsklasse (typische Subass. bis typische Variante der *Carex-fusca*-Subass. des Holcetums) eingenommen. Der Boden ist ein typischer Gley mit stark humushaltigem bis anmoorigem Oberboden über Sand. Die Mächtigkeit des A-Horizontes ist unterschiedlich, z. T. nicht größer als in trockeneren Ackertypen, doch bewirkt der hohe Humusgehalt eine gute Wasserhaltekraft und ausreichende Wasserversorgung auch bei stärkerem Absinken des Grundwassers. Im allgemeinen behält jedoch das Grundwasser während der ganzen Vegetationsperiode Einfluß auf die Vegetation, im Winter und Frühjahr steigt es bis nahe unter Flur, kann die Standorte sogar überstauen, oder das Niederschlagswasser dringt in den vom aufsteigenden Kapillarwasser gesättigten Boden kaum ein und vernäßt den Acker oberflächlich. Die Grundwasserstände schwankten in der MS 25 zwischen etwa 30—80 cm. Wie die Bodenfeuchtemessungen an der gleichen MS zeigten, blieb namentlich die obere Schicht des A-Horizontes ständig frisch. In den tiefergelegenen Schichten und in der Übergangszone zum G_{ox}-Horizont wurde dagegen zeitweise ein stärkerer Wasserentzug durch die Vegetation bemerkbar, dennoch sanken die Werte kaum unter 20%. Der eigentliche G_{ox}-Horizont stand meist bereits unter dem Einfluß des kapillar aufsteigenden Grundwassers und blieb so ständig gleichmäßig durchfeuchtet (33—34%). In nassen Jahren, wie z. B. 1965, sind die Flächen vernässungsgefährdet, so daß es notwendig war, die Flächen, auf denen z. B. die Kartoffeln nicht aufgegangen waren, nachträglich mit Futterpflanzen anzusäen. Die feuchten Senken fielen in diesem Jahr schon von weitem durch geringe Wuchsleistungen der Feldfrüchte auf. Besonders empfindlich sind offenbar die Kartoffel und die Getreidearten, wobei unter letzteren der Hafer noch die besten Leistungen zeigte. Bei günstigem Witterungsverlauf liefern die Äcker mit dieser Gesellschaft jedoch die besten Erträge im UG. Feuchteunterschiede machen sich im Gebiet nur in einem quantitativ höheren Anteil an Feuchtezeigern (*Rorippa islandica*, *Bidens tripartitus*, *Juncus bufonius*), jedoch nicht durch Ausbildung floristisch differenzierter Varianten bemerkbar.

4.4.2. Sandmohn-Gesellschaft [Papaveretum argemone (Libb. 32) Krusem. & Vlieg. 39],

Hühnerhirse-Gesellschaft [Panicetum crus-gallii Krusem. & Vlieg. (39) 40]

Papaveretum		Panicetum	
Scleranthus-annuus-Subass.		*Spergula-arvensis*-Subass.	
Stachys-palustris-Var.	F	*Stachys-palustris*-Var.	F
Typische Var.	F	Typische Var.	F
Juncus-bufonius-Subvar.	f	*Juncus-bufonis*-Subvar.	f
Typische Subvar.	f	Typische Subvar.	f
Arnoseris-minima-Var.	F	*Digitaria-ischaemum*-Var.	F

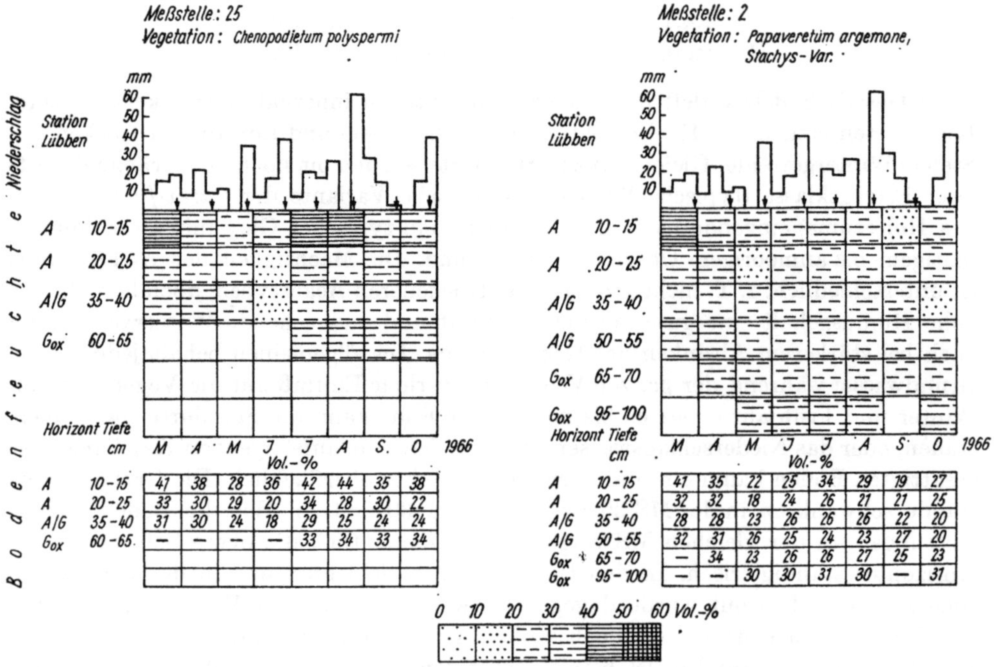

Abb. 6. Bodenfeuchtewerte der Meßstellen 25 und 2

Die beiden Gesellschaften nehmen die mittleren Ackerstandorte mit guten bis mittleren Erträgen ein, das Papaveretum unter Wintergetreide, das Panicetum unter Sommerfrüchten (Hackfrüchte, Sommergetreide), wobei ihre Standorte einander entsprechen und sie sich bei Fruchtwechsel ablösen können. Parallele Untereinheiten kennzeichnen gleiche Grundwasserstufen, geringe Unterschiede bestehen in der jahreszeitlichen Ausnutzung der Bodenfeuchtevorräte, bedingt durch den unterschiedlichen Entwicklungsrhythmus der Winter- und Sommerfruchtkulturarten.

Die am stärksten grundfeuchten, zum Chenopodietum polyspermi tendierenden Bestände stellen die *Stachys*-Varianten mit zahlreichen Grund- und Krumenfeuchtezeigern dar, wobei die Unterschiede zwischen beiden Gesellschaften noch weniger ausgeprägt sind — eine noch stärkere Durchfeuchtung und zeitweise Vernässung im Winter und zeitigen Frühjahr läßt die für die Winterfruchtäcker typischen Herbst- und Frühjahrskeimer schlecht aufkommen, und die „Hackfruchtunkräuter“ können sich ausbreiten. Hier ist der Grundwassereinfluß noch über das ganze Jahr wirksam, die Höchststände lagen bei etwa 30—50 cm, die Tiefststände bei etwa 100—115 cm. Zudem garantiert das noch relativ hohe Wasserhaltevermögen des gut humosen Oberbodens (Bodentyp: typischer Gley) eine ausreichende Wasserversorgung während der Vegetationsperiode. Die mittleren Bodenfeuchtewerte lagen in der MS 2 unter Winterroggen bei etwa 25% in den vom Grundwasser nicht beeinflußten Schichten. Nur in der starken Wachstumsphase bei geringen Niederschlägen im Mai und nach Ernte und Umbruch im September sanken die Werte in den oberen 40 cm deutlich tiefer. In dem Kartoffelacker der MS 15 erfolgte bei sonst etwa ähnlichen

Verhältnissen vom Juli bis September eine offensichtlich stärkere Austrocknung im G_{ox}-Horizont, so daß es bis zu Tiefstwerten von 16% kam. Auch nach der Ernte im Oktober waren diese Verluste noch nicht wieder ausgeglichen. Auffallend sind dagegen die vergleichsweise höheren Werte im Mai (25—30%) in den oberen 50 cm (gegenüber 18—23% in MS 2), also zu einem Zeitpunkt, als die Kartoffeln erst im Austreiben begriffen waren.

In den typischen Varianten ohne eigentliche Grundfeuchtezeiger ist jeweils eine krumenfeuchte *Juncus*-Subvariante neben einer trockeneren typischen Subvariante ausgebildet. Hierbei ist Anlaß für das Auftreten der Krumenfeuchte zwar der noch relativ hohe Humusgehalt des A-Horizontes, doch wird zugleich ein etwas höherer Grundwasserstand angezeigt, da die reichlichere Humusbildung auch erst durch sein Wirksamwerden möglich war. Die Grundwasserstände liegen mit Höchstwerten von etwa 50—70 cm und Tiefstwerten zwischen 100 und 130 cm v. a. in der ersten Hälfte der Vegetationsperiode noch hoch genug, um über den Saugsaum einen merklichen Einfluß auf die Vegetation zu nehmen (vgl. MS 1). In der typischen Subvariante muß bei Grundwasserhöchstständen von 70—90 cm und Tiefstständen von etwa 130—150 cm bereits ein Großteil der Wasserversorgung aus den Bodenwasservorräten erfolgen, doch blieb der Boden meist noch mittelmäßig durchfeuchtet (MS 17 u. 18), was bei normalen Niederschlägen zu keinen größeren Ertragsausfällen führt. Sämtliche Werte (bis auf eine Ausnahme) lagen über 15%. In der *Arnoseris*- bzw. *Digitaria*-Variante tritt jedoch bereits eine Trockenheitsgefährdung in niederschlagsarmen Sommern ein, da die Grundwasserstände nur mit ihren Höchstwerten noch bis etwa 90—110 cm im Winter und Frühjahr ansteigen, während die Tiefststände schon bei etwa 150—170 cm liegen.

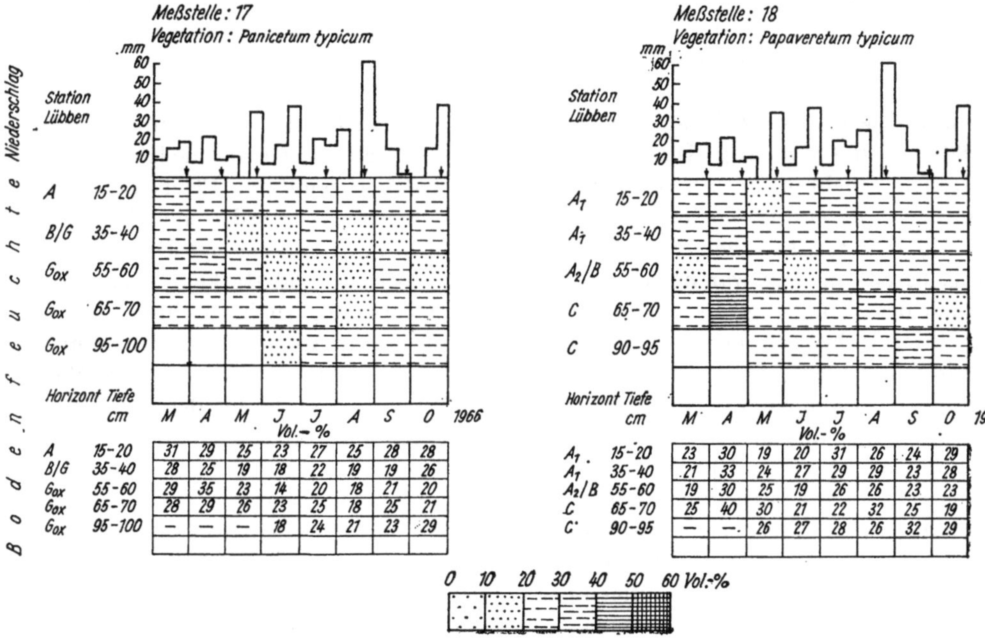

Horizont	Tiefe cm	M	A	M	J	J	A	S	O
A	15-20	31	29	25	23	27	25	28	28
B/G	35-40	28	25	19	18	22	19	19	26
G_{ox}	55-60	29	35	23	14	20	18	21	20
G_{ox}	65-70	28	29	26	23	25	18	25	21
G_{ox}	95-100	—	—	—	18	24	21	23	29

Horizont	Tiefe cm	M	A	M	J	J	A	S	O
A_1	15-20	23	30	19	20	31	26	24	29
A_1	35-40	21	33	24	27	29	29	23	28
A_2/B	55-60	19	30	25	19	26	26	23	23
C	65-70	25	40	30	21	22	32	25	19
C	90-95	—	—	26	27	28	26	32	29

Abb. 7. Bodenfeuchtewerte der Meßstellen 17 und 18

4.4.3. Lammkraut-Gesellschaft [Arnoseretum minimae (All. 22) Tx. 37]
Artenarme Gänsefuß-Gesellschaft [Chenopodietum albi Pass. (55) 64]

Arnoseretum		Chenopodietum, *Rumex-acetosella*-Subass.	
Myosotis-arvensis-Subass.	F	*Echinochloa-crus-galli*-Var.	F
Juncus bufonius-Var.	F	*Juncus-bufonius*-Subvar.	f
Typische Var.	F	Typische Subvar.	f
Typische Subass.	F	Typische Var.	F
Corynephorus-canescens-Subass.	F	—	

Die Gesellschaft besiedelt die humusärmsten, trockensten Sandstandorte innerhalb der Ackerflächen unter Winterroggen (Arnoseretum) bzw. Hackfrüchten — fast ausschließlich Kartoffeln — (Chenopodietum). Nicht so sehr das Grundwasser, das kaum noch Einfluß auf die Vegetation hat, als die Nährstoff- und Humusarmut, verbunden mit äußerst geringer Wasserhaltekraft, sind die begrenzenden Standortsfaktoren gegenüber den trockenen Papaveretum- und Panicetum-AF. Während in der *Myosotis*-Subass. bzw. *Echinochloa*-Var. noch ähnliche Verhältnisse wie in den trockensten Varianten des Papaveretums bzw. Panicetums herrschen und das Grundwasser Höchststände bis etwa 85—100 cm und Tiefststände bei etwa 140—160 cm hat, steigt es in der typischen Subass. bzw. Variante schon kaum mehr über 110 cm und sinkt bis etwa 170 cm. In der *Corynephorus*-Subass. des Arnoseretums — auf dessen Standorten Kartoffeln nicht mehr gedeihen, so daß eine Paralleleinheit des Chenopodietums nicht zu finden war — liegen die Werte noch ungünstiger, ein Grundwassereinfluß während der Vegetationsperiode ist nicht mehr vorhanden. Bei den tiefen sommerlichen Grundwasserständen kommt es in den Gesellschaften zu einem Abreißen des Saugsaums vom Wurzelraum, so daß der Grundwassergang nur in der ersten Hälfte der Vegetationsperiode z. T. von den Kulturarten beeinflußt wird. Die Schwankungswerte lagen hier auch mit Werten um 40—50 (—60) cm relativ niedrig. Die Böden der Gesellschaften wiesen die niedrigsten gemessenen Feuchtewerte innerhalb der landwirtschaftlich genutzten Flächen des UG auf, so daß deren Standorte in Trockenzeiten stark ertragsgefährdet sind. MS 20 in der *Myosotis*-Subass. zeigte nur im März und April noch eine mäßige Durchfeuchtung, bereits im Mai war nach trockenerer Wetterperiode und starkem Wasserverbrauch des Winterroggens eine tiefgreifende Austrocknung zu beobachten, die im Juni ihren Höhepunkt erreichte. Bezeichnend ist die starke Austrocknung des humusarmen Oberbodens, die sogar bei der Julimessung mit starken Regenfällen am Vortage noch vorhanden war, denn der oft staubtrockene Boden leistet der Benetzung Widerstand, so daß nur langanhaltende Niederschläge tiefer eindringen können. Erst in 1 m Tiefe blieb die Durchfeuchtung relativ unverändert (25—28%), wenn nicht durch Einfluß des Saugsaums zeitweise noch höhere Werte auftraten. In den krumenfeuchten *Juncus*-AF liegen die Grund- und Bodenfeuchteverhältnisse noch am günstigsten. Zusätzlich kommt es in ihnen auch zu einer zeitweisen Durchfeuchtung des Oberbodens infolge ihrer häufigen Lage in flachen Senken, in die das Regenwasser, das in den trockenen humusarmen Sandboden der höhergelegenen umliegenden Flächen kaum eindringt, teilweise abfließt. Vereinzelt finden sich auch gering stauende Schichten (anlehmige, feinsandreiche, wenige cm starke Sandstreifen oder humusreicherer Boden bei künstlicher oder natürlicher Übersandung, letztere durch Flugsand) im oberflächennahen Untergrund an den entsprechenden Standorten. Die Bodenfeuchtewerte unter der typischen Subass. (MS 21 im Arnoseretum unter Winterroggen), wo nur einmal 20% erreicht wurden, oft jedoch bis in eine Tiefe von etwa 1 m hinab die Werte unter 10% fielen, lassen die Ernteerträge meist unter die Rentabilitätsgrenze sinken, und die Standorte der *Corynephorus*-Subass. mit sehr lückigen, kaum meterhohen Winterroggenbeständen sind für die Ackernutzung nicht mehr geeignet.

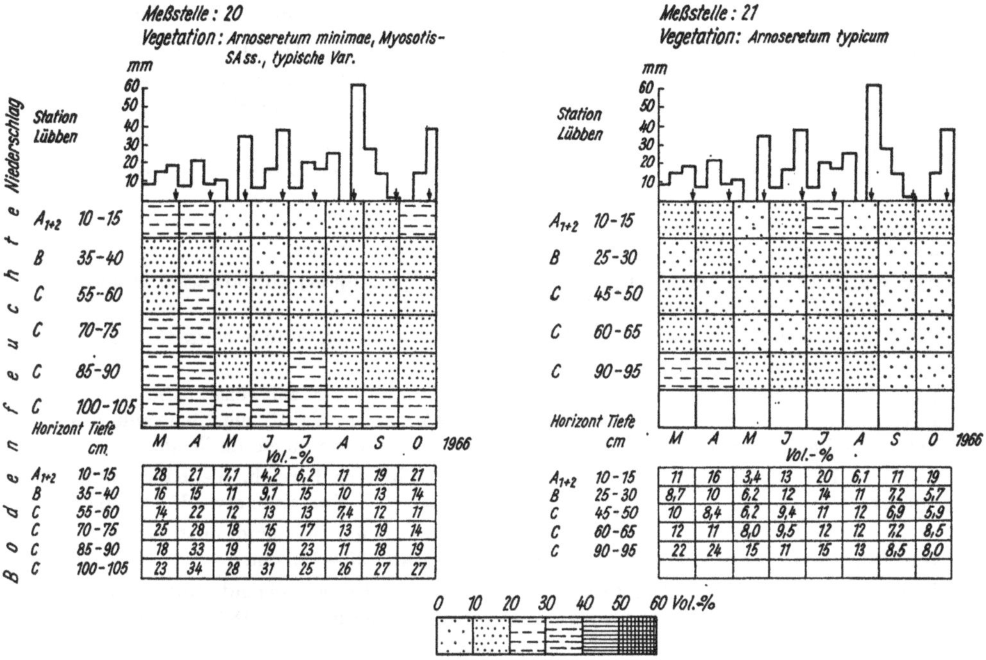

Meßstelle 20 (Vol.-%)

Horizont	Tiefe cm	M	A	M	J	J	A	S	O
A_{1+2}	10-15	28	21	7,1	4,2	6,2	11	19	21
B	35-40	16	15	11	9,1	15	10	13	14
C	55-60	14	22	12	13	13	7,4	12	11
C	70-75	25	28	18	15	17	13	19	14
C	85-90	18	33	19	19	23	11	18	19
C	100-105	23	34	28	31	25	26	27	27

Meßstelle 21 (Vol.-%)

Horizont	Tiefe cm	M	A	M	J	J	A	S	O
A_{1+2}	10-15	11	16	3,4	13	20	6,1	11	19
B	25-30	8,7	10	6,2	12	14	11	7,2	5,7
C	45-50	10	8,4	6,2	9,4	11	12	6,9	5,9
C	60-65	12	11	8,0	9,5	12	12	7,2	8,5
C	90-95	22	24	15	11	15	13	8,5	8,0

Abb. 8. Bodenfeuchtewerte der Meßstellen 20 und 21

5. *Die Wasserstufen*

Die ausgeprägte Abhängigkeit der einzelnen Vegetationseinheiten vom Wasserfaktor innerhalb des UG machte es möglich, die Pflanzengesellschaften bestimmten Wasserstufen zuzuordnen und diese kartenmäßig durch Ableitung von der Vegetationskarte darzustellen (Klemm 1967). Namentlich in der Grünlandkartierung sind von verschiedenen Autoren derartige Stufen aufgestellt und benutzt worden. Prinzipiell muß diabei zwischen zwei Methoden unterschieden werden. Besonders bei der Charakterisierung einzelner Bewirtschaftungsformen (wie vor allem bei reiner Grünlandkartierung) hat sich die Darstellung der Güte- und Ertragsbeeinflussung durch die Wasserstufen, also die Darstellung optimaler, zu trockener und zu feuchter Standorte für die entsprechende Nutzungsform, bewährt (Wagner 1955, Hundt 1957, 1964 etc.).

Seltener wurde diese Methode auch bei der Beurteilung von Grünland-, Acker- und Waldgesellschaften angewandt (MEISEL 1954), was aber besonders im Wald zu einer ungenügenden Differenzierung führen kann. Im UG wurde deshalb die Methode der Grundwasserstufen innerhalb aller Vegetationseinheiten angewandt, wie sie in vergleichbarer Form ebenfalls von mehreren Autoren benutzt wurde (TÜXEN 1954, BUCHWALD 1954, WAGNER 1955, HUNDT 1957, BOCHERT 1958, NIEMANN 1963 u. a.). Durch Kombination der Wasserstufendarstellung mit der der Bewirtschaftungsformen lassen sich aus entsprechenden Karten die Entwicklungstendenzen der Vegetation und die landeskulturellen Möglichkeiten bei zukünftigen Grundwasseränderungen in Gebieten mit Boden- und Wasserverhältnissen ähnlich wie im UG ablesen.

Ausgehend von den bei der Grundwassermessung gewonnenen Ergebnissen wurden für das UG 8 Wasserstufen aufgestellt. Innerhalb dieser Stufen ist im frischen bis mäßig trockenen Bereich eine meist noch weitere Differenzierung an Hand der Vegetation möglich. Dagegen ist im trockeneren Bereich nicht immer eine völlige Übereinstimmung zwischen Grundwasserverhältnissen und Vegetation vorhanden, so daß einige Gesellschaften in zwei Wasserstufen erscheinen. Die Wasserstufen sind:

Stufe 1: Sehr naß. Längere Überstauung im Winterhalbjahr, Grundwasser ständig dicht unter Flur, geringe Schwankungen, Absinken selten unter 20 cm.

Stufe 2: Naß. Regelmäßige Überstauung im Winterhalbjahr, Grundwasser ständig nahe unter Flur, Tiefststände etwa 30—50 cm.

Stufe 3: Feucht. Nur selten Überstauung, Grundwassertiefststände etwa 45—65 cm, zeitweise sehr dicht unter Flur.

Stufe 4: Frisch. Keine Überstauung mehr, Grundwasser zeitweise nahe unter Flur, Höchststände etwa 15—40 cm, Tiefststände etwa 70—110 (Wald!) cm. Schwankungswerte unterschiedlich, etwa 50—60 cm in Acker und Grünland, etwa 80 cm im Wald.

Stufe 5: Frisch bis trocken. Grundwasser im mittleren Bereich, nie nahe unter Flur, mittlere Schwankungen. Höchststände etwa 40—70 cm, Tiefststände etwa 110—140 cm.

Stufe 6: Mäßig trocken. Grundwasser oft noch über 1 m ansteigend, mittlere Schwankungen. Höchststände etwa 80—120 cm, Tiefststände etwa 130—180 cm.

Stufe 7: Sehr trocken. Grundwasser nicht mehr über 1 m ansteigend, mittlere bis geringe Schwankungen. Höchststände etwa 110—150 cm, Tiefststände etwa 180—220 cm.

Stufe 8: Extrem trocken. Grundwasser nicht mehr über 150 cm ansteigend, Tiefststände bis mehrere Meter. Die Stufe ist meist nicht mehr an der Vegetation ablesbar, sondern nur an der Gelände-Morphologie.

Die Verteilung der Wald- und Forst-, Grünland- und Ackerunkrautgesellschaften auf die einzelnen Wasserstufen ist den Abb. 9—11 zu entnehmen.

Die Eignung der einzelnen Wasserstufen für die Bewirtschaftungsformen ist folgendermaßen einzuschätzen: Innerhalb des Waldes sind bei Anbau standortsgerechter Gehölze im (naß-)feuchten bis mäßig trockenen Bereich [Stufen (1)2—6] befriedigende Erträge zu erzielen. Die Kiefer zeigt auch in der trockeneren Stufe 7 und an den tiefergelegenen Stellen innerhalb der Stufe 8 noch mittlere bis mäßige Wuchsleistungen. Dagegen wächst sie auf den Dünenstandorten der Stufe 8 nur sehr langsam und unbefriedigend, bei Moränensanden ist offensichtlich die Grundwassernähe nicht von gleicher entscheidender Bedeutung. Wesentlich stärker ist die Abhängigkeit der Erträge vom Grundwasser in den Acker- und Grünlandbeständen. Die wiesenwirtschaftlich genutzten Bestände des Schlankseggen-Riedes der Stufe 1 ebenso wie die

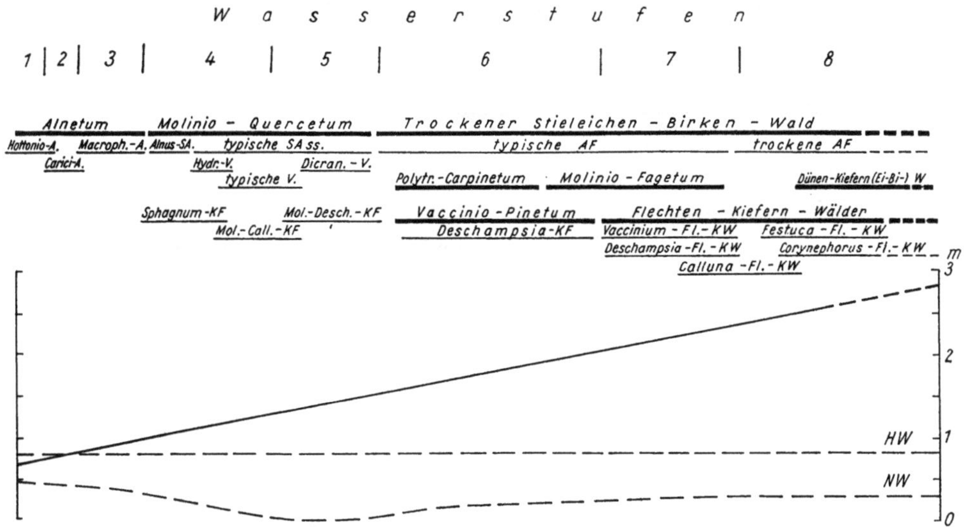

Abb. 9. Wasserstufen-Verteilung der Wald- und Forstgesellschaften

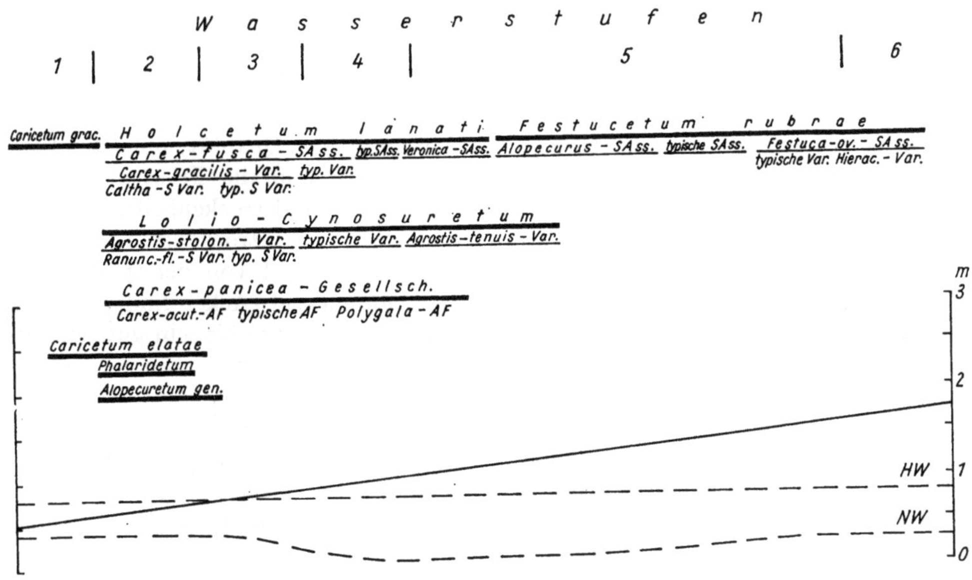

Abb. 10. Wasserstufen-Verteilung der Grünlandgesellschaften

seggenreichen Feuchtwiesen der Stufe 2 bringen zwar meist mittlere bis hohe Erträge, jedoch von minderer Qualität. Hier wäre die Förderung von Rohrglanzgras- und Wiesenfuchsschwanzbeständen durch gute Düngung und Melioration eine Möglichkeit der Qualitätsverbesserung. Die teilweise durchgeführte Weidenutzung auf den Standorten der Stufe 2 oder gar der Stufe 1 ist völlig unrentabel. Die Standorte der Stufe 3 und noch mehr die der Stufe 4 sind die für Wiesen- und Weidenutzung günstigsten im

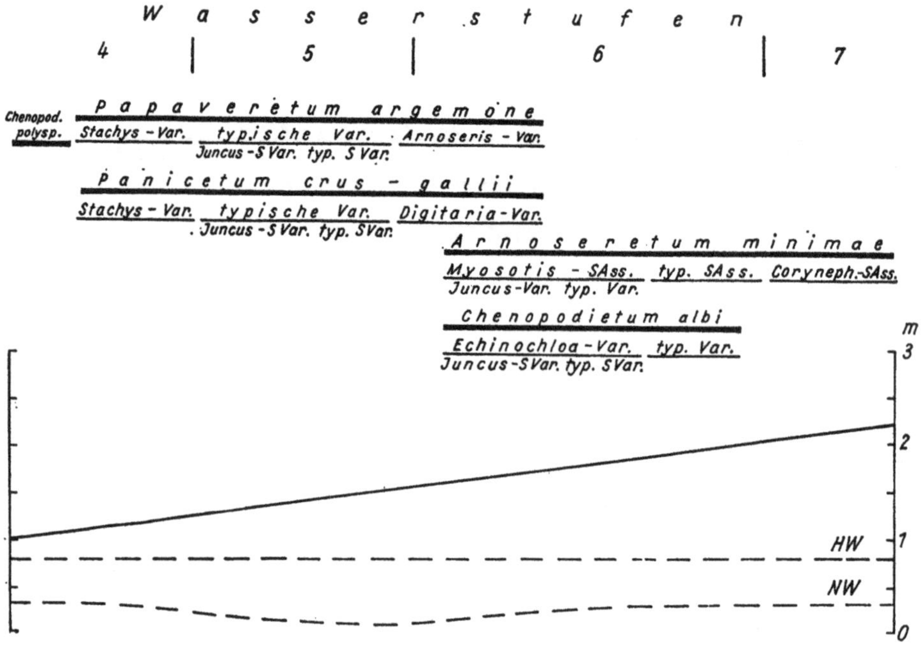

Abb. 11. Wasserstufen-Verteilung der Ackerunkrautgesellschaften

UG, sie liefern mittlere bis gute Erträge von meist mittlerer Qualität, ohne trockenheitsgefährdet zu sein. Die trockeneren Feuchtwiesen und die Frischwiesen der Stufe 5 sind bereits in ihren Ertragsleistungen schwächer und trockenheitsgefährdet, obwohl ihre Futtergüte nicht geringer zu bewerten ist. Die trockensten, noch als Grünland genutzten Flächen der Stufe 6 dagegen sind sowohl von der Qualität als auch von der Quantität ihrer Erträge her völlig unrentabel. Äcker finden wir ab Stufe 4, wo bei günstiger, nicht zu feuchter Witterung zwar gute bis sehr gute (Hackfrucht!) Erträge erzielt werden, die aber in niederschlagsreichen Jahren oft vernässungsgefährdet sind. Die günstigsten Ackerstandorte sind daher die der Stufe 5, ihre Erträge sind von mittlerer Höhe und fallen in feuchteren Jahren günstiger aus als in trockeneren. Die Äcker im frischeren Bereich der Stufe 6 (namentlich bei Krumenfeuchte) können mäßige Erträge bringen und sind im UG noch häufig in Nutzung. Aber schon die trockeneren Äcker innerhalb dieser Stufe sind in nicht allzu feuchten Jahren ausgesprochen ertragsschwach, während eine Ackernutzung auf Standorten der Stufe 7 völlig unrentabel ist.

6. Die möglichen Auswirkungen einer oberflächennahen Grundwasseranreicherung auf die Vegetationsverhältnisse

Ebenso wie die Grundwasserabsenkung in den vergangenen Jahrzehnten würde (in entgegengesetzter Richtung) eine Grundwasseranreicherung nicht ohne Folgen für die Land- und Forstwirtschaft bleiben. Nach den bisher vorliegenden Untersuchungen

(Herrmann et al. 1967) könnten die Versickerungsbecken für das aus dem Neuendorfer See entnommene Oberflächenwasser, also das Zentrum der Grundwasseranreicherung, in dem Moränenausläufer 1,2 km nordöstlich Neu-Schadow angelegt werden. Das so von der Grundwassererhöhung in erster Linie betroffene Gebiet wird im W und N vom Tal der Hauptspree bzw. dem Neuendorfer See, im E und SE vom Tal der Pretschener Spree und im S von einem fast geschlossen durchgehenden Dünenzug, der als natürlicher Grundwasserstauer zu wirken scheint, begrenzt. Die derzeitige, noch ungestörte Grundwasserfließrichtung biegt fast überall vor diesem Zug um und verläuft dann parallel zu ihm. Demzufolge wäre im S-Teil des UG nur eine wesentlich abgeschwächte Grundwassererhöhung zu erwarten.

Die vorhergesehene Grundwasseraufhöhung um etwa 50 cm würde ungefähr die Verhältnisse, wie sie vor der Absenkung bestanden, wiederherstellen. Freilich ist nur eine ungefähre Voraussage über den Einfluß der Anreicherung auf die Vegetation möglich. In Abhängigkeit von der Morphologie, dem Untergrund, den Bodenverhältnissen, der Entfernung der Standorte vom Versickerungszentrum und der bei jahreszeitlich unterschiedlicher Versickerung und Wasserentnahme aus dem Speicherraum wechselnden Höhe der Anreicherung sind die 50 cm Erhöhung nur als mittlerer Richtwert zu betrachten. So kann auch verallgemeinernd angenommen werden, daß sich die Wasserstufen 3—7 jeweils zur nächstfeuchten Stufe verschieben würden. Ihr mittlerer Grundwasserniveau-Unterschied beträgt etwa 40—50 cm. Die Stufe 3 würde allerdings stellenweise zur Stufe 1 werden. Im nassen Bereich der Stufen 1 und 2 käme es wahrscheinlich gleichermaßen zu einer langandauernden Überstauung und damit zur weitgehend nicht mehr möglichen landwirtschaftlichen Nutzung. Die extrem trockene Stufe 8 dagegen, deren Grundwasserstände meist mehrere Meter tief liegen, würde nur sehr kleinflächig an ihren tiefergelegenen Stellen in den Bereich der Stufe 7 gelangen, sonst aber weitgehend unverändert bleiben.

Die nördliche Hälfte des UG, die von den Veränderungen am stärksten getroffen würde, wird flächenmäßig zum weitaus größten Teil von trockenen Flechten-Kiefern-Wäldern der Wasserstufen 7 und 8 eingenommen. Nur im Gebiet östlich Alt-Schadow sind auf größerer Fläche auch frischere Kiefern-Forste der Stufen 4—6 und Laubwaldbestände geringer Ausdehnung (Stufen 1 und 4) anzutreffen. All diese Bestände würden von einer Grundwasser-Erhöhung kaum in Mitleidenschaft gezogen werden, da die Baumartenkombination in den feuchteren Bereichen ohnehin noch weitgehend den vor der Grundwasser-Absenkung herrschenden Verhältnissen entspricht und die trockeneren Bestände durch ein größeres Wasserangebot nur in ihrer Produktivität gesteigert werden könnten.

Die Auswirkungen auf die Ackerflächen sind unterschiedlich zu beurteilen. Die Äcker im nördlichen Teilgebiet (im Bereich der Ortschaften Alt-Schadow, Neu-Schadow, Hohenbrück) gehören bis auf wenige Ausnahmen zu den Wasserstufen 5—7. Vielfach sind sie zu trocken und ertragsgefährdet, was zu einer weitreichenden Aufgabe der Äcker und nachfolgender Kiefern-Aufforstung führte. Sie würden in ihrer Ertragsleistung gefördert werden. Die Flächen der Stufe 5 blieben zum Großteil auch noch ohne allzu starke Vernässungserscheinungen. Einige Flächen der Stufe liegen jedoch bereits so tief, daß sie zeitweise zu stark vernässungsgefährdet wären und in Grünland überführt werden müßten, was auch für die nur kleinflächig verbreiteten Äcker der Stufe 4 gilt. In den meisten Fällen stehen die feuchteren Äcker schon jetzt mit Grünlandgesellschaften in Kontakt, so daß die veränderte Bewirtschaftung keine größeren Umstände machen dürfte. Bei ständig genügend hohem Grundwasserstand ist auch auf extrem humusarmen Sanden eine Grünlandwirtschaft möglich, wie das der Bestand an der MS 24 zeigt. Außerdem könnten einige jetzt zu trockene Ödlandflächen — soweit sie noch nicht aufgeforstet sind — wieder in Ackernutzung genommen werden.

Ausgedehnte Grünlandkomplexe fehlen dem nördlichen Teil des UG. Einige Flächen, die sich in moorigen bis anmoorigen Senken befinden (im Bereich der Ortschaften Neu-Schadow und Hohenbrück), werden von Schlankseggen-Riedern, Feuchtwiesen, Weiden und Knickfuchsschwanz-Rasen eingenommen. Eine Schädigung durch zu hohen Grundwasserstand wäre in den Schlankseggen-Riedern und Knickfuchsschwanz-Rasen der Wasserstufen 1 und 2 mit Sicherheit zu erwarten, wahrscheinlich wird auf diesen Flächen eine Grünlandnutzung kaum noch möglich sein. In den Feuchtwiesen der Stufe 3 ist mit einer starken Ausbreitung der bereits vorhandenen Sauergräser (v. a. *Carex gracilis*) und dem Zurückbleiben der besseren Wiesenarten zu rechnen, was zur Ausbildung von Schlankseggen-Riedern führen würde. Auch hier träte also eine Verschlechterung ein, v. a. zuungunsten der Futterqualität. Die Wiesenflächen der Stufe 4 hingegen würden nur wenig in Mitleidenschaft gezogen werden, es käme zu einer geringen Futterverschlechterung bei zu erwartender Ertragssteigerung. Obwohl die Schädigung der Grünlandflächen infolge ihrer geringen Ausdehnung nicht allzu stark ins Gewicht fiele, kann hier durch grünliche Melioration das Ausmaß des Schadens von vornherein eingedämmt werden.

Insgesamt sind also die zu erwartenden Veränderungen innerhalb der Vegetation und der land- und forstwirtschaftlichen Nutzung bei einer oberflächennahen Grundwasseranreicherung in der nördlichen Hälfte des UG als vorwiegend positiv zu bezeichnen. Inwieweit sich der veränderte Grundwasserstand auf das südliche Teilgebiet auswirken würde, hängt davon ab, wie vollkommen sich der Grundwasserstau entlang des begrenzenden Dünenzuges bemerkbar macht. Grundsätzlich gelten jedoch auch hier die gleichen Schlußfolgerungen.

7. *Zusammenfassung*

Am Beispiel des nordöstlichen, durch Grundwasserabsenkung charakterisierten Unterspreewald-Randgebietes wurden die Beziehungen zwischen Pflanzengesellschaften und Bodenwasserverhältnissen eines grundwassernahen Talsandgebiets untersucht. Die Klärung des jetzigen Abhängigkeitsverhältnisses der Vegetation insbesondere vom Grundwasser sollte dazu dienen, die möglichen Folgen einer Grundwasseranreicherung für die land- und forstwirtschaftlichen Nutzflächen in diesem Gebiet abzuschätzen. Es werden die Feuchteverhältnisse, beruhend auf Grundwassergang- und Bodenfeuchteuntersuchungen, in den Wald- und Forst-, Grünland- und Ackergesellschaften des Gebiets beschrieben. Auf Grund einer relativ einheitlichen Bodenzusammensetzung in den einzelnen Teilen des Untersuchungsgebiets aus Sanden weitgehend ohne stauende Schichten und der daraus folgenden engen Abhängigkeit der Vegetation vom Grundwasser konnten 8 vegetationsdifferenzierende Grundwasserstufen aufgestellt und diesen die einzelnen Pflanzengesellschaften zugeordnet werden. Die möglichen Auswirkungen einer künftigen Grundwasseranreicherung mit einer mittleren Spiegelaufhöhung von etwa 50 cm im Untersuchungsgebiet sind sowohl für die Land- als auch für die Forstwirtschaft als überwiegend positiv vorauszusagen.

Summary

The relations between plant associations and soil water conditions of a region with alluvial sands (external district of the Lower Spreewald), which are influenced by ground water, were studied. This region is characterized by a lowering of the ground water level. The present dependence of the vegetation especially of the factor ground water was analysed, in order to estimate the eventual consequences of an artificial increase of the ground water level for agriculture and forestry. The moisture conditions (ground water, soil moisture) of the wood, forest, grassland and field associations are described.

The soil structure in the different parts of the whole region is relatively homogeneous (sands without stagnant moisture). By this way it was possible to produce eight ground water scales with variant dependence of the vegetation from the ground water level. The plant associations were attributed to these scales.

The eventual consequences of an increase of the ground water level (averagely about 50 cm) in the region can be predominantly positive predicted as for agriculture as for the forestry.

8. Literatur

ANDREAE, H., 1956: Grundwasserabsenkungen und landeskulturelle Änderungen. — Berlin.

BOCHERT, H., 1958: Standörtliche Gliederung der Nutheniederungen am südwestlichen Fläming-rand auf Grund vegetationskundlicher, hydrologischer und bodenkundlicher Untersuchungen. — Diss., Martin-Luther-Univ. Halle—Wittenberg.

BUCHWALD, K., 1954: Wasserhaushaltsstufen und -zonen des Wassermangelgebietes Unteres Illertal. — Angew. Pflanzensoz. **8**, 37—55.

ESKUCHE, U., 1962: Herkunft, Bewegung und Verbleib des Wassers in den Böden verschiedener Pflanzengesellschaften des Erfttales. — Arb. Bundesanst. Veg.-Kart. Düsseldorf.

FREITAG, H., 1957a: Die Grünlandgesellschaften des Oberspreewaldes. — Diss., Päd. Hochsch. Potsdam.

— 1957b: Vegetationskundliche Untersuchungen an Gründlandgesellschaften des Nieder-Oderbruchs. — Wiss. Z. Päd. Hochsch. Potsdam, Math.-nat. R. **3**, 125—139.

HERRMANN, R., 1963: Zur Bedeutung der unterirdischen Wasserspeicherung für die Volkswirtschaft. — Ber. geol. Ges. DDR **8**, 38—48.

— 1968: Hydrologische Untersuchungsmethoden in Sanden des Baruther Urstromtales (Raum Neu-Schadow—Alt-Schadow—Pretschen) unter besonderer Berücksichtigung der Beziehungen von Kornlagerung, Kornform und Korngrößenverteilung auf die Grundwasserbewegung. — Diss., Humboldt-Univ. Berlin.

HERRMANN, R., K. BAUER, G. PESCHKE & G. KLEMM, 1967: Erkundungen der Grundwasserverhältnisse in einem Teileinzugsgebiet der Spree (Raum Neu-Schadow—Alt-Schadow—Pretschen) unter besonderer Berücksichtigung der Erweiterung des Grundwasserdargebots und seiner Auswirkungen auf die Landeskultur. — Unveröff. Forsch.-Ber., Berlin.

HUNDT, R., 1957: Pflanzensoziologische Methoden zur Beurteilung der Grünlandwasserstufen und des Ertragswertes. — Dtsch. Landw. **8**, 333—338.

— 1964: Ein vegetationskundliches Verfahren zur Bestimmung der Grünlandwasserstufen. — Z. Landeskd. **5**, 161—186.

KLEMM, G., 1967: Vegetationskundliche Untersuchungen im nordöstlichen Unterspreewald-Randgebiet zur Einschätzung möglicher Auswirkungen einer oberflächennahen Grundwasseranreicherung auf die Landeskultur. — Diss., Martin-Luther-Univ. Halle—Wittenberg.

— 1969/70: Die Pflanzengesellschaften des nordöstlichen Unterspreewald-Randgebietes. — Verh. Bot. Ver. Brandenbg. **106**, 24—62; **107**, 3—28.

KUBIENA, W., 1953: Bestimmungsbuch und Systematik der Böden Europas. — Stuttgart.

MEISEL, K., 1954: Anwendung der Pflanzensoziologie zur Beurteilung von Wasserschäden in der Landwirtschaft. — Angew. Pflanzensoz. **8**, 137—150.

MÜLLER, A. v., 1956: Über die Bodenwasserbewegung unter einigen Grünlandgesellschaften des mittleren Wesertales und seiner Randgebiete. — Angew. Pflanzensoz. **12**.

NIEMANN, E., 1963: Beziehungen zwischen Vegetation und Grundwasser. — Arch. Natursch. u. Landschaftsforsch. 3, 3—36.
— 1968: Zur Rolle des Grundwasserfaktors im Vegetationsgefüge. — Hab.-Schr., TU Dresden.
PASSARGE, H., 1955: Die Pflanzengesellschaften der Wiesenlandschaft des Lübbenauer Spreewaldes. — Fedd. Repert., Beih. 135.
— 1964: Pflanzengesellschaften des nordostdeutschen Flachlandes I. (Pflanzensoz. 13). — Jena.
PESCHKE, G., 1968: Spezielle hydraulische Untersuchungen über die Grundwasserbewegung und die sie beeinflussenden Faktoren in Mittelsanden aus dem Bereich des Baruther Urstromtales. — Diss., Humboldt-Univ. Berlin.
ROSENTRAEGER, H., 1962: Über die Schwankungen und Absenkungen des Grundwassers und landeskulturelle Änderungen im mittleren Spreegebiet. — Unveröff. Forsch.-Ber., Berlin.
SCAMONI, A., 1955: Die Wiesen, Grasfluren und Heiden des Unterspreewaldes. — Wiss. Z. Humboldt-Univ. Berlin, Math.-nat. R. 5, 253—267.
— 1960: Waldgesellschaften und Waldstandorte. — Berlin.
SCHLÜTER, H., 1969: Vegetationskundlich-synökologische Untersuchungen zum Wasserhaushalt eines hochmontanen Quellgebietes im Thüringer Wald. — Hab.-Schr., Martin-Luther-Univ. Halle—Wittenberg.
TÜXEN, R., 1954: Pflanzengesellschaften und Grundwasserganglinien. — Angew. Pflanzensoz. 8, 64—98.
WAGNER, H., 1955: Die Bewertung der Wasserstufen in der Bodenschätzung des Grünlandes. — Bodenkultur 8, 133—150.
WALTER, H., 1960: Einführung in die Phytologie 3, 1. — Stuttgart.
Klimaatlas für das Gebiet der DDR. — Berlin 1953.
Meteorologischer Dienst der DDR, 1964—1966: Monatliche Witterungsberichte für das Gebiet der DDR, 18—20.
Erstaufnahmen der Meßtischblätter (1846) der Königl. Preuß. Landesaufnahme. Blatt Alt-Schadow, Blatt Schlepzig. — Kartograph. Abt. Deutsch. Staatsbibl. Berlin.
Geologische Übersichtskarte von Deutschland (1:200000) Nr. 90, Blatt Berlin-Süd (K. KEILHACK). — Berlin 1923.

Eingegangen am 15. 6. 1969.

Anschrift des Verfassers:

Dr. GUNTHER KLEMM
DDR — 116 Berlin
Parsevalstr. 6

Limnologica (Berlin)	11	1	1976	161–174

FRIEDERIKE WAWRIK

Wetterbedingte Phytoplanktonentwicklungen in den Teichen des niederösterreichischen Waldviertels

Mit 9 Abbildungen und einer Tabelle

Abstract

Base water temperature and heavy rainfall during summer and autumn 1974 influenced the phytoplankton growth in the fishponds of the Waldviertel very strongly. 42 ponds were explored and those algae studied, whose appearance under the abnormal weather conditions was striking: *Actinastrum hantzschii, Chlamydomonas cingulata, Ch. bichlora, Chlorogonium maximum* var. *oogamum* nova var., *Chlorogonium maximum, Colacium ciclopicola, C. simplex, Hemitoma meandrocystis, Lambertia gracilipes, Mallomonas schwemmlei* (Hologamie), *Microcystis parasitica* f. *dinobryonis* n. f., *Nephroselmis angulata, Rhizosolenia eriensis, Stephanodiscus hantzschii.*

1. Einleitung

BEHRE & WEHRLE (1944) haben in einer größeren Studie jene Faktoren diskutiert, „welche die Zusammensetzung von Algengesellschaften entscheiden". Für den Ausdruck „entscheiden" möge an dieser Stelle der Begriff „beeinflussen" stehen, denn es gibt außer jenen 12 Faktoren, die behandelt worden sind, noch viele andere Einflüsse, deren Wirkungsgrade sich kaum erfassen lassen. Daß Licht und Wärme die Kardinalpunkte des sehr komplexen Geschehens sind, resultiert aus klaren Erfahrungstatsachen. Wesentliche Faktoren, die weder bei BEHRE & WEHRLE (1944) noch bei UHLMANN (1966) ins Gewicht fielen, sind die Funktionen des Pegelstandes: Stagnation und Wassererneuerung, Durchspülung, Ausschwemmung und im Zusammenhang damit die Nährstoffkonzentration (WAWRIK 1952).

2. Das Beobachtungsgebiet und die Witterungsverhältnisse

Das Beobachtungsgebiet (Abb. 1) erstreckt sich auf einem breiten Streifen, südlich entlang des Donauraumes beginnend, in nördlicher Richtung bis an die Staatsgrenze zur ČSSR:

a) Gebiet III/IV Gutenbrunn-Bärnkopf (siehe WAWRIK 1962), auf Seehöhen von 800–900 m; es wurden 12 Teiche untersucht. Um den 5. VII. 1974 lagen die Oberflächentemperaturen in den Teichen bei 12,8°–13,2°. Die Nachttemperaturen der Luft sanken auf 6 °C.

b) Gebiet XVII Kirchberg am Walde, auf Seehöhen um 560 m; es wurden 11 Teiche untersucht. Um den 21. VII. lagen die Oberflächentemperaturen bei 16,5°–17,0°.

c) Gebiet XVIII Schrems, auf Seehöhen zwischen 500 und 600 m; es wurden 19 Teiche untersucht. Um den 6. VII. lagen die Oberflächentemperaturen bei 17,0°–17,6°.

Da es sich in der Hauptsache um dystrophe Kleingewässer (Braunwasserteiche) handelt, stiegen beispielsweise im Gebiet Schrems im Juli nach mehrstündiger Sonneneinstrahlung die Oberflächentemperaturen bald über 20 °C an, waren aber wegen der kurzfristigen Dauer der Schönwetterlage nicht von Bestand.

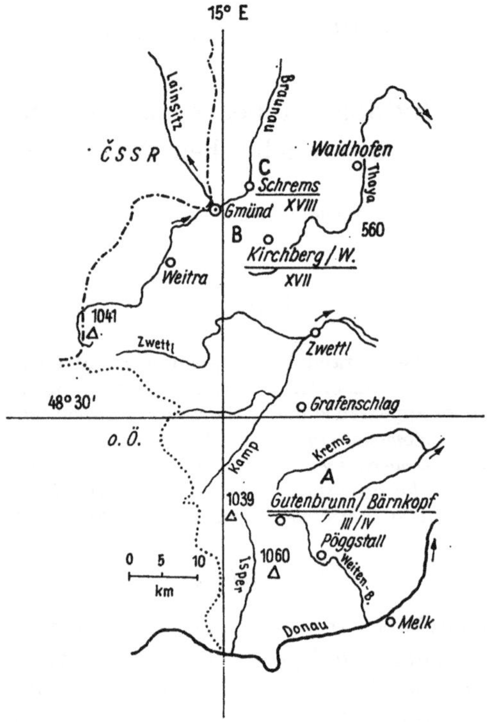

Abb. 1. Lageplan

Die Witterungsverhältnisse der Monate V, VI und VII des Jahres 1974 haben die Zusammensetzung der Phytoplankton-Gesellschaften unserer wenig oder nicht meliorierten und durch Abwässer nicht wesentlich belasteten Waldteiche beispielhaft beeinflußt.

Die Schneeschmelze bewirkt häufig eine Ausschwemmung der bisweilen nicht unbeträchtlichen Wintervegetationen. Nach Eisbruch erwärmen sich die Braunwasserteiche rasch. Sie erreichen im Frühjahr Planktondichten bis zu 200000 Z/ml; es treten Vegetationsfärbungen auf, seltener Wasserblüten (WAWRIK 1966, 1972). Im Sommer wird das Plankton durch Fraß dezimiert; im Herbst erholt es sich wieder durch Wassererneuerung und Einschwemmung. Es gibt einen Teich im südlichen Waldviertel, den Edlesbergerteich, dessen Spezialität spätherbstliche Wasserblüten sind, die sogar unter Eis noch andauern können (WAWRIK 1969, 1971).

Jene Teiche des Waldviertels, die sich noch eine weitgehend natürliche Physiognomie erhalten haben, sind eine Fundgrube seltener und interessanter Algen. Durch Witterungseinflüsse war die sommerlicheEntwicklung in den Teichen im Jahre 1974 offensichtlich gestört. Bei der Beurteilung einer Algenvegetation ist die der Beobachtung vorangegangene Wetterlage wichtig.

Es ergibt sich, daß in den Beobachtungsgebieten gegenüber dem 60- bzw. 30jährigen Durchschnitt die Niederschlagsmengen um 2,2—62,7% höher, die Temperaturminima um 0,1—1,5 °C und die Temperaturmaxima um 1,5—3,2 °C niedriger waren.

Tab. 1. Niederschlagsmengen und Temperaturen der dem Untersuchungsgebiet nächstgelegenen Stationen

	RR	DRR	TMAX	DTMAX	TMIN	DTMIN
Weitra:						
Mai	129,0	78	14,8	18,0	5,5	5,8
Juni	126,9	94	17,7	20,5	8,4	9,2
Juli	124,0	115	20,1	22,3	10,7	11,2
August	85,6	90	22,9	21,5	11,8	10,3
St. Zwettl:						
Mai	116,8	70	15,8	17,7	5,2	5,3
Juni	102,7	86	18,6	20,8	8,1	8,8
Juli	108,7	102	20,8	22,8	10,0	10,7
August	79,4	85	23,6	22,1	11,4	10,1
Martinsberg-Gutenbrunn:						
Mai	152,1	88	13,2	15,2	4,1	4,3
Juni	162,5	100	15,9	18,2	6,9	7,5
Juli	132,5	128	18,4	20,1	9,0	9,6
August	58,1	103	21,4	19,7	10,5	9,0

Monatliche Niederschlagsmengen in mm (RR), durchschnittliche Niederschlagsmengen (1901–1960) in mm (DRR), durchschnittliches Temperaturmaximum in °C (TMAX), langjähriges Temperaturmaximum in °C (DTMAX), durchschnittliches Temperaturminimum in °C (TMIN), langjähriges Temperaturminimum in °C (DTMIN) Zeitraum 1931–1960

Vorstehende Meßwerte hat in entgegenkommender Weise die Klimaabteilung der Zentralanstalt für Meteorologie und Geodynamik, Wien, Hohe Warte, zur Verfügung gestellt.

3. *Die Phytoplanktonentwicklung*

Die Planktonentwicklung war nicht nur von den niedrigen Sommertemperaturen, sondern noch viel mehr von den hohen Niederschlagsmengen beeinflußt. Im Gebiet Kirchberg gingen beispielsweise drei Hochwässer durch. Die Zapfen wurden gezogen, das Plankton ausgeschwemmt. Die Abflüsse traten über das Ufer und führten das Plankton nicht dem Unterlieger zu, sondern spülten es in die Wiesen hinaus.

Über die sommerlichen Planktondichten in den Teichen (bis zu 180000 Z/ml) geben Planktonzählungen Auskunft, die in den Jahren 1967/68 durchgeführt worden sind (Wawrik 1972). Dagegen wurden 1974 stichprobenweise nur einige hundert bis wenige tausend Zellen gezählt, ein Ergebnis, das nichts mit einem „grazing effect" zu tun hat, denn das Zooplankton wurde gleichermaßen ausgeschwemmt.

Allerdings gab es auch Ausnahmen: in kaum oder wenig durchflossenen Himmelsteichen, in zu- und abflußlosen Löschteichen und in Grundwasserbecken. Beispielsweise bestand in einem Straßenteich nächst Gebharts (XVIII), der im Frühjahr ausgebaggert und frisch bespannt worden war, eine Vegetationsfärbung von Protococcalen, offenbar bedingt durch die „jungfräulichen" ökologischen Verhältnisse in diesem Lebensraum. In einem Löschteich gab es eine Vegetationsfärbung durch *Pediastrum boryanum* (Turp.) Meneg. Ein kleines Wasserbecken nächst Schrems (XVIII) wies eine dichte Reinkultur von *Asterionella formosa* Hass. auf, der Dorfteich in Groß Haselbach (XVIII) eine solche von *Nephroselmis angulata* (KORSCH.) SKJ., auf die wir noch zurückkommen. In diesen Fällen handelt es sich offenbar um Plankter, denen steigende Pegelstände besonders zusagen.

Auffallend waren die qualitativen Verhältnisse. Viele Teiche sind Fundgruben seltener Arten. Ich kenne die Standorte und spüre ihnen von Jahr zu Jahr nach. Im Sommer 1974 war nichts davon da. Lediglich im Spielbergerteich (XVII) gab es ein einziges Exemplar von *Pachycladon umbrinus* G. M. SMITH, erstmalig mit 5 statt der bisher beobachteten 4 Borsten.

Es stellte sich heraus, daß es unter den „Wald- und Wiesenformen" mit langgestreckten Optimumskurven viele gibt, denen schwankende Pegelstände bzw. eine Wassererneuerung nichts ausmachen, weil sie in mehreren Teichen Dominanz erreichten oder durch ihre Häufigkeit auffielen. Es könnte sich dabei um Algen handeln, die befähigt sind, den durch wiederholte Wassererneuerung bedingten Schwankungen der Nährstoffkonzentration mit einem entsprechenden Permeabilitätsverhalten zu begegnen. Es ist bekannt, daß Ubiquisten nur äußerste Extreme meiden. Die Pegelschwankungen des Sommers 1974 haben offensichtlich in vielen Teichen extreme Umweltverhältnisse verursacht.

4. *Beschreibung interessanter Sippen*

Nachfolgend werden einige Sippen besprochen, die in den Algengesellschaften der Teiche im Sommer 1974 aufgefallen sind.

1. *Microcystis parasitica* KÜTZ. f. *dinobryonis* nova forma (Abb. 2)

Unter den Chrysophyceen des Mitterteiches (XVII) dominierten *Synura petersenii* KORSCH., *S. uvella* EHRENBG., *Dinobryon divergens* IMHOF und *D. sertularia* EHRENBG. Zahlreiche Becher der *Dinobryon*-Kolonien enthielten eine – offenbar raumparasitäre – *Microcystis*. Die hellblauen, kugeligen Zellen maßen $D = 1-2\,\mu m$ und können wohl am besten zu *M. parasitica* KÜTZ. gestellt werden. Die Alge füllte die Becher von unten nach oben fortschreitend, nicht nur in der Kolonie, sondern auch im Einzelbecher auf und siedelte auch unterhalb gesunder Protoplasten, die sie bei rasanter Vermehrung aus den Bechern drängte. Auch im Edlesbergerteich (III) gab es *Dinobryon*-Kolonien, die von derselben Alge befallen waren. Sie dürfte demnach auf *Dinobryon* spezialisiert sein, was die Beschreibung einer forma *dinobryonis* bedingt:

Microcystis parasitica **KÜTZ f. *dinobryonis* nova forma, differt a typo habitatione in calycibus *Dinobryonis sertulariae*.**

2. *Nephroselmis angulata* (KORSCH.) SKJ. (Abb. 3a–i)

Diese Polyblepharide – siehe SKUJA (1956) – lebt in den Teichen des Waldviertels verbreitet, eingestreut, ganzjährig. Im Winter 1966 war sie im Edlauteich (XVII) bei 5,5 °C, pH 7,0, SBV 0,64 mval/l, El_{20} 129 μS, 38% Bikarbonat und 105% O_2-Sättigung mit 20000 Z/ml dominant (WAWRIK 1968). Im Amtwiesteich (XVII) trat sie im März 1968 unter Eis bei pH 5,6, SBV 0,64 mval/l, El_{20} 130 μS, 43 % Bikarbonat und 69% O_2-Sättigung mit 2500 Z/ml auf; es gab Zellteilungen und Geißelkontakte.

Abb. 2. *Microcystis parasitica* KÜTZ. f. *dinobryonis* n. f. in Bechern von *Dinobryon sertularia* EHRENBG.

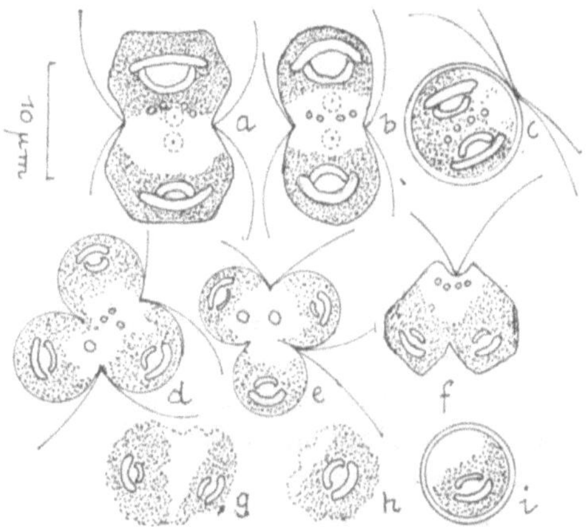

Abb. 3. *Nephroselmis angulata* (KORSCH.) SKJ.

a, b – isogame Hologamie; c – Zygote; d, e – Drillingskopulation; f – Zellteilung; g–i – Encystierung

Im Juli 1974 erreichte sie bei 17,5 °C, pH 8,2 und SBV 1,14 mval/l aspektbeherrschend 80000 Z/ml. Sie war mit *Planctomyces bekefii* GIMESI, *Lagerheimia ciliata* (LAGERH.) CHOD., *Synedra acus* KÜTZ. und verschiedenen Trachelomonaden vergesellschaftet. Das Milieu im Gr. Haselbacher Dorfteich (XVIII) war saprob. Unter diesen ökologischen Verhältnissen war ihre rasante Reproduktion besonders bemerkenswert: Etwa $^1/_3$ der Zellen befand sich in vegetativer Teilung, ein weiteres Drittel in isogamer Zellverschmelzung; es gab viele Zysten und Zygoten. Für die junge Zelle gibt es drei Entwicklungsmöglichkeiten:

(1) Vegetative Teilung: Sie verläuft median sehr langsam. Vom Beginn bis zur Trennung der Tochterzellen vergehen einige Stunden.

(2) Sexuelle Reproduktion — Isogamie: Die Tochterzellen wachsen nicht heran, sondern suchen sofort nach der Trennung mit vehementer Kopulationsbereitschaft Geißelkontakte, das heißt, sie funktionieren sofort nach der Trennung als Gameten. Man sieht nicht nur 2, sondern nicht selten auch 3—4 Gameten im Geißelkontakt. Doch werden die überzähligen Gameten wieder abgestoßen. Es vereinigen sich 2 Gameten zu einer kugeligen Zygote (D = 12—13 µm; Abb. 3a, b, c). Es gab aber auch 3 oder 4 Fälle je Präparat, bei denen bereits vor der Trennung der Tochterzellen ein Gamet apikal Kontakt gewann und die Zellumina miteinander verschmolzen. Dabei entstanden Gebilde ähnlich einem Dreipaß, von dem 2 Geißelpaare abstanden. „Drillingszygoten" wurden nicht gesehen (Abb. 3d, e). Wollte man diese „Drillingsgebilde" dahin deuten, daß bei der vegetativen Teilung bereits vor der Trennung der Tochterzellen eine davon schon wieder zur Teilung geschritten wäre, so fände man für die beiden Geißelpaare keine Erklärung.

(3) Enzystierung (Abb. 3f—i): Sie wurde häufig beobachtet. Während der Zellteilung löste sich der Periplast völlig auf. Es entstand ein mehr oder weniger „formloses" Stadium. Geißeln wurden nicht ausgebildet. Nach der Trennung der Protoplasten entwickelte sich jeder zu einer kugeligen Zyste (D = 9—10 μm).

Das Material vom Haselbachteich konnte eine Woche hindurch — es wurde zweimal frisches Material geschöpft — studiert werden. Während dieser Zeit änderte sich die Artdichte von *N. angulata* kaum. Zellteilungen, Isogamie und Enzystierungen hielten uneingeschränkt an.

Zusammenfassend kann gesagt werden, daß die Art in unseren Teichen perenniert, bei einem pH-Spektrum von 6,6—8,2 in oligo- bis eutrophen Gewässern, in denen sie Sommer- und Wintermaxima ausbildet. Die größte Individuendichte erreichte sie nach heftigen Niederschlägen. Sie scheint damit Wassererneuerungen zu bevorzugen.

3. *Chlamydomonas cingulata* Pasch.

4. *Stephanodiscus hantzschii* Grun.

Im Kotwiesteich (XVII) gab es eine deutlich wetterbedingte Algenvegetation. Dieser Teich empfängt von den Siedlungshäusern Abwasser, das am Hang in saure Wiesen versickert. Alljährlich zur Zeit der Schneeschmelze, wenn der Teich durch Einschwemmungen vom Hang her mächtig saprobiert wird, hat er eine Massenproduktion von *Stephanodiscus hantzschii* aufzuweisen. Im Juli 1974, bei 19,2 °C, pH 7,0 und SBV 1,12 mval/l dominierte *Chlamydomonas cingulata*. Skuja (1956) bezeichnete diesen Flagellat als charakteristisch für das Frühjahrsplankton des Stadtkanales von Riga. Im Kotwiesteich nutzte er die wetterbedingten polysaproben Verhältnisse des Sommers und vermehrte sich rasant. Es gab viele vegetative Teilungsstadien, Heterogameten im Geißelkontakt und zahlreiche Zysten. Sie maßen im Durchmesser 29 bis 30 μm einschließlich der 5 μm dicken, ungeschichteten Gallerthülle. Auch *Stephanodiscus hantzschii* nutzte die sommerliche Saprobität. Er lebte subdominant, solitär und mit vielen 4- oder 5zelligen Ketten (Trennungsverzug der Tochterzellen!) in rasanter Vermehrung.

5. *Chlamydomonas bichlora* Pascher & Jahoda

Bemerkenswert war das Vorkommen dieser recht seltenen Art in den Algengesellschaften des Pfaffenstegteiches (IV). Sie wurde aus dem Lunzer Gebiet beschrieben und im Waldviertel bisher erst einmal im Herbst 1968 (Wawrik 1972) im Pointgrabenteich (XVII) beobachtet. Nun fand sich dieser Flagellat unter ähnlichen ökologischen Verhältnissen neuerdings im Herbstplankton. Es gab Gameten im Geißelkontakt.

6. *Chlorogonium maximum* Skuja (Abb. 4a—v)

Die abnormen Witterungsverhältnisse des Sommers 1974 setzten sich mit kurzen Unterbrechungen im Oktober und November fort: Es war unterdurchschnittlich kühl und naß. Als ich am 7. XI. zur gewohnten Herbstexkursion in das Gebiet Gutenbrunn (III)/Bärnkopf (IV) kam, lag bereits auf den Höhen Schnee; die Teiche waren eisbedeckt (Wawrik 1966, 1972c). Es wurden 12 Teichgewässer untersucht. Das Materialstudium enttäuschte, denn die Besiedlung war einförmig und spärlich: *Gymnodinium pascheri* (SUCHL.) SCHILLER, ein paar Desmidiaceen und Diatomeen. Nur der Pfaffenstegteich (IV) machte eine Ausnahme. Er war im September 1974 abgelassen und ausgefischt worden (Forellenteich). Bald hernach wieder gesteckt, floß er langsam ein und war, als ich in der 2. Novemberhälfte erstmals Material entnahm, bis zur Hälfte vollgelaufen und eisbedeckt. Die phytoplanktische Besiedlung erwies sich überraschend individuenreich und interessant. Diese Entwicklung war eindeutig auf radikale Wassererneuerung zurückzuführen.

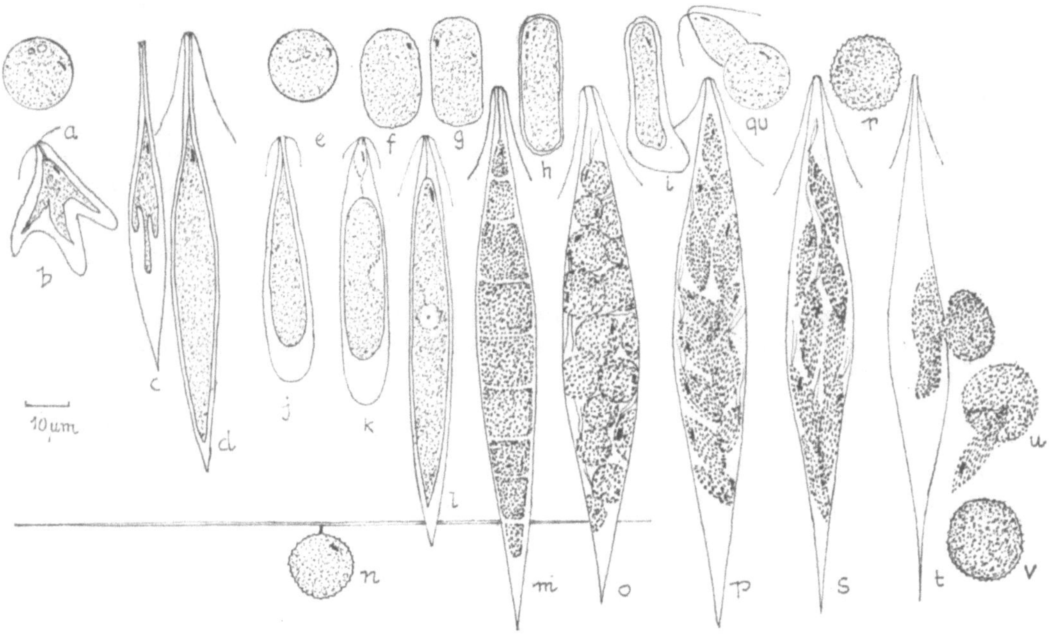

Abb. 4. *Chlorogonium maximum* SKJ.

a – d – Frühentwicklung I; e – k – Frühentwicklung II; l, m – vegetative Stadien und Protoplastenteilung; n – Parthenocyste; o, p – Heterogameten; qu – Heterogamie; r – Zygote; s – Spermatozoide; t – Macrogamet; u – Oogamie; v – Zygote

So war *Chlorogonium maximum* SKUJA, wenn auch nur mit 10—12 Individuen im 10-ml-Zentrifugat vertreten, bemerkenswert, weil es bisher (WAWRIK 1963) nur als Frühjahrsform beobachtet worden ist. Auch SKUJA (1956) beschrieb es als solche. Nun lebte es zur Zeit des Eisschließens bei pH-Werten von 6,2—6,4 und 0,16 mval/l SBV im Edlesberger- und Pfaffenstegteich in folgender Algengesellschaft: *Gymnodinium pascheri*, *Phytodinium* sp., *Chrysococcus cordiformis* NAUM., *C. minutus* (FRITSCH) NYG., *C. rufescens* KLEBS, *Mallomonas akrokomos* RUTT., *M. coronifera* MATV., *M. elongata* REV., *Chlamydomonas bichlora* PASCH. & JAHODA. Es gab nur vegetative Stadien, in einem Fall beginnende Protoplastenteilung. Das Materialstudium dauerte 10 Tage.

Wieder konnte die „Frühentwicklung II" dieses Flagellaten beobachtet werden. Die „Frühentwicklung I" läuft über eine spindelförmige Jugendform, bei welcher der Protoplast Fortsätze ausbildet, die eine genetische Verwandtschaft zur Gattung

Haematococcus (AGH.) FLOT. dokumentieren (WAWRIK 1972a). Die „Frühentwicklung II" aber führt über ein keulenförmiges Frühstadium zur Vollzelle (damit ist der ausgewachsene Flagellat gemeint). Beim Aufstieg vom Grund wirft die Zyste die scrobiculierte Hülle ab. Zysten von *Ch. maximum* werden sonst im zeitigen Frühjahr vor Eisbruch beobachtet. Sie hatten daher im Spätherbst längstens eine 7monatige Ruhezeit hinter sich. Die kugelige Zelle ($D = 20\ \mu m$) besitzt einen zusammengesetzten sattgrünen Chromatophor, vorn 2 träge Vakuolen, körperlange Geißeln, selten schon ein Stigma. Die Gestalt wird breitellipsoidisch, kurz- und langzylindrisch, erweitert sich basal keulenförmig und läuft vorn spitz zusammen. Geißeln und Stigma können auch erst im Laufe dieser Entwicklung zur Ausbildung kommen, während der Plasmafortsätze nicht beobachtet worden sind.[1])

Seit mehr als 10 Jahren wird dieser Flagellat studiert und immer noch sind Fragen offen: Welche Form der Frühentwicklung führt über die Vollzelle zur Heterogamie, welche zur Oogamie (WAWRIK 1969, 1972a)?. Diese Fragen könnten wohl nur Reinkulturen beantworten. Zweimalige Versuche in dieser Richtung (WAWRIK 1972a) schlugen bisher fehl. SKUJA (1956) kannte keine Frühentwicklung, beobachtete wohl Heterogameten, aber nicht Heterogamie. Ich sah beide Formen der sexuellen Reproduktion und beide Formen der Frühentwicklung, kenne aber die Zusammenhänge nicht. Gesprächsweise bedeutete mir Prof. BOURRELLY, daß eine Art keinesfalls heterogam und oogam sein könne. Diese Tatsache bedingt die Beschreibung einer neuen Varietät:

Chlorogonium maximum SKUJA var. *oogamum* nova var. differt a typo: reproductione oogama.

Es ist klar, daß der Wert dieser Beschreibung gering ist, denn Stadien der Frühentwicklung und sexuelle Vermehrung werden nur selten im Freiland beobachtet. Die Algologen können den Flagellaten weiterhin nur nach seinen morphologischen Merkmalen und ökologischen Bedingtheiten — die in beiden Fällen gleich sind — als *Chlorogonium maximum* bestimmen.

7. *Hemitoma meandrocystis* SKUJA

Im Jahre 1939 stellte SKUJA als Übergang von den Coccomonadineen zu den Phacoteen die Gattung *Hemitoma* auf. Die bisher einzige Art der Gattung, *Hemitoma meandrocystis*, lebt in unseren Teichen zu allen Jahreszeiten eingestreut. Als Begleiter von *Chlorogonium maximum* trat sie im Pfaffenstegteich aber recht häufig auf; es gab zweigeißelige bewegliche Stadien und alle Entwicklungsphasen. Es wurde beobachtet, wie aus der mehr oder weniger aequatorial gesprengten Mutterzelle ein sattgrüner Protoplast austrat, aus dem sich innerhalb der hyalinen Blase zwischen den beiden Schalenhälften durch Schrägteilung 4 Schwärmer entwickelten: kugelig, $D = 8-10\ \mu m$, mit farbloser Membran, sattgrünem Chromatophor und kleinem mehr oder weniger zentralem Stigma. Die endgültige feinstrukturierte Membran färbt sich durch Fe-Einlagerung langsam braun. SKUJA (1956) beobachtet schon, daß diese 4 Schwärmer nicht unbedingt frei werden müssen, sondern sich bereits innerhalb der Blase enzy-

---

[1]) Wahrscheinlich kam ich nie zur richtigen Zeit an das Mikroskop; denn ich bin der Auffassung, daß diese Fortsätze kennzeichnend für die Frühentwicklung sind. Wege der Naturerkenntnis sind Wege der Geduld!

stieren können. Daraus ergibt sich die auffallende Tatsache, daß das bewegliche Schwärmerstadium auf die Blase beschränkt bleiben oder überhaupt ausfallen (?) kann.

Nach meinen Beobachtungen ist *Hemitoma meandrocystis* nicht als „stenotherme Frühjahrsform“ anzusehen, als die sie HUBER-PESTALOZZI (1961) auswies. Es bestätigt sich hier, wie auch in anderen Fällen, daß ökologische Einordnungen nicht nach einmaliger Beobachtung oder auf Grund vereinzelter Feststellungen getroffen werden sollten. Gültige Aussagen bedürfen mehrjähriger Erfahrung. Die Anpassungsfähigkeit vieler Arten ist meist weitaus größer als allgemein vermutet wird. Die Zahl der gesicherten stenothermen Arten dürfte verhältnismäßig klein sein.

8. *Lambertia gracilipes* (LAMBERT) KORSCHIKOV (Abb. 5a)

In mehreren Teichen des Gebietes XVII trat diese Art besonders gehäuft auf. Es wurde Zystenbildung beobachtet, Die kugeligen Dauerstadien waren gelbgrün mit glatter Membran: $D = 7{,}5\ \mu m$. Für das Waldviertel handelt es sich um eine erstmalige Beobachtung (Abb. 5a). Im selben Material lebte mit c Frequenz — etwas ungewöhnlich für die Jahreszeit — die Kieselalge *Rhizosolenia eriensis* H. L. SMITH, ebenfalls mit Dauersporen (Abb. 5b). Innerhalb einiger Frusteln siedelten, offenbar raumparasitär, 4 µm kleine Flagellaten.

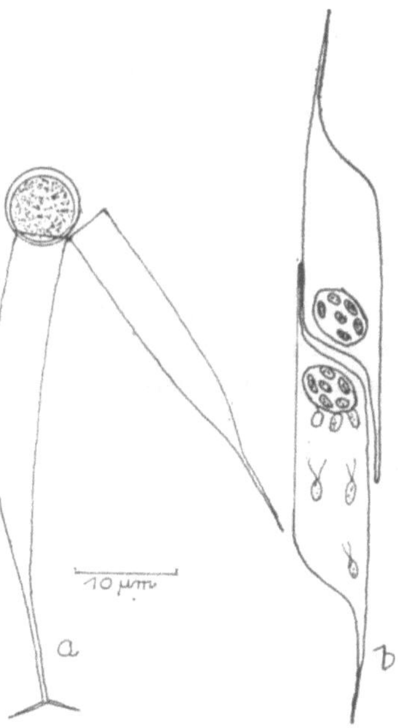

Abb. 5a — *Lambertia gracilipes* (LAMBERT) KORSCH., Zystenbildung

Abb. 5b — *Rhizosolenia eriensis* H, L. SMITH, mit Dauersporen und raumparasitären Flagellaten

9. *Actinastrum hantzschii* LAGERH. (Abb. 6a, b)

Diese sehr polymorphe Art war in mehreren Teichen des obigen Gebietes ebenfalls sehr häufig. Sie lebte im Juli 1974 im Mitterteich, einem Braungewässer, bei 16,5 °C, pH 6,8 und SBV 0,16 mval/l in zwei verschiedenen Ausbildungen zwischen einer Chrysophyceen-Gesellschaft, die für niedere Wassertemperaturen und sonnenarme Monate kennzeichnend ist. Zahlreiche Kolonien waren von einem Parasit befallen, von dem nur Dauerstadien vorhanden waren: ellipsoidisch, farblos, mit derber Membran, 8 × 4 µm groß. Mehrere vegetative Zellen der Alge enthielten Karotinausfällungen, die auf Nährsalzmangel im Substrat hindeuten könnten.

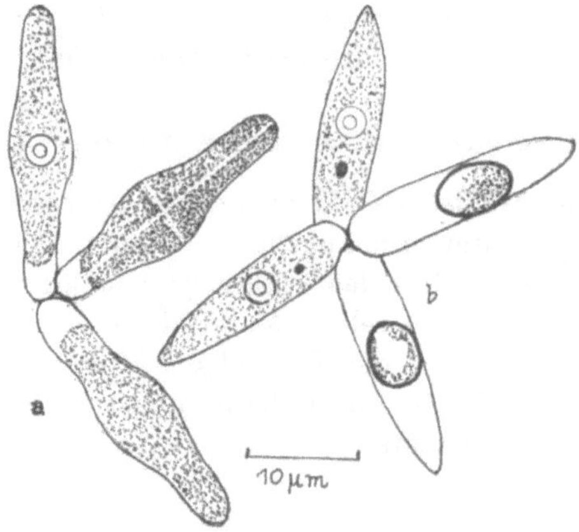

Abb. 6. *Actinastrum hantzschii* LAGERH.

a – spindelförmige Zellform; b – Zellen mit parasitären Dauersporen und Karotinausfällungen

10. *Colacium simplex* HUBER-PESTALOZZI (Abb. 7a—c)

Im Löschteich des Dorfes Hirschbach bestand im Juli eine Vegetationsfärbung von *Colacium simplex*. Diese Art ist wegen ihrer weitgespannten vertikalen Verbreitung interessant. HUBER-PESTALOZZI (1955) beschrieb sie aus einem Almtümpel in 2222 m Seehöhe, später fand man sie auch in kleinen Teichen in der Umgebung von Prag. Im Löschteich lebte sie zusammen mit *Chroomonas acuta* UTERMÖHL bei 19,0 °C,

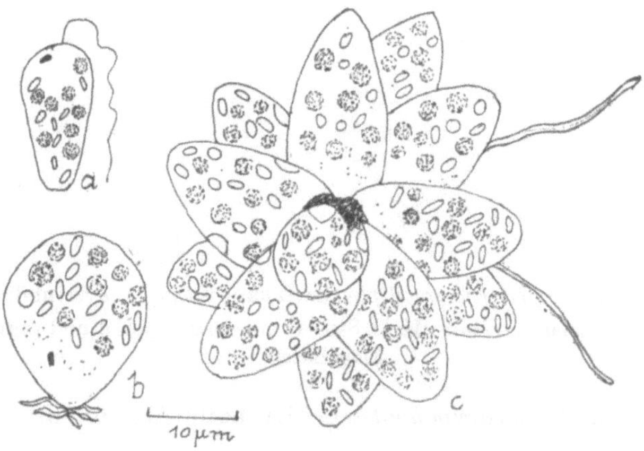

Abb. 7. *Colacium simplex* HUB.-PESTAL.

a – monadoid, freischwimmend; b, c – epibiont auf organischem Detritus pH 7,4 und SBV 0,64 mval/l monadoid freischwimmend und epibiont auf organischem Detritus. Die Art entsprach völlig der Beschreibung. Von einigem Interesse ist die rosettenförmige Anordnung der epibionten Exemplare, die im Wasser flotierten (euphotisch ?).

11. *Colacium cyclopicola* (GICKELH.) BOURR. (Abb. 8a—c)

Diese Art wurde schwärmend und epibiont auf *Volvox globator* (L.) EHRENBERG mit cc Frequenz beobachtet. Im Juli bei 16,2 °C, pH 7,0 und SBV 0,48 mval/l war sie über 2 Wochen mit unveränderter Individuendichte im Otterteich (XVII) anzutreffen. Die freischwimmenden Exemplare maßen 20—30 × 10 µm und waren walzlich breit gerundet; die derbe Geißel war körperlang und stets rasch bewegt, darunter 2 pulsierende Vakuolen, nahe daneben das kleine Stigma; 8—10 hellgrüne parietale Chromatophoren, wenig Paramylon; Kern basal. In keinem Fall konnte eine Streifung des Periplasten festgestellt werden, die BOURELLY (1947) gesehen hat. Die sessile Form war mit einem kreisringförmigen, unregelmäßig begrenzten, eiseninkrustierten Scheibchen an der Oberfläche der Volvoxkugel[2]) festgeheftet. Die Zellen saßen gehäuft-mosaikartig dicht nebeneinander. Bei HUBER-PESTALOZZI (1955) ist die Art aus Frankreich und der ČSSR gemeldet.

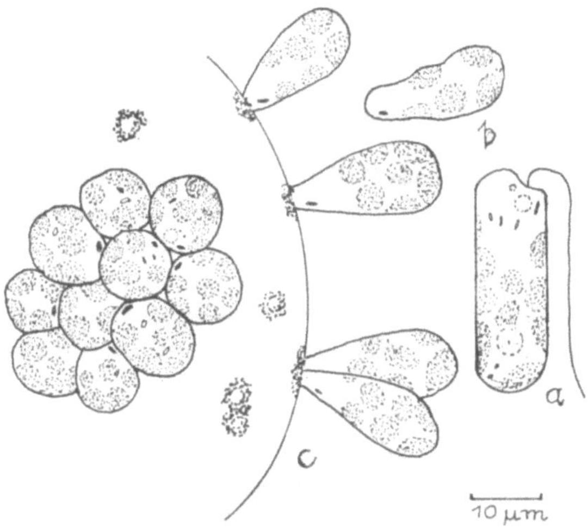

Abb. 8. *Colacium cyclopicola* (GICKELH.) BOURR.

a, b — monadoid; c — epibiont auf *Volvox globator* (L.) EHRENBG.

12. *Mallomonas schwemmlei* GLENK (Abb. 9a—h)

Das Auftreten dieser, 1956 aus einem fränkischen Fischteich beschriebenen Mallomonade ist offensichtlich von Wasser- bzw. Nährstofferneuerung bedingt. Nach bisherigen Beobachtungen (GLENK & FOTT 1971) erreichte sie als Frühjahrsform nach Eisbruch große Volksdichten, sogar Vegetationsfärbungen.

---

[2]) Bei *V. globator* konnten mehrfach Kolonienteilungen beobachtet werden.

Meine erste Bekanntschaft mit diesem Protisten machte ich im Februar 1971 im kleinen Ferenczyteich im nördlichen Waldviertel. Damals lebte unter Eis mit einer Frequenz von 50000 Z/ml *Mallomonas coronifera* MATVIENKO. Sehr vereinzelt gab es dazwischen eine Mallomonade ohne Kaudalborsten, die sich im Präparat von der häufigen Art nicht weiter zu unterscheiden schien. Ich signalisierte sie mit der Feststellung (WAWRIK 1972b, S. 356): Kaudalborsten 0—5 innerhalb der Art.

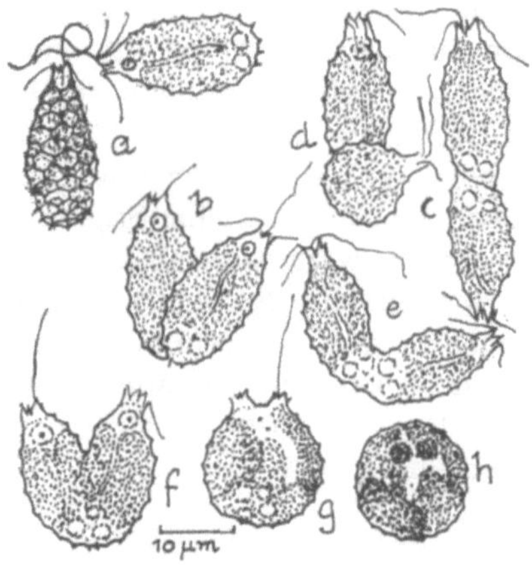

Abb. 9. *Mallomonas schwemmlei* GLENK.
a — Geißelkontakt; b — Kaudalkontakt der Isogameten; c — Strecklage; d, e — V-Lage; f, g — Gametenfusion; h — Zygote, zweikernig (Karmin-Essigsäure-Färbung).

Nun trat diese kaudalborstenlose Mallomonade zur Zeit des Eisschließens, während der Pfaffenstegteich einlief, bei pH 6,4, SBV 0,16 mval/l und rasanter Wassererneuerung mit 800 Z/ml nahezu rein auf. Ich sah die basalen „Zähnchen" und dachte an eine Varietät von *M. coronifera*, die nur sehr vereinzelt in viel größeren Exemplaren im Material vorkam. Erst das Studium konservierten und präparierten Materials ergab ein Herbstvorkommen von *M. schwemmlei*. Auf diese verwandte Art beziehen sich alle nachfolgenden Beobachtungen.

In einer Schöpfprobe, am Morgen des 19. XI. ufernah entnommen, sah ich erstmalig Geißelkontakt in der Gattung *Mallomonas*. Die Gestalt der Gameten war mehr oder weniger ellipsoidisch und sie lagen einander unter einem Winkel von 90° genähert. Die Geißeln bewegten sich kräftig. Die beiden basal gelagerten Vakuolen, die in meinem Material für die Art kennzeichnend zu sein scheinen, waren aktiv. Die Beobachtung wurde nach 16 Minuten von einem Rotator gestört. Später wurde auch noch ein zweites Paar gleicherweise im Geißelkontakt gesehen. Es gibt also in der Gattung *Mallomonas*, obschon die Zellfusion kaudal erfolgt, zweifellos vorher auch Geißelkontakte (Abb. 7a). Um 9.15 Uhr schwamm ein Gametenpaar in das Gesichtsfeld; der Kaudalkontakt war auf Zellbreite vollzogen. In der 25. Beobachtungsminute begann die Drehung aus der Strecklage in die V-Lage. Dabei wurden einige Schuppen abgestoßen, was offenbar der Grund dafür gewesen ist, daß das Gametenpaar am Deckglas haften blieb: Eine große Erleichterung für die weitere Beobachtung, die insgesamt 210 min währte! Die Verschmelzung der Isogameten vollzog sich sehr langsam, wie in allen bisher in der Gattung beobachteten Fällen. Während der lateralen Fusion lagen die Gameten nahezu parallel. Im Endstadium wurde eine Geißel abgeworfen. Sehr langsam erfolgte auch die Abkugelung der Zygote, die im Durchmesser 18,5 µm maß. Eine Kernfärbung mit Karmesin-Essigsäure gelang gut: Die beiden Zellkerne lagen einander genähert unter den beiden Apices, die verhältnismäßig lange deutlich sichtbar geblieben waren. Das Materialstudium von Schöpfproben aus dem Pfaffenstegteich dauerte bis zum 27. Nov. 1974. Eine wesentliche Zunahme der Volksdichte von *Mallomonas schwemmlei* wurde nicht festgestellt, doch hielt die Sexualperiode der Art die Woche hindurch an.

Der Winter 1974/75 war mild und niederschlagsarm. Förster Angerer, Gutenbrunn, hatte die Freundlichkeit, für mich am 2. Februar eine Schöpfprobe zu entnehmen. Die planktische Besiedlung unter Eis war geringfügig. Im Zentrifugat von 10 ml fanden sich 6—8 Exemplare von *Mallomonas schwemmlei*, durchweg junge, frischfarbene und rasch bewegliche Individuen, keineswegs Überwinterer. Ein Beweis dafür ist die Tatsache, daß sich auch ein Isogametenpaar in Strecklage darunter befand. Es wurden 250 ml Probenwasser zentrifugiert, und es blieb bei dieser Einzelbeobachtung: Eine beginnende Sexualperiode? Die Ruhezeit der Mallomonade hat demnach nur etwa 6 Wochen gedauert. Begleiter waren *Phytodinium* sp. und *Chlamydomonas* sp., beide rr. Eine Schöpfprobe, am 1. März 1975 unter 40 cm Eis mit 8 cm Schneeauflage entnommen, enthielt keinerlei autotrophe Produktion: Der Lichtfaktor war ins Minimum gekommen.

Bei einer neuerlichen Probeentnahme am 7. IV. 1975 entsprachen die Verhältnisse am Teich weitgehend jenen vom 2. Februar: Der Teichrand war eisfrei; die maximale Eisdicke maß 5 cm; es gab keine Schneeauflage. Der Teichfrühling hatte seinen Anfang genommen: Beherrschend lebte *Chromulina sphaeridia* Schiller ohne Stigma mit einer 1—$1^1/_2$mal durchmesserlangen Geißel. *Chlorogonium maximum* Skj. war mächtig im Aufkommen. Es gab nur vegetative Stadien und noch keine Protoplastenteilungen. Zwischen *Synura pertersenii* Korsch. r, *S. sphagnicola* Korsch. r, *Chloromonas* sp. *Cryptomonas* sp. rr und reichlich karotinhaltigen Exemplaren von *Gymnodinium pascheri* (Suchl.) Schiller gab es auch vereinzelt wieder *Mallomonas schwemmlei*.

Am 18. IV. war der Teich nur mehr zu einem Drittel eisbedeckt (Luft 6 °C, 0 m 3,3 °C, pH 5,6, SBV 0,10 mval/l). *Mallomonas schwemmlei* war in fast reinem Aspekt mit 3800 Z/ml anwesend. Seine Maße entsprechen den Angaben bei Glenk: 12—20 × 6—10 µm. Es besteht eine mäßige Sexualperiode: Kaudalkontakte, Streck- und V-Lagen, keine Zysten oder Zygoten. Bei der Untersuchung im geheizten Raum erweist sich der Organismus nicht als besonders wärmeempfindlich, aber in der kaltgestellten Rohkultur ist er am 3. Tag nicht mehr da.

Wegen der weiten Entfernung des Pfaffenstegteiches wurde mit der nächsten Probeentnahme zugewartet. Glenk & Fott (1971) geben eine mehr als 6wöchige Vegetationsdauer an und eine Vegetationsfärbung nach dem Abschmelzen des Eises! Am 24. IV. war der Teich eisfrei (Luft 8°, Wasser 6,2°). Den Planktonaspekt beherrschte ein kleiner (5 µm) Flagellat cc, dazwischen *Chlorogonium maximum* rr. *Eunotia rhomboidea* Hust. rr, *Synura sphagnicola* rr, *Mallomonas schwemmlei* 6 Z/ml! Keine Sexualperiode, keine Zysten und Zygoten. Möglicherweise hatte ein Schmelzwasserstoß den Aspekt dermaßen verändert und die Vegetationsdauer der Mallomonade auf kaum 2 Wochen begrenzt. Durch die Einstrahlung war die Temperatur allerdings etwas über 6° angestiegen. In ihren ökologischen Ansprüchen stimmt *M. schwemmlei* im Pfaffenstegteich (860 m) mit den bisherigen Beobachtungen überein; neu sind ihr Herbstvorkommen und die Sexualperiode bei Eisschluß und Eisbruch. Auch lebte sie in einer völlig anderen Algengesellschaft als beispielsweise FOTT & ETTL in der Talsperre Sedlice (ČSSR) festgestellt haben. Ihr Vorkommen ist bislang aus der BRD, der ČSSR und nun auch aus Österreich gemeldet.

5. *Literatur*

BEHRE, K., & E. WEHRLE, 1944: Welche Faktoren entscheiden über die Zusammensetzung von Algengesellschaften? — Arch. Hydrobiol. **39**, 1—23.

BOURELLY, P., 1957: Recherches sur les Chrysophycees. — Rev. Algol. Mém. Hors-Sér. No. 1, 409 S.

FOTT, B., 1959: Algenkunde. — VEB Gustav Fischer Verlag Jena, 482 S.

GLENK, H. O., & B. FOTT, 1971: Zur Frage der Validität der Art *Mallomonas schwemmlei* GLENK und ihre Unterscheidung von *Mallomonas coronifera* MATVIENKO. — Arch. Protistenkd. **113**, 220—229.

HUBER-PESTALOZZI, G., 1955: Euglenophyceen. — 4. Teil. In HUBER-PESTALOZZI; Phytoplankton des Süßwassers. — Schweizerbart'sche Verlagsbuchhandlung Stuttgart, 606 S.

— 1961: Chlorophyceae. — 5. Teil. — Ibid., 744 S.

SKUJA, H., 1956: Taxonomische und biologische Studien über das Phytoplankton schwedischer Binnengewässer. — Nov. Act. Reg. Soc. Scient. Upsl. IV, 16/3, 1—400.

UHLMANN, D., 1966: Artendichte. — Limnologica (Berlin) **4** (2), 221—233.

WAWRIK, F., 1952a: Nährstoffkonzentration und Produktionsmaxima als Funktionen des Pegelstandes. — Wetter u. Leben **4** (1), 38—40.

— 1952b: Beziehungen zwischen Kieselsäurehaushalt und Diatomeenblüte in der Seebachlacke bei Kienberg-Gaming, N. Ö. — Österr. Bot. Z. **99** (2/3), 286—294.

— 1963: Planktonorganismen aus Waldviertler Fischteichen II. — Arch. Protistenkd. **106**, 369 bis 377.

— 1966: Die Waldviertler Fischteiche und ihre Entomostrakenfauna auf oekologischer Grundlage. — Hydrobiologia **28** (4), 385—552.

— 1968: Winterschichtung 1965/66 in 12 Waldviertler Fischteichen. — Ibid. **31** (3/4), 516—544.

— 1969: Sexualität bei *Cryptomonas* sp. und *Chlorogonium maximum* SKUJA. — Nov. Hedw. **8**, 283—292.

— 1971: Vegetationsfärbungen und Wasserblüten. — Wetter u. Leben **23**, 203—210.

— 1972a: Oogamie bei *Chlorogonium maximum* SKUJA. — Arch. Protistenkd. **114**, 489—494.

— 1972b: Isogame Hologamie in der Gattung *Mallomonas* PERTY. — Nov. Hedw. **23**, 353—362.

— 1972c: Vergleichende Braunwasser-Teichstudien im niederösterreichischen Waldviertel. — Hydrobiologia **39** (1), 17—82.

Eingegangen am 2. 9. 1975.

Anschrift des Verfassers:

Dr. FRIEDERIKE WAWRIK
Biologisches Labor Waldviertel
A-3943 Schrems/Gebharts
Österreich

Limnologica (Berlin)	11	1	1976	175—176

DIETRICH UHLMANN

Bericht über das Internationale Biologische Programm, Sektion „Produktivität der Binnengewässer"

Im September 1974 beendete die Sektion „Produktivität der Binnengewässer" ihre Arbeit. Von der Sektion wurden 260 Untersuchungsprojekte in 39 Ländern koordiniert. Nach Einschätzung des Sekretariates (Leitung: Dr. J. RZOSKA) hat dieses Programm in sehr starkem Maße dazu beigetragen, die junge Forschergeneration auf dem Gebiet der limnischen Produktionsbiologie auszubilden. Eine Auswertung der Zwischenergebnisse erfolgte auf Symposien 1970 in Polen und 1974 in Großbritannien. Die wesentlichsten Ergebnisse der letzten Jahre werden unter dem Titel „Die biologische Produktivität von limnischen Ökosystemen" voraussichtlich noch 1976 im Verlag Cambridge Univ. Press als Sammelband erscheinen. Als sehr wichtiges Ergebnis kann die Herausgabe der fünf methodischen Handbücher (siehe Literaturzusammenstellung) gewertet werden, die alle strukturellen Ebenen der limnischen Ökosysteme, vom chemischen Milieu zum Fischertrag, erfassen. Diese Handbücher waren das Ergebnis von Spezialistentreffen in Italien, Großbritannien, der ČSSR, der UdSSR und den Niederlanden. Weitere Publikationen über Makrophyten, Huminstoffe, Detritus, Wetlands und Photosynthese wurden gemeinsam mit anderen IBP-Sektionen herausgebracht. Als Erfolg kann auch die durch das IBP stark intensivierte Bearbeitung der Binnengewässer in Lateinamerika, Afrika und Südostasien bezeichnet werden, die zur Entwicklung einer wissenschaftlich betriebenen Binnenfischerei in den betreffenden Ländern beitragen soll.

Eine Auswahl der neueren IBP-Publikationen (bei Blackwell Scientific Publications, Oxford) sei hier angeführt:

Handbook	Editor(s)	Title	Year
No. 3	W. E. RICKER	Methods for Assessment of Fish Production in Fresh Waters	1968
No. 3	W. E. RICKER	Second (revised) edition	1971
No. 8	H. L. GOLTERMANN with assistance of R. S. CLYMO	Methods for Chemical Analysis of Fresh Waters. Revised third printing	1971
No. 12	R. A. VOLLENWEIDER with collab. of J. TALLING & D. WESTLAKE	A Manual on Methods of Measuring Primary Production in Aquatic Environments. Second revised edition	1974
No. 17	W. T. EDMONDSON & G. G. WINBERG	A Manual on Methods for the Assessment of Secondary Productivity in Fresh Waters	1971

Handbook	Editor(s)	Title	Year
No. 21	H. LUTHER & J. RZÓSKA	Projekt Aqua, a source book of inland waters proposed for conservation. Revised edition	1971
No. 23	Y. I. SOROKIN & H. KADOTA	Techniques for the Assessment of Microbial Production and Decomposition in Fresh Waters	1972
	G. G. WINBERG et al.	Symbols, Units and Conversion factors in studies of freshwater productivity. IBP Central Office — Private print	1971
1972	Z. KAJAK & A. HILLBRICHT-LKOWSKA (Eds.)	Productivity problems of freshwaters. Proc. of the IBP/UNESCO symposium 6—12 May 1970 Warszawa-Kraków. PWN Polish Scientific Publishers Warsaw	1972

Publikationen in Zusammenarbeit mit anderen Sektionen des IBP

	Romanian Academy of Sciences	Proc. of an IBP/UNESCO-IDH meeting on Macrophytes (1—20 Sept. 1970) Hidrobiologia vol. 12, 1971 — Academie Republicii Socialiste Romania, București	1971
	A. SZCZEPÁNSKI	Proc. of IBP Wetlands symposium 11—18 Juni 1972 at Mikolajki, Poland. Polish Archives of Hydrobiology. Vol. 20, no. 1. PWN Polish Scientific Publishers Warsaw	1973

Anschrift des Verfassers:
Prof. Dr. DIETRICH UHLMANN
Technische Universität Dresden
Sektion Wasserwesen, Bereich Hydrobiologie
DDR-8027 Dresden
Mommsenstr. 13

Limnologica (Berlin)	11	1	1976	177–224

Rezensionen

ANDRZEJ BATKO: Zarys Hydromikologii. Warszawa: Państwowe Wydawnictwo Naukowe 1975. 478 S., 63 Taf. Ln. zł 95,—

Im allgemeinen denkt man, wenn von Pilzen die Rede ist, an die „Großpilze" in unseren Wäldern oder auf unseren Wiesen und Weiden. Nur wenige Eingeweihte wissen, welch große Rolle diese Organismengruppe in den Gewässern spielt. Eine moderne zusammenfassende Darstellung der Gewässerpilze in Mitteleuropa fehlt. Der geplante Pilzband der PASCHERschen „Süßwasserflora von Mitteleuropa" ist nie erschienen. Hier schließt das vorliegende Werk eine empfindliche Lücke.

A. BATKO hat sich der schwierigen Aufgabe unterzogen, einen Überblick über die Wasserpilze zu vermitteln. Er führt Chytridio-, Hyphochytrio-, Plasmodiophoro-, Oo-, Zygo-, Tricho-, Endo-, Asco-, Basidio- und Deuteromycetes mit insgesamt 47 Familien und 276 Gattungen vor, wobei der Löwenanteil auf Chytridio- und Oomycetes entfällt. Alle Gattungen und die 1067 Arten werden aufgeschlüsselt, viele von ihnen auf 63 Tafeln durch Strichzeichnungen dargestellt. Die Tilletiaceae und Laboulbeniales hat T. MAJEWSKI bearbeitet. In vier einführenden bzw. allgemeinen Kapiteln werden Fortpflanzung und Entwicklung sowie die Ökologie der Wasserpilze behandelt. Methodische Hinweise auf die Kultur und Untersuchung sind sehr förderlich. Der technische Apparat besteht aus einem umfangreichen Literaturverzeichnis und einem ausreichenden Sach- und Sippenindex. Die Darstellung enthält auch einige nomenklatorisch und taxonomisch wichtige Neukombinationen. Es ist zu befürchten, daß diese von der mykologischen Wissenschaft kaum bemerkt werden.

Es ist dem Rezensenten nicht möglich, in das Werk tiefer einzudringen, da ihm die sprachlichen Voraussetzungen dazu fehlen. Er hat aber den Eindruck gewonnen, als ob mit diesem Bestimmungsbuch den polnischen Mykologen und Hydrobiologen ein wertvolles Hilfsmittel in die Hand gegeben worden ist, von dem zu hoffen ist, daß es in absehbarer Zeit in einer der Weltsprachen erscheinen möge.

S. J. CASPER

WALTER HOPPE & GERD SEIDEL (Hrsg.): Geologie von Thüringen (Bezirke Erfurt, Gera, Suhl). Gotha, Leipzig: VEB Hermann Haack 1974. 1000 S., 200 Abb., 124 Tab., 99 Photos, 1 geol. Kt. Ln. M 140,—

Das von einem großen Autorenkollektiv (genannt sind 36 Bearbeiter) geschaffene Werk ist in doppelter Hinsicht „gewichtig". Der graphisch und typographisch hervorragend gestaltete Band wird für Jahre, wahrscheinlich sogar Jahrzehnte, die Grundlage geologischer Arbeiten in diesem Gebiet darstellen. Beim Vergleich etwa mit der „Einführung in die Geologie Thüringens" von H. WEBER (1955) wird deutlich, welcher gewaltige Kenntniszuwachs in den letzten Jahrzehnten erreicht wurde, an dem die Bearbeiter weit mehr beteiligt waren, als es die Literaturverzeichnisse ausweisen, denn die unveröffentlichten Berichte dieser und anderer Autoren sind zwar meist nicht genannt, aber mit ausgewertet worden.

Das Buch gliedert sich in sieben Kapitel: 1. Geschichte der geologischen Erforschung (13 S.), 2. Physisch-geographischer Überblick (17 S.), 3. Regionalgeologische Gliederung und Stellung (26 S.), 4. Baueinheiten (713 S.), 5. Geologische Entwicklungsgeschichte (71 S.), 6. Bodenschätze (98 S.), 7. Böden (19 Seiten). Den Abschluß bilden die neu bearbeitete Geologische Karte von

Thüringen 1:500000 und ein Ortsregister (14 S.). Jedes Kapitel bzw. Abschnitt endet mit einer Zusammenfassung und Literaturangaben.

Den Hauptumfang des Werkes nimmt das Kapitel „Baueinheiten“ ein, in dem die Untergliederung nicht nach den regionalen Einheiten, sondern im wesentlichen nach der geochronologischen Skala bzw. nach tektonischen Stockwerken erfolgt. Stratigraphie und Fazies, Petrographie, Fossilien, Paläogeographie und Genese, Tektonik und Magmatismus werden behandelt. Eine Vielzahl von Fakten wird übersichtlich dargestellt, wobei aber teilweise auffällt, daß noch mehr Koordinierung nötig gewesen wäre, um einzelne Unstimmigkeiten und Wiederholungen vor allem in den Kap. 3—5 sowie den etwas ungleichwertigen Umfang der Abschnitte zu vermeiden. Nach diesen Detailanalysen folgt im nächsten Kapitel die Synthese, die Eingliederung des thüringischen Raumes in die geologische Entwicklung Mitteleuropas.

Für die „Nutzer und Anwender“ geologischer Ergebnisse sind besonders die Kapitel über das Quartär, die Bodenschätze und die Böden wichtig, weil vorwiegend hier die für sie relevanten Fakten zu finden sind. So interessieren den Limnologen z. B. die Angaben über die Entwicklung des heutigen Gewässernetzes, die hydrogeologischen Parameter und die Böden.

Dieses Werk braucht nicht empfohlen zu werden, es wirbt für sich selbst. Schade ist eigentlich nur, daß die Herausgabe erst so spät erfolgte, denn die Manuskripte stammen aus dem Jahre 1968. D. Storch

Helen I. Alston: Aquatic plants of Australia. A guide to the identification of the aquatic ferns and flowering plants of Australia, both native and naturalized. Melbourne: University Press 1973. XV + 368 S., 138 Abb. Kunstleder. £ 18.00

Dieses Wasserpflanzenbuch ist eine bemerkenswerte Publikation, stellt es doch den ersten Versuch dar, die in aquatischen Ökosystemen des australischen Kontinents vorkommenden höheren Pflanzen zu erfassen, zu beschreiben und abzubilden. Es ist dem Fachmann klar, daß mit diesem Werk nicht nur eine bestimmte ökologisch umrissene Pflanzengruppe in einem begrenzten Raum vorgestellt wird, sondern taxonomische und pflanzengeographische Fragen aufgeworfen und so weit wie möglich beantwortet werden, die die Einordnung und Verbreitung der Wasserpflanzen der ganzen Erde berühren. Hier muß der in Mitteleuropa arbeitende Taxonom eine ganze Menge liebgewordener Vorstellungen aufgeben: Die Wasserpflanzen sind gar nicht so weltweit „gleichartig“ verbreitet. Viele der bislang auf dem australischen Kontinent als beheimatet angegebenen Sippen kommen dort nicht vor, sondern werden durch andere, oft verwandte, ersetzt, z.B. *Sagittaria sagittifolia* L. durch *S. montevidensis* Cham. & Schlecht. und *Typha angustifolia* L. durch *T. domingensis* Pers. An einigen Stellen wird die Unsicherheit in der taxonomischen Beurteilung der Sippen besonders deutlich, so etwa in den Familien Lemnaceae, Potamogetonaceae, Elatinaceae, Callitrichaceae und Najadaceae, Familien, die auch in Europa eine „crux botanicorum“ darstellen. Besonderen Aufschluß gewährt die Behandlung der Aponogetonaceae, Hydrocharitaceae, Onograceae, Menyanthaceae und Haloragaceae, deren tropisch-subtropische Vertreter hier und da in Europa adventiv auftreten oder von Aquarienfreunden gehalten werden.

Insgesamt werden rund 220 Arten ausführlich vorgestellt, von denen 109 in Victoria zu Hause sind. Zu beachten ist, daß Characeen und Ricciaceen als „Makrophyten“ behandelt werden. Der klare Text wird durch eine große Zahl einfacher, detailreicher, meist originaler Abbildungen wirkungsvoll unterstützt. Punktkarten vermitteln eine Vorstellung von der Verbreitung der Sippen in Victoria. Der technische Apparat besteht aus einem Glossarium der botanischen Fachausdrücke, Benutzungshinweisen für das Buch überhaupt und die Bestimmungsschlüssel, einem Verzeichnis der gebrauchten Abkürzungen und Symbole, einer Bibliographie mit einer Einführung, einer Liste metrischer Umrechnungsfaktoren, Karten der im Text erwähnten Örtlichkeiten in Australien und insbesondere im Staate Victoria sowie einem ausführlichen Sippenindex. Im Anhang finden sich Angaben über *Eichhornia crassipes* (mit Quellenverzeichnis), über die sogenannten Seegräser (marine Hydrocharitaceae, Posidoniaceae, Zannicheliaceae, Zosteraceae) sowie eine Verbreitungsliste der Wasserpflanzen, aufgeschlüsselt nach ihrem gegenwärtigen oder früheren Vorkommen in den einzelnen Staaten des Kontinents.

Alles in allem ein wichtiges, weit über den australischen Raum hinaus Impulse ausstrahlendes, sehr brauchbares Kompendium, gut gedruckt (wenn auch nicht ohne, manchmal ärgerliche Druckfehler) und ausgestattet. S. J. Casper

Norman L. Allinger & Janet Allinger: Strukturen organischer Moleküle. Übersetzt von Helga Schreider. Stuttgart: Georg Thieme Verlag 1974. IX + 161 S., 24 Abb., 8 Tab. Flex. Taschenb. DM 9,80

Die 1865 von Kekule von Stradonitz aufgestellte Strukturformel für Benzol wird heute noch aus anschaulichen Gründen verwendet, obwohl wir wissen, daß diese Art der Darstellung nicht in der Lage ist, die Besonderheiten des aromatischen Bindungszustandes exakt wiederzugeben. Die Autoren des vorliegenden Bandes verlassen die alte zweidimensionale Darstellungsmethode und versuchen, ausgehend vom atomar-räumlichen und elektronischen Bau der organischen Moleküle, chemische Bindungen und Reaktionen zu erklären. Der Inhalt des Buches führt über Atommodelle zu den Molekülstrukturen, über ungesättigte und zyklische Kohlenwasserstoffe, Verbindungen mit funktionellen Gruppen, Rotationsisomerie und andere Ringsysteme bis zu chemischen Reaktionen mit Thermodynamik und Kinetik. Die Methodik zur Untersuchung der Struktur mit UV-, IR- sowie Kernresonanz-Spektren wird an Beispielen erläutert. An Hand einiger Aufgaben am Ende der 7 Kapitel kann sich der Leser prüfen, ob der Inhalt erfaßt wurde.

Der dargelegte Stoff wird in kurzer, straffer, aber trotzdem leicht verständlicher Form geboten, so daß das Büchlein für einen breiten Leserkreis als Einführung geeignet ist. G. Proft

Paul Duvigneaud: La synthèse écologique. Populations, communautés, écosystèmes, biosphère, noosphère. Paris: Doin éditeurs 1974. 296 S., 94 Abb., 17 Farbtaf. Geb. fr. 145,—

Die Ökologie, die Wissenschaft von den Beziehungen der Lebewesen zu ihrer Umwelt, ist aus ihrem Dornröschenschlaf erwacht. Die in allen hochindustrialisierten Ländern entstandenen und viel diskutierten Probleme des Umweltschutzes haben ihr zu einer Popularität verholfen, die die Schöpfer dieser wissenschaftlichen Disziplin vielleicht wünschen mochten, aber wohl kaum ahnen konnten. Die Verfeinerung der methodischen Grundlagen und die Verbindung mit den exakten naturwissenschaftlichen Disziplinen der Physik und Chemie sowie der Mathematik haben die Ökologie aus der beschreibenden Phase ihrer Entwicklung in die experimentelle hinübergeführt. Als in hohem Grade interdisziplinäre, vorwiegend biologische Wissenschaft schlägt sie die Brücke zur Geographie und Soziologie und läßt sich umgekehrt von daher befruchten. Freilich hat dieser Aufschwung der Ökologie auch manchen Nachteil mit sich gebracht: Unter Ökologie werden hier und da Dinge verstanden, die mit ihr wenig oder gar nichts zu tun haben.

Am Anfang ging die ökologische Forschung vom Einzelorganismus aus und war daher überwiegend autökologisch orientiert. Alsbald wendete sie sich der Analyse ganzer Populationen oder Gesellschaften von Organismen zu, „entdeckte“ die Biozönose und wurde so zur Synökologie. Der vorliegende Band bringt in seinem Titel diese „Synthese“ zum Ausdruck, und die im Untertitel genannten Begriffe Populationen, Gesellschaften, Ökosysteme, Biosphäre und Noosphäre spiegeln ein Programm wider. In wesentlichen Teilen seines Textes und seiner Illustrationen geht das Werk auf eine Broschüre zurück, die unter dem Titel „Ecosystèmes et biosphère“ zum ersten Male 1962 vom Ministerium für nationale Erziehung und Kultur in Belgien veröffentlicht worden ist. Auf ihrer Grundlage wurde weitergearbeitet und durch Zufügung eines völlig neuen Teils „Populations et Communautés“ dem Buche eine dem Gesamtgebiete der Ökologie entsprechende Abrundung gegeben. Die Einbeziehung des Menschen und der menschlichen Gesellschaft in die Gesamtschau vermittelt der 1921 von Vernadsky geprägte Terminus Noosphäre.

Der Verfasser, Copräsident der Gruppe Wälder im Internationalen Biologischen Programm, hat unter Mithilfe kompetenter Fachleute eine Darstellung geschaffen, die sowohl wissenschaftlichen Ansprüchen genügt, als auch und vor allem Laien und Interessierte mit der Ökologie als Wissenschaft und ihrer Bedeutung für die menschliche Gesellschaft vertraut macht. Im Mittelpunkt steht das Konzept des Ökosystems. Binnengewässer (Limnologie!), Ozean, Wald, Wiese und Acker und die in ihnen sich vollziehenden Energie- und Stoffkreisläufe sowie ihre Produktivität beherrschen die Darlegungen, die zwar den Einzelorganismus und die Gesellschaft nicht vergessen, sie aber als Glieder übergeordneter autoregulativer Einheiten begreifen. Mit der Konzeption der Biosphäre und ihrer Gliederung in eine Litho- und Hydrosphäre wird jeder regionale Aspekt überwunden und der Gesamtbereich unserer Erde erfaßt. Drei umfangreiche Kapitel beschäftigen sich mit der Stellung des Menschen in der Biosphäre, seiner Aktivität und Rolle in Gegenwart und Zukunft. Dabei werden futurologische Aspekte westlicher Provenienz und ihre pessimistischen

Aussagen zwar behandelt, aber als überwindbar erkannt. Der Autor verspricht sich von der internationalen Solidarität bei der Aus- und Verwertung der natürlichen Ressourcen die Sicherung der menschlichen Existenz und die Lösung dringender Probleme, wie Unterernährung und Unterentwicklung in weiten Bereichen vor allem der Dritten Welt.

Der Text ist übersichtlich und klar, zeichnet sich durch eine Fülle aussagekräftiger schematischer Abbildungen aus und ist wissenschaftlich fundiert. Direkte Hinweise auf die verwendeten Quellen erleichtern den Zugang zu ihnen. Ein umfangreiches, nach Sachgebieten geordnetes Literaturverzeichnis, das neben Büchern auch Zeitschriften, Symposiumsberichte und ökologische Bibliographien erfaßt, sowie ein umfangreicher Index erhöhen die Brauchbarkeit des vorzüglich gedruckten und gut ausgestatteten Buches beträchtlich. Insgesamt eine beachtliche Leistung.

S. J. Casper

Erhard Eschler: Verhaltensforschung und Evolution. Jena: VEB Gustav Fischer Verlag 1974. 130 S. Br. M 15,20

Der vorliegende, aus einer Habilitationsschrift hervorgegangene schmale Band aus der Reihe „Philosophie und Biowissenschaften" setzt sich aus der Sicht des dialektischen Materialismus mit den ethologischen Aspekten der Evolution des Menschen auseinander, wobei die Widerlegung gewisser scholastischer Positionen bürgerlicher Autoren Hauptanliegen der Ausführungen des Autors ist. Insofern ist der anspruchsvolle Titel der Schrift irreführend, bei dem der Leser eher eine zusammenfassende Behandlung der Bedeutung der Verhaltensforschung für die Entwicklung der synthetischen Evolutionstheorie erwartet.

Nach einem einleitenden Kapitel, in dem die marxistisch-leninistischen Grundpositionen des behandelten Themas herausgearbeitet werden, folgt in zwei Hauptabschnitten die Darstellung der natürlichen Entstehung des menschlichen Bewußtseins und der Bedeutung der Geselligkeit in der Evolution. Die Ausführungen schließen mit einer philosophischen Abgrenzung des Themenkomplexes von verschiedenen theologisch beeinflußten Deutungsversuchen der Evolution des Menschen und einer kurzen Zusammenfassung der sich aus der Verhaltenskunde ergebenden weltanschaulichen Konsequenzen für die Theorie der Anthropogenese.

Obwohl das Grundanliegen des Autors, die Soziogenese als das wesentliche Merkmal der Evolution des Menschen zu betrachten, klar herausgearbeitet worden ist, haften der Darstellung nicht selten unklare Formulierungen, falsche Vorstellungen und stilistische Mängel an, die dem Verständnis des Stoffes nicht gerade förderlich sind. Einige Beispiele seien herausgegriffen. Der Theologe Albert Schweitzer dient mit seiner Ansicht vom Leid der Kreatur einmal „als Beispiel einer letztlich anthropomorphistischen Tendenz religiöser Ansichten" (S. 16), zum anderen aber scheint die Überzeugung Schweitzers dem Autor als Hinweis auf mögliche Vorstufen des Bewußtseins durchaus berechtigt (S. 18). Bei Tieren finden sich nicht nur Gruppenbildungen, sondern auch eine Vielfalt von kommunikativen Beziehungen der Gruppenmitglieder untereinander (S. 23). Es trifft nicht zu, daß „bei der Fortpflanzung ein längerer oder kürzerer Kontakt zwischen den Partnern der gleichen Art bei allen Metazoen die Regel" ist, ganz zu schweigen von der angeblich generellen Verbreitung der Brutpflege bei Metazoen (S. 23). Sprachlich falsch ist die Bezeichnung „Züge" für die periodischen Wanderungen der Vögel, die im übrigen durchaus nicht nur im Schwarm unternommen werden (S. 26). Von der „Teilung der Arbeit im Geschlechtsakt" zu sprechen, klingt zwar sehr wissenschaftlich, ist aber trotzdem eine seltsame Stilblüte (S. 23). Besondere Schwierigkeiten scheint dem Autor der Begriff der Fauna zu bereiten, denn wie anders lassen sich die folgenden, teilweise grotesken Formulierungen erklären: „Der Gedanke der Ablehnung sozialer Aspekte im Bereich der Fauna ..." (S. 30), „... bestimmter Grundlagen der Geselligkeit im Reich der Fauna ..." (S. 33), „... da die verschiedenen Arten in der Systematik der Fauna ..." (S. 104) und „... daß soziale Lebensäußerungen ... in der Fauna vorhanden sind" (S. 111).

Von solchen Klippen und Unebenheiten der Diktion abgesehen, die auch eine mangelnde redaktionelle Sorgfalt verraten, kann der vorliegende Beitrag als erste Einführung in die philosophische Bewältigung des Problemkreises Verhalten und Menschwerdung Lehrern, Studenten und interessierten Oberschülern empfohlen werden.

D. Flössner

RUTHERFORD PLATT: Wasser. Die Substanz des Lebens. (Aus dem Amerikanischen von RICHARD KAUFMANN). Tübingen, Stuttgart: Rainer Wunderlich Verlag Hermann Leins 1973. 318 S. Lw. DM 26,—

Ausgehend von der Physik und Chemie des Wassers werden die Entstehung des Lebens im Urozean, die Entwicklung der Tier- und Pflanzenwelt im Wasser, die Landbesiedlung, die Evolutionsmechanismen und vieles andere behandelt. Dabei hat der Verfasser immer die Bedeutung des Wassers für alle Erscheinungen des Lebens im Auge. Die letzten Kapitel des Buches sind der Umweltkrise gewidmet.

So originell die Grundkonzeption des Buches sein mag, so schlecht ist ihre Ausführung. Was dabei zu Lasten des Übersetzers geht, vermag der Referent nicht zu sagen. Der Stil ist lax und häufig unklar. Die wohl poetisch gemeinten Verzerrungen führen zu einer groben Entstellung der Sachverhalte und lassen das Buch zu einer Sammlung unfreiwilliger Komik werden. Auch das „Reißerische" ist nicht zu übersehen: Sex aus den Tiefen des Urozeans bis zum „Amnion-See" in der menschlichen Gebärmutter. Hier einige Kostproben:

„Primär ist nämlich das Ozeanwasser ein Gemisch starker Lösungen von Wasserstoff, Sauerstoff, Stickstoff und Kohlenstoff. Zusätzlich gibt es im Meerwasser energiereiche Elemente ... Eisen ist dabei von unerhörter Wichtigkeit, ohne dieses magnetische Element gäbe es unseres Wissens überhaupt kein Leben. Eisen ließ die Eisenfresser unter den Bakterien gedeihen, Eisenlösungen im Meerwasser waren zur Hand, als es galt, die roten Blutzellen kommender Zeiten zu bilden ... Überlegen Sie bitte einmal, wie einige Elemente, die im Plankton zu finden sind — herumtreibend, jahrmillionenlang vom Sonnenlicht gebacken —, irgendwann später wieder im Strom des Lebens auftauchten ... Oder, noch seltsamer, in der Frage des Sex scheint der Brauntang ein wichtiger Vorläufer des menschlichen Tieres zu sein." (S. 80 und 83). „Im Verlauf der zeitlosen Ozeanzeit waren den Haaren, Haarbürstchen und falschen Füßen der Amöben schließlich Beine, die mit Gelenken versehen sind, gefolgt. Obwohl plump wie bei einer stelzfüßigen Krabbe, konnte man sich mit ihnen doch schon über festen Grund bewegen. Andere der Auserwählten hatten Schwimmblasen entwickelt. . ." (S. 107).

Populärwissenschaftliche Literatur, wie sie nicht sein soll und nicht sein darf. W. SCHÖNBORN

BRIAN E. S. GUNNING & MARTIN W. STEER: Ultrastructure and the Biology of Plant Cells. London: Edward Arnold (Publishers) Ltd. 1975. VIII u. 312 S., 12 Abb., 49 Taf. Ln. £ 27.50

Es ist etwa 12 Jahre her, daß mit Hilfe verfeinerter Methoden der elektronenmikroskopischen Forschung der endgültige Durchbruch in der Biologie gelang. Seither ist die Zahl der einschlägigen Untersuchungen so groß geworden, daß ihre Ergebnisse kaum mehr vollständig überschaubar sind. Auf dem Gebiete der Zellforschung sind wesentliche Fortschritte gemacht worden. Zwar ist das von der Lichtmikroskopie entworfene Bild der Zelle nicht gänzlich verworfen, jedoch in fast jeder Hinsicht ergänzt, verfeinert und vertieft worden.

Der vorliegende Band führt an Hand von mehr als 200 ausgezeichneten, vorwiegend mit dem Transmissionselektronenmikroskop gewonnenen Mikrophotographien die Beziehungen zwischen Struktur und Funktion pflanzlicher Zellen vor. Hier und da dienen licht- und rasterelektronenmikroskopische Mikrophotos als Orientierungshilfe. Zellwand und Plasmamembran, pflanzliche Gewebe, Vakuolen und Tonoplast, Kern und Endoplasmatisches Retikulum, Golgiapparat, Mitochondrien und Plastiden, Microbodies und Sphärosomen, Mikrotubuli und Mikrofilamente, Zellteilung sowie strukturelle und funktionelle Integration der Zellen in Geweben und Organen werden auf 49 Bildtafeln und anhand von etwa 130 biologischen Objekten dargestellt. Dem Bildteil geht ein ausführlicher Textteil voraus, der so gestaltet ist, daß der direkte Bezug zum Tafelteil stets herstellbar ist. Im Tafelteil steht jeweils eine Bildseite einer erläuternden Textseite gegenüber, wobei die Erläuterungen auf die entsprechenden Stellen des Textteiles verweisen.

Die Behandlung des Stoffes selbst steht absolut auf der Höhe der Zeit. Der Text ist übersichtlich gegliedert, klar und einfach geschrieben und kommt mit einem Minimum wissenschaftlicher Termini aus, ohne dabei den Charakter strenger Wissenschaftlichkeit einzubüßen. An botanischen Objekten gewonnene Erkenntnisse werden vielfach mit entsprechenden Ergebnissen zoologischer Provenienz verglichen. Kormophyten und Thallophyten finden vor allem Berücksichtigung, auf Bakterien wird hier und da verwiesen. Insgesamt läßt die Darstellung nur wenige Wünsche offen.

Vielleicht hätte der Bewegungsapparat der Zoosporen, Gameten und der Phytoflagellaten eine Darstellung verdient, das um so mehr, als die Autoren ihr Werk Frau Prof. I. MANTON, einer bedeutenden Förderin der Ultrastrukturforschung und Entdeckerin des „9 + 2-Musters“ der Geißeln, gewidmet haben.

Die Photographien sind durchweg ausgezeichnet, Druck und Ausstattung hervorragend. Die Bibliographie umfaßt rund 940 Einzeltitel. Insgesamt gesehen ein Buch, vorzüglich geeignet, Botaniker und Biologen in die Prinzipien der pflanzlichen Zellbiologie einzuführen. S. J. CASPER

BRIAN, E. S. GUNNING & NARTIN W. STEER: Plant Cell Biology an ultrastructural approach. London: Edward Arnold (Publishers) Ltd. 49 Taf. Brosch. £ 2.75

In gewisser Weise ist der Titel des Buches irreführend, stellt es doch nichts anderes dar, als den „Sonderdruck“ des Tafelteils des von den beiden Autoren verfaßten größeren Werkes „Ultrastructure and the Biology of Plant Cells“. Der veränderte Titel weckt neue Erwartungen. Von einer kurzen, vierseitigen Einführung und einem neuen Bild auf dem Umschlag abgesehen, ist das Heft vollständig mit dem Tafelteil des ebenerwähnten Werkes identisch. Der Verlag begründet sein Vorgehen damit, daß den Lehrern und Studenten der Biologie in Schulen oder auf Universitäten bisher kein Buch zur Verfügung gestanden hätte, daß bei einem vertretbaren Preis eine dem heutigen Stand der Wissenschaft widerspiegelnde illustrierte Darstellung der Ultrastruktur der pflanzlichen Zelle geboten hätte. Das mag richtig sein. Dennoch hätte man den Titel nicht ändern dürfen. Vom Inhalt gilt das, was über das Gesamtwerk gesagt worden ist. Das Heft wird seinen Zweck erfüllen. S. J. CASPER

Beiträge zur Meereskunde, Heft 23 (Hrsg.: ERICH BRUNS unter Mitwirkung von O. KOLP, G. SAGER, K. VOIGT). Berlin: Akademie-Verlag 1969. 94 S., 26 Abb., 10 Tab. Brosch. M 28,—

Das Heft beinhaltet 4 Beiträge aus dem Institut für Meereskunde der AdW der DDR. D. NEHRING, S. SCHULZ & K.-H. ROHDE berichten im Rahmen von Untersuchungen zur Produktivität der Ostsee über stichprobenartige (April/Mai 1967) hydrographische, chemische und primärproduktionsbiologische Messungen in der mittleren Ostsee und in der Boddensee. Es folgt ein interessanter Beitrag von W. MATTHÄUS „Zur Entdeckungsgeschichte des Äquatorialen Unterstromes im Atlantischen Ozean“. Die Arbeiten von G. SAGER über „Amplitudenkurven bei Gezeiten und Surf beats“ und von D. BÖHL über „Fluchtlinientafeln für spezielle Schwebeerscheinungen“ vervollständigen das insgesamt sehr heterogen angelegte Heft. R. KOSCHEL

Beiträge zur Meereskunde, Heft 24/25 (Hrsg.: ERICH BRUNS unter Mitwirkung von O. KOLP, G. SAGER, K. VOIGT). Berlin: Akademie-Verlag 1969. 192 S., 69 Abb., 24 Tab., 20 Taf. Brosch. M 92,—

Das Heft steht unter dem Thema: Meeresgrund- und Küstenforschung im Bereich der Ost- und Nordsee III. Sediment und Benthos. GUDELIS demonstriert Oszillationen der litauischen Küstenlinie im Postglazial, die deutlich im Abklingen begriffen sind. SCHULZ fand 80 Makrozoobenthosarten in der Mecklenburger Bucht, differenziert einzelne Gemeinschaften nach den Sedimenttypen (Sande und Schlicke) und findet eine signifikante Faunenverarmung im Bereich der Lübecker Bucht in den letzten 40 Jahren. Schwierigkeiten bei der Interpretation des gewonnenen Probenmaterials können sehr schön durch autonome Tauchuntersuchungen ausgeglichen werden. Die *Abra-alba*-Zönose erscheint in 3, die *Macoma-baltica*-Zönose in 2 Varianten. KOSLER untersuchte das Makrozoobenthos des höchstens 0,5 m tiefen Vierendehlgrundes bei Hiddensee. Von den 66 festgestellten Arten gehören $^2/_3$ in den marinen Bereich. Die Individuendichten sind hoch. Die Flachwasserzonen besiedeln jedoch nur wenige Arten. Mit zunehmender Tiefe steigt die Artenzahl. NEUMANN & BUBLITZ beschreiben die Zusammensetzung der Sedimentoberfläche im südlichen Ostseebereich zwischen Rügen, der Oderbank und dem Festland.

LANGE untersuchte Ablagerungen, die sich an der Grenze zwischen Kreide und Tertiär auf Seeland gebildet haben und fand 18 Arten von Dinoflagellaten und Hystrichosphaerideen. 3 Sippen werden neu beschrieben. Die kurzen Diagnosen ergänzen nur Photos. BEHRE & MENKE zeigen die pollenanalytische Auswertung eines Bohrkernes von der südlichen Doggerbank. Danach war auf einem Niveau von —46 m NN vor etwa 10000 Jahren noch trockenes Land. Die weiter zurückreichenden Proben entsprechen den Bedingungen, wie wir sie parallel in England vorfinden. Die Absenkung in das Meeresniveau führte zur Erosion der oberen Torfschichten, Überdeckung mit Sand, Ansiedelung von *Mytilus* und weitere Entwicklung zum typischen Meeresboden, wie wir ihn heute kennen. Die Entwicklung lief nicht kontinuierlich ab. PAZOTKA v. LIPINSKI & WIEGANK stellen die Foraminiferen eines Stechprofiles von der Doggerbank vor (Holozän). Der taxonomische Teil enthält Kurzdiagnosen, Maße sowie Angaben zur geschichtlichen wie geographischen Verbreitung. Der ökologische Teil befaßt sich mit der Temperatur- und Salinitätsentwicklung im Gebiet der Doggerbank. SCHMIDT, EYCKE & KOLP demonstrieren ein neu entwickeltes Vibrationsstechrohr in Bau- und Funktionsweise. Abgeschlossen wird das Heft mit einem Bericht über die 3. Arbeitstagung der INQUA-Subkommission „Baltic Shorelines“ in Warnemünde sowie der Erinnerung an Geburtstage von Prof. Dr. NEEF und Dr. HORN. G. MOTHES

Beiträge zur Meereskunde, Heft 26 (Hrsg.: ERICH BRUNS unter Mitwirkung von O. KOLP, G. SAGER, K. VOIGT). Berlin: Akademie-Verlag 1969. 95 S., 26 Abb., 16 Tab. Brosch. M 28,—

Das vorliegende Heft enthält 5 Beiträge. U. KREMSER & H. J. BROSIN stellen Ergebnisse über vertikale Austauschkoeffizienten und RICHARDSONsche Zahlen aus der mittleren Ostsee vor. In einer weiteren Arbeit zur Produktivität der Ostsee (vgl. Rezension Heft 23) werden von W. SCHNESE quantitative Angaben zu den wichtigsten Phytoplanktonarten in der mittleren Ostsee und in der Beltsee im Frühjahr (April/Mai 1967) gemacht. S. SCHULZ berichtet über das Makrobenthos der Mecklenburger Bucht und angrenzender Seegebiete. Es werden Angaben zur Verbreitung und Häufigkeit von insgesamt 75 Arten (Gastropoden, Bivalven, Polychaeten, Priapuliden, Krustazeen, Echinodermen und Tunikaten) vermittelt. H. BERGE & L. BRÜGMANN behandeln „Möglichkeiten zur polarographischen Bestimmung einiger Hauptkomponenten im Meerwasser“ und W. MATTHÄUS teilt in einer umfangreichen Arbeit (37 S.) Ergebnisse experimenteller Untersuchungen der Durchflußkoeffizienten von Rohrpegelanlagen mit. R. KOSCHEL

JOHN GILBERT & RAYMOND LEGGE: Das Große Aquarienbuch. Die schönsten tropischen Süßwasserfische. Deutsche Bearbeitung von GOTTFRIED SCHUBERT. Stuttgart: Verlag Eugen Ulmer 1975 (2. Aufl.). 251 S., 430 Farbfotos u. Zeichnungen. Balacron. DM 68,—

Das verlängerte Wochenende und die dadurch gewonnene Freizeit für die Beschäftigung mit vielen schönen Nebensächlichkeiten unseres Daseins hat der Aqaurienkunde wie der Ornithologie, Mykologie, Philatelie oder dem Sport neue Freunde gewonnen, Freunde, die in einem wissenschaftlichen Hobby ernsthafte, geistige Beschäftigung suchen. Die weite Verbreitung etwa der Zeitschrift „Aquarien und Terrarien“ oder der inzwischen so beliebt gewordenen zweibändigen Aquarienkunde des Urania-Verlages legen davon Zeugnis ab. Es entspricht daher durchaus dem Zuge der Zeit, wenn überall eine reiche Aquarienliteratur, darunter Gutes und weniger Gutes, angeboten wird. Das vorliegende Werk nimmt aber zweifellos innerhalb dieser Literaturgattung eine Sonderstellung ein: Es ist in erster Linie ein prächtiges Bilderbuch, zusammengestellt aus meisterlichen Farbaufnahmen hervorragender englischer Photographen, und spricht zunächst durch seine Schönheit den Aquarianer sowie überhaupt jeden Freund schöner Bilder an. Nun ist allerdings Schönheit allein nicht geeignet, dem Anspruch auf eine geistig befriedigende Freizeitbeschäftigung zu genügen. Die Verfasser haben sich erfolgreich bemüht, den Bildern kurze, aber adäquate Beschreibungen an die Seite zu stellen, so daß die Mehrzahl der gängigen Aquarienfische und auch manche Seltenheit in Bild und Wort vorgestellt werden.

Der systematisch ausgerichtete Teil, der in die Abschnitte Salmler, Eierlegende Zahnkarpfen, Lebendgebärende Zahnkarpfen, Karpfenartige, Labyrinthfische, Buntbarsche, Welse und sonstige Arten gegliedert ist, führt fast 300 der beliebtesten Formen der Teiche, Flüsse und Ströme Asiens, Afrikas und Amerikas an. Er spricht außerdem die biologischen und ökologischen Verhältnisse an und enthält fachkundigen Rat und eine Fülle zuverlässiger Informationen über die Aquarienpflanzen, die Wahl, Vorbereitung, Einrichtung und Gestaltung des Aquariums, die Auswahl der Fische, Futter und Fütterung, Paarung und Zucht, Fischkrankheiten, Aquariendesinfektion und Medikamente. Zweifellos ist es mit Hilfe deses Werkes möglich, die Aquarienfische leicht zu erkennen und zu identifizieren. Besonders hervorgehoben sei, daß konsequent stets die wissenschaftliche Bezeichnung einer jeden Art angeführt wird, der volkstümliche Name dann, wenn einer vorhanden ist. In jedem Falle haben sich die Autoren bemüht, nomenklatorisch den Regeln der Internationalen Codes zu folgen. Dieses Bemühen wird besonders deutlich bei den Aquarienpflanzen, wo die Darstellung im Vergleich mit manchem Wasserpflanzenbuch exaktere Information bietet.

Die Farbaufnahmen sind hervorragend, die drucktechnische Wiedergabe von höchster Qualität, die Gesamtausstattung und der Druck vorzüglich: Insgesamt ein ungewöhnliches Buch, das zwar die bekannten und bewährten Aquarienbücher nicht ersetzen kann, das aber besonders geeignet ist, der Aquarienkunde neue Freunde zuzuführen. S. J. CASPER

MAX VON DEM BORNE: Die Angelfischerei. Hrsg.: WOLFGANG QUINT. 14. vollständig neu bearb. Aufl. Hamburg, Berlin: Verlag Paul Parey 1974. XIX u. 348 S., 401 Abb. im Text u. auf 40 Taf., davon 1 Farbtaf. Lw. DM 38,—

Die „Angelfischerei“, 1875 von M. v. D. BORNE begründet und seitdem vom Parey-Verlag betreut, liegt jetzt in 14. Auflage vor. Die völlig neu bearbeitete Auflage ist zum ersten Male ein Gemeinschaftswerk, an dem 16 Autoren, einschließlich des Herausgebers, beteiligt sind. Obwohl die einzelnen Kapitel und Abschnitte von verschiedenen Autoren bearbeitet werden, bleibt die Einheitlichkeit des Buches durchaus gewahrt. An neuen Abschnitten sind hinzugekommen: Sportfischerei und Gewässerschutz, Angelboote, Wettfischen, Castingsport.

Das Buch umfaßt alle Bereiche des Angelsports. Es ist mit vielen Abbildungen versehen (bemerkenswert die sowohl motivisch als auch technisch hervorragenden Fotos). Sein Hauptteil ist der Angelfischerei im Süßwasser gewidmet. Ein zweiter Teil behandelt das „Angeln im Meer“ und ein dritter Teil, „Verschiedenes“ betitelt, informiert in kurzen Kapiteln über allerlei Wissenswertes für den Angler. Der Schwerpunkt liegt auf der Darstellung der Angeltechnik mit ausführlichen Beschreibungen der Angelgeräte, Köder und Angelmethoden. Dabei werden auch alle für den Angelsport bedeutsamen Fische systematisch beschrieben und wiederum in hervorragenden Schwarz-Weiß-Abbildungen vorgestellt. Im Kapitel „Die Gewässer“ wird dem Angler limnologisches Grundwissen geboten, das ihm die fischereiliche Einordnung des von ihm beangelten Gewässers gestattet und ihm die günstigsten Angelstellen herleitet. Auch wird an dieser Stelle im Rahmen des Angelsports deutlich und energisch auf die Verschmutzung der Gewässer eingegangen. Der Angler wird in den Umweltschutz einbezogen, ein folgerichtiger und aktueller Gedanke! Relativ ausführlich werden die Probleme der Fischhege besprochen. Fischgerechtes Angeln und Fischhege sind die Grundlagen eines gesunden Angelsports. Dabei werden auch kritische Worte über die eigene Zunft gefunden.

Auf die Nennung von Mindestmaßen und Schonzeiten wurde verzichtet, dafür wird ihre Bedeutung für die Fischhege diskutiert und somit dem Angler das Verständnis für diese Festlegungen erschlossen. Die Einhaltung der Mindestmaße und Schonzeiten machen im allgemeinen nicht die Fischhege aus. Die Möglichkeiten eines Gewässers (Nahrungangebot, Einstände usw.) bestimmen Qualität und Dichte der Fischbestände. Dem daraus abgeleiteten Qualitäts- und Dichteziel müssen die Besatz- und Entnahmemaßnahmen angepaßt sein. Diese moderne Auffassung von der Fischhege wird dem Angler aus ökologischen Gesichtspunkten heraus entwickelt. Weitere Abschnitte informieren über „Fischereirechtliches“, „Markierte Fische“, „Altersbestimmungen an Fischen“, „Präparieren von Fischen“ und bringen viele praktische Winke für den Angler. Das Werk schließt mit einem kurzen historischen Überblick über die Sportfischerei. W. SCHÖNBORN

ARTHUR D. HASLER (Hrsg.): Coupling of Land and Water Systems. (Ecological Studies 10). Berlin, Heidelberg, New York: Springer-Verlag 1975. XVII u. 309 S., 95 Abb., 41 Tab. Ln. DM 60,90 (US $ 26.20)

Es ist schon lange kein Geheimnis mehr, daß Ökosysteme trotz eines beachtliches Grades autoregulativer Eigenschaften nicht im „luftleeren Raum schweben“. Ob in der Hierarchie der Ökosysteme hoch oder niedrig „angebunden“, ob „Biosphäre“ oder „Wald“, „See“ oder „Meer“ — überall wetteifern autochthone und allochthone Kräfte miteinander und formen das jeweilige „System“. So untersucht wohl die Limnologie als die Lehre von den Binnengewässern primär entsprechende Ökosysteme wie Quelle, Bach, Fluß, See, Teich, Weiher oder Talsperre, sie sieht sie aber stets im Zusammenhang mit ihrer jeweiligen „Umwelt“, d. h., in ihrer Kopplung bzw. Rückkopplung mit den umgebenden terrestrischen, subterranen oder aerischen Ökosystemen. Der heute in den Industrieländern so gefürchtete Prozeß der rasanten Seeneutrophierung ist ein typisches Produkt der „Einflußnahme von außen" auf den Wasserkörper, das Resultat zunehmender Nährstoffeinschwemmung aus intensiv gedüngten land- oder forstwirtschaftlichen Flächen, begleitet von Erosion und Zufuhr von Abwasser. Unter dem Einfluß der menschlich-gesellschaftlichen Tätigkeit wandelt sich in unserem Beispiel der Prozeß der natürlichen Seenalterung, der natürlichen Eutrophierung, abhängig von Nährstoffreichtum und Größe des Einzugsgebiets im Verhältnis zur Größe des Seebeckens, von Schichtung und Erneuerungszeit des Wassers, in einen mehr oder weniger irreversiblen Prozeß, der ein natürliches „recycling" nicht mehr vollziehen kann. In- und Output-Bilanzen spielen daher eine große Rolle bei der Beurteilung des Gewässerzustandes und bei den Bemühungen um eine effektive Gewässerprognose.

Der vorliegende Band wendet sich dieser Problematik von vielen Seiten und unter Berücksichtigung verschiedener Objekte zu. Gestützt auf das UNESCO-Programm „Man and the Biosphere (MAB), das Internationale Biologische Programm (IBP) und die Internationale Hydrologische Dekade (IHD) brachte die Internationale Vereinigung für Ökologie (INTECOL) die 25 Autoren auf einem Symposium anläßlich des XVIII. Internationalen Limnologischen Kongresses in Leningrad zusammen, dessen Ergebnis sie in 18 Einzelbeiträgen nunmehr vorlegen.

In 12 Kapiteln werden u. a. behandelt: Zusammenhänge zwischen Nährstoff und Hydrologie am Beispiel nordamerikanischer Gewässer; Stickstoff-, Phosphor- und Eisenbilanz in Talsperren der ČSSR; der Einfluß von Mooren auf die Landschaft; der Einfluß der Marschen und Mangroven auf Wasserqualität und -produktivität; Beziehungen zwischen Nahrungsangebot und Verhalten bei Salmoniden; Beziehungen zwischen Fluß und Landschaft am Beispiele von Paraná und Amazonas; Trophiezustand italienischer Seen als Folge menschlicher Tätigkeit; Mineralversorgung tropischer Flüsse und Seen (Uganda) sowie die ökologischen Beziehungen von Land und Uferzone masurischer Seen.

Die meisten Beiträge sind klar gegliedert und gut illustriert; Ausstattung und Druck sind von hohem, die Ökologie-Reihe kennzeichnenden Niveau. Ein Buch, das den Stand der Ökosystemforschung auf einem speziellen Teilgebiet der Limnologie in hervorragender Weise repräsentiert.

S. J. Casper

S. Jost Casper: Grundzüge eines natürlichen Systems der Mikroorganismen. Jena: VEB Gustav Fischer Verl. 1974. 232 S., 42 Abb., 32 Tab., 2 Taf. M 45,— (Sonderpreis für die DDR: M 35,—)

Der Titel dieses Buches kann falsche Vorstellungen wecken. Wer unter Mikroorganismen hauptsächlich Bakterien versteht, wird in seinen Erwartungen ebenso enttäuscht werden wie derjenige Leser, der von einem „System" eine Hierarchie von Klassen, Ordnungen, Familien etc. erwartet.

Das Anliegen dieses eigenwilligen, in gewisser Weise „unbequemen" Buches ist vielmehr, die Großgliederung der eukaryotischen Mikroorganismen (Phytoflagellaten, Algen, Pilze, Protozoen) zu diskutieren; Bakterien und Cyanophyten werden nur kurz und summarisch charakterisiert. Casper wendet sich gegen einen weit verbreiteten, aber durch den gegenwärtigen Kenntnisstand in keiner Weise mehr gerechtfertigten Schematismus, insbesondere gegen die sogenannte Farbensystematik der Algen. Als systematische Kriterien zieht Casper die Organisationsstufen, die Assimilationspigmente und deren Trägerstrukturen, die Reservestoffe, vor allem aber die Begeißelung der Mikroorganismen heran. Eine umfassende und kritische Literaturauswertung (über 700 Zitate) führt zu der Schlußfolgerung, daß die Mannigfaltigkeit der existierenden Merkmalskombinationen („Muster") weit größer ist, als es uns die gängigen Lehr- und Handbücher glauben machen.

Allein die sehr komprimierte, aber durch Tabellen übersichtlich gestaltete Darstellung der Pigmentmuster- und -träger auf 41 S., die aus dem Gestrüpp zahlreicher einander widersprechender und korrigierender Publikationen herausgeschält wurde, macht das Buch lesenswert. Die in diesem Zusammenhang sicherlich interessante Endosymbiontentheorie wird leider von Casper nicht diskutiert, sondern mit einem Nebensatz abgetan. Den Hauptteil bildet eine 87 S. lange Analyse der Geißelapparate, wobei der Feinstruktur der Geißelbasis und -wurzel vorrangige Bedeutung beigemessen wird. Neben der Literatur wurden hier auch eigene Untersuchungen an *Uroglena* und *Codonosiga* verwertet. Im Ergebnis werden 33 Begeißelungsmuster beschrieben und, soweit genauer bekannt, durch Schemazeichnungen illustriert; Belegfotos müssen in der Originalliteratur aufgesucht werden. Neben der Ultrastruktur sind auch Anzahl, Längenverhältnis, Insertionsart,

Dynamik und Beflimmerung der Geißeln durch sogenannte Begeißelungsformeln ausgedrückt, in das „Muster" einbezogen. Diese mehr „klassischen" Geißelmerkmale erscheinen aber neben der Ultrastruktur von sekundärer Bedeutung, z. B. stimmen die verschieden langen, verschieden beflimmerten etc. Geißeln der Chrysophyceen, Oomyceten u. a. im Bau ihrer Basis- und Wurzelregion überein.

Als Konsequenz dieser Erörterungen erfolgt dann die Systemdarstellung auf nur 8 Seiten. Die traditionelle Einteilung der Organismen in das Tier- und Pflanzenreich wird verworfen, stattdessen werden die Reiche der Protokaryota (mit den nicht weiter gegliederten Phyla der Bakterien und Blaualgen) und der Karyota (mit allen eukaryotischen Organismen) unterschieden, wobei in diesen 2 Reichen weniger phyletische Einheiten als zytologische Organisationsstufen gesehen werden. Die Karyota werden in 31 Hauptgruppen oder Phyla und 91 Klassen gegliedert; erstere sowie die da und dort akzeptierten Subphyla bzw. Superclasses werden kurz charakterisiert, die Klassen nur aufgezählt. Die hohe Zahl von Phyla resultiert offenbar aus der Prämisse, daß keine evolutionäre Abwandlung der Geißelstruktur innerhalb eines Phylums möglich ist. Die nicht behandelten Meta-, Para- und Mesozoen sind in den o. g. Zahlen nicht enthalten; ihr Anschluß an irgendwelche Gruppen von Mikroorganismen bleibt offen, während die Cormophyta klar auf das Phylum „Volvocicea" zurückgeführt werden.

Die Pilze präsentieren sich in diesem Systemvorschlag polyphyletisch und unabhängig von den Algen. Die Begeißelungsmuster scheinen es nicht zu gestatten, irgendeine Gruppe der niederen Pilze auf irgendeine Algengruppe zurückzuführen oder umgekehrt. Auf die höheren Pilze, denen Begeißelung und Assimilationspigmente fehlen, lassen sich die systematischen Kriterien des Verf. kaum anwenden. Um so mehr überrascht es, daß CASPER etlichen Pilzgruppen einen sehr hohen systematischen Stellenwert einräumt, wofür er eine Begründung schuldig bleibt: Die geißellosen Pilzgruppen verteilen sich auf die 6 (anscheinend voneinander unabhängigen?) Phyla der Endo-, Zygo-, Asco- und Basidiomyceten, der Rost- und der Brandpilze; die früher als Familie (Tremellaceae), heute meist als Ordnung bewerteten Gallertpilze nehmen in diesem System den Rang einer Superclassis ein; dies alles erscheint befremdlich und unmotiviert, man könnte es als einen Verschleiß systematischer Kategorien bezeichnen.

Befremdlich ist auch die hier angewendete Nomenklatur. CASPER verwendet nämlich „neutrale" Endungen, um nicht die Zugehörigkeit einer Gruppe zum Tier- oder Pflanzenreich zu suggerieren; dadurch kommen Namen wie „Volvocicea" (statt Chlorophyta), „Sclerodermaticea" (statt Basidiomycota) oder „Russulatae" (statt Hymenomycetes) zustande, die das Studium des Buches nicht gerade erleichtern. In der Praxis werden sich derartige Namensformen kaum jemals einbürgern, sie sind wohl auch nicht so gemeint. Algologen und Mykologen betrachten jedenfalls den botanischen Nomenklaturcode als verbindlich.

Insgesamt ist dieses Buch ein interessanter, durch die Berücksichtigung reichen Faktenmaterials wertvoller Beitrag zur Frage der Klassifizierung niederer Organismen. Es weckt Verständnis für die oft ungern akzeptierte Tatsache, daß die „Welt der Mikroben" ein buntes Nebeneinander mannigfacher Organisationstypen ist und nicht etwa nur eine schmale, ungegliederte Wurzel des Tier- und Pflanzenreiches.

Dank gebührt dem Verf. auch für das Vorwort, in welchem das Anliegen biologischer Systematik in klaren Worten, frei von modernistischen Verkrampfungen, umrissen wird. H. KREISEL

FREDERICK COULSTON & FRIEDHELM KORTE (eds.): Environmental Quality and Safety. Global Aspects of Chemistry, Toxicology and Technology as Applied to the Environment. Vol. 2. Stuttgart: Georg Thieme Verlag u. Academic Press, New York, London 1973. XVIII u. 333 S., 64 Abb., 92 Tab. Ln. DM 58,—

Im Rahmen des Umweltschutzes nehmen die Wasserprobleme eine zentrale Stelle ein. Die Hauptzahl der Umweltgifte und Schadstoffe werden im Wasser gefunden, transportiert und gelangen über das Trinkwasser in die Nahrung.

In den 35 Originalarbeiten von EQS Vol. 2 werden häufiger Wassersprobleme angesprochen, als aus den Titeln ersichtlich ist. Speziell mit Wasserproblemen beschäftigt sich C. MENDIA: „Drinking Water and Waste Water Problems". Abwasser, biologische Abbaubarkeit und die Trinkwasserbeeinflussung sind Hauptprobleme dieses Beitrages. Aber auch hygienische und bakteriologische Fragen werden diskutiert. Über die Wirkung von Nitriten im Tierversuch, bei chronischer Verabreichung von Natriumnitrit im Trinkwasser, berichten N. GRUENER & H. I. SHUVAL (Studies on the Toxicology of Nitrites).

T. JAAKKOLA, H. TAKAHASHI & J. K. MIETTIENEN behandeln den „Cadmium Content in Sea Water, Bottom Sediment, Fish, Lichen, and Elk in Finland". Nach einem Überblick über das Kadmiumproblem, unter besonderer Berücksichtigung der Erkrankungen in Japan, folgt eine Untersuchung von Cd-Anreicherungen in Finnland, welche im Zusammenhang mit dem dortigen Zinkbergbau stehen. Die Anreicherungszone und die Anreicherungsmechanismen werden bis in den Finnischen Meerbusen verfolgt.

J. J. WENT beschäftigt sich mit „Power Generation Chemicals". Zur Erzeugung von 1000 MW Elektroenergie werden 40—50 m^3/s Kühlwasser benötigt. Der Energiebedarf verdoppelt sich etwa alle 10 Jahre. Das Luftverschmutzungsproblem kann durch die künftigen Kernkraftwerke gelöst werden. Das Kühlproblem unter Ausnutzung der natürlichen Gewässer ist Inhalt des Beitrages. Selbst bei Einhaltung einer Maximaltemperatur von 30 °C steigern die Kühlwässer die Eutrophierung und schädigen die Ökologie. Bei Flüssen kann die Temperaturbeeinflussung bis zu 100 km nachweisbar sein.

In den weiteren Beiträgen werden Probleme der Insektizide und anderer Gifte in Umwelt und Nahrung, die Luftverschmutzung, das Kunstdünger- und Müllproblem sowie Schutzmaßnahmen von Vereinigungen und Ländern behandelt.

Die Darlegungen geben einen guten Überblick über den Untersuchungsstand und eine fundierte Extrapolation in die Zukunft. Dem Verlag muß für die Aufmachung und den guten Druck gedankt werden.

G. PROFT

FREDERICK COULSTON & FRIEDHELM KORTE (eds.): Environmental Quality and Safety. Global Aspects of Chemistry, Toxicology and Technology as Applied to the Environment. Vol. 3. Stuttgart: Georg Thieme Publishers and Academic Press, New York, London 1974. VIII u. 245 S., 41 Abb., 39 Tab. Ln. DM 58,—

Die Frage „How Safe is Safe" wird im vorliegenden Band EQS Nr. 3 aus verschiedener Sicht beleuchtet: J. S. TURNER: „A Consumer's Viewpoint", O. H. LOWRY: „A Scientist's Viewpoint", W. C. WESCOE: „A Producer's Viewpoint", P. B. HUTT: „A Regulator's Viewpoint".

Das Hauptanliegen der Beiträge ist die Luftverunreinigung (Internat. Symp. on the Establishment of Air Quality Standards, Paris 1972).

P. CHOVIN berichtet über automatische physikalische Methoden zur Bestimmung von Schwefeldioxid, Stickoxiden und Kohlenwasserstoffen und in einem Spezialbeitrag über die Bestimmung des Kohlenstoffmonoxids sowie dessen Wirkung auf den Menschen (Carboxyhämoglobin-Bildung).

Biochemische Diagnosetests werden zur Analyse von Luftverschmutzungsstoffen eingesetzt, spielen aber keine Rolle bei der Festlegung von Luftqualitätsrichtlinien, die Gründe werden von J. C. GAGE diskutiert. Gleichfalls mit biochemisch-enzymatischen Reaktionen beschäftigt sich F. KORTE in „Fundamental Biochemical Aspects of Air Pollution". Aus klinischer Sicht beleuchtet das Problem J. H. KNELSON in seinem Beitrag „The Role of Clinical Research in Establishing Air Quality Criteria and Standard".

Weitere 5 Beiträge sind den Insektiziden, Pestiziden und radioaktiven Substanzen gewidmet. Den Limnologen wird besonders die Arbeit von B. W. HALSTEAD „Marine Biotoxicology" interessieren. Einer kurzen Einteilung in marine Phytotoxikologie und marine Zootoxikologie folgen Klassifikation sowie ausführlichere Beschreibungen der Arten, Giftwirkung und Behandlung.

Die Beiträge weisen ein auserlesenes Niveau auf, der Druck ist hervorragend.

G. PROFT

ALFONS THEODOR CZAJA: Einführung in die praktische Polarisations-Mikroskopie. Zum Gebrauch bei Untersuchungen von Lebensmitteln, Drogen, pflanzlichen Textilfasern und botanischen Objekten. Stuttgart: Gustav Fischer Verlag 1974. 78 Seiten, 12 Abb., 9 Tab., 21 Taf. mit 96 schwarz-weißen u. 10 farb. Mikrofotos. Kart. DM 22,—

Das vielseitige Polarisationsmikroskop wird hier als Hilfsmittel zur Erkennung von pflanzlichen Zellbestandteilen vorgestellt, wie sie in zerkleinertem oder andersartig deformiertem Zustand in Mehlen, Pulvern und Fasern vorliegen und mit anderen Methoden nicht oder nur schwer diagnostiziert werden können.

Vor allem die Strukturen der Zellwände, aber auch einige Zellinhaltsstoffe wie z. B. Stärkekörner, zeigen im polarisierten Licht ein charakteristisches Verhalten.

Wie im Titel vermerkt, liegt das Schwergewicht auf der praktischen Arbeitsanleitung. Sie besteht in der Gegenüberstellung von Beschreibung und Abbildung des polarisationsoptischen Erscheinungsbildes der wichtigsten Getreidemehle, Leguminosenprodukte, Alginate, Früchte, Samen und deren Zerkleinerungsprodukten (Epidermen, Steinzellen, Fasern etc.) sowie einigen ausgewählten botanischen Objekten. Außer pflanzlichen Textilfasern werden auch halb- und vollsynthetische vorgestellt. Im Kapitel der Drogen helfen Bestimmungsschlüssel bei der Diagnose.

Die Grundlagen der Polarisationsmikroskopie werden dem Zweck des Buches entsprechend nur kurz erörtert. Auf 21 S. werden die Erzeugung polarisierten Lichtes, die Doppelbrechung und Interferenz des polarisierten Lichtes als Ursache der zu beobachtenden Erscheinungen, das Instrumentarium und die Untersuchungstechnik beschrieben. Der Benutzer wird hier sicherlich Lehrbücher und Firmendruckschriften als Ergänzung benötigen.

Der Druck und die Ausstattung sind sehr gut. Man wünschte sich eine bessere Koordinierung von Text und Bildern. In dem so farbenfreudigen Gebiet der Mikroskopie wäre im Kapitel Interferenzfarben eine Farbtafel angebracht gewesen, um das Verständnis für diese Erscheinungen zu fördern. Botaniker, Pharmazeuten, Lebensmittel- und Textilfachleute können dieses Buch und die beschriebene Methodik zu ihrem Vorteil nutzen. CHR. SCHÖNFELD

OTAKAR HASIK: Vodohospodářská výstavba a životni prostředi člověka (Wasserbau und Umwelt des Menschen). — Praha: Academia 1974. 381 S., 109 Abb., 24 Tab. Geb. Kčs 65,— [tschech.; dtsch. u. russ. Zusammenfassung]

Das Buch ist für Wasserwirtschaftler und all jene bestimmt, die sich mit Fragen des Umwelt- und Naturschutzes, Gesundheitsschutzes, der Territorialplanung und des Städtebaus befassen. Der Autor betrachtet die Wasserbauten vom ökologischen Gesichtspunkt und untersucht ihre Beziehungen zur Natur, zu Siedlungen und Städten, zur Industrie und zu den Erholungsstätten. Das Inhaltsverzeichnis mag die Vielfalt der Thematik darstellen: 1. Einführung, 2. Grundlagen und Definitionen, 3. Wasserbauten und Umwelt, 4. Konzeption und Ziele der Publikation, 5. Staudämme, Staubecken und Umwelt, 6. Teiche umd Umwelt, 7. Zierwasserflächen, Zierbauten und Kunstwerke, 8. Regulierungen von Wasserläufen, Wehre und Umwelt, 9. Binnenwasserwege und Umwelt, 10. Melioration und Umwelt, 11. Badeanstalten und Umwelt, 12. Wasserversorgung und Umwelt, 13. Kanalisation und Reinigung der Abwässer und Umwelt, 14. Schlußfolgerungen.

Den Band beschließen eine deutsche und eine russische Zusammenfassung (jeweils die Übersetzung von Kapitel 14), ein Literaturverzeichnis und ein Sachregister. Knapp ein Drittel aller Abbildungen, die den Text gut unterstützen, stammen aus dem Ausland. Die ca. 70 Photos (meist vom Verfasser) hätten beim Druck auf Kunstdruckpapier sicher an Aussagekraft gewonnen. Beim 235 Titel umfassenden Literaturverzeichnis fällt auf, daß die ausländische Literatur wenig berücksichtigt ist (nur ein Fünftel der Zitate).

Das Buch bringt sowohl negative als auch positive Beispiele aus Vergangenheit (u. a. aus dem Römischen Reich) und Gegenwart der wasserbaulichen Praxis mit Ausblicken bis zum Jahr 2000. Sein Autor ist Ingenieur und Kandidat der Wissenschaften; er appelliert an alle Verantwortlichen, mit sämtlichen Wasserbauten einschließlich Wasserversorgungen und Abwasserbehandlung einen Beitrag zum Umweltschutz zu leisten, um ein gesundes Gleichgewicht zwischen den Anforderungen des Menschen und der Erhaltung der natürlichen Ökosysteme herzustellen bzw. wiederzugewinnen.

D. STORCH

RUPERT RIEDL: Die Ordnung des Lebendigen. Systembedingungen der Evolution. Hamburg, Berlin: Verlag Paul Parey 1975. 372 S., 317 Abb., 7 Tab. Geb. DM 98,—

Es gilt heute vielfach als „unerfreulich“, dem Leser eines Buches „Mühen zu bereiten“, d. h. zum Denken oder besser Nachdenken zu zwingen. Beliebter sind Bücher, die Althergebrachtes in leicht eingänglicher, gefälliger Form verpackt bieten und optisch durch „schöne“, oft auch, dem Zuge der Zeit entsprechend, „plakative“ Abbildungen auffallen. Der Autor aber will die Mitarbeit des Lesers, und die zahlreichen, gutgelungenen und oft die Sachlage wirklich erhellenden Strich-Strichzeichnungen (etwa die Figuren I1, VI1 oder VII1) dienen eben zuallererst der „Sache“ und dann dem (ästhetischen) „Wohlbefinden“ des Betrachters.

Riedl will mehr als das bekannte und anerkannte Gebäude der Evolutionstheorie vor uns aufbauen. Zwar zweifelt er nicht an dem fundamentalen Erklärungswert der Mechanismen der Selektion, Mutation und Populationsdynamik für den Ablauf der Evolution, erkennt folglich die Bedeutung der „Synthetischen Theorie" an, fragt aber nach dem „ordnenden" Mechanismus der transspezifischen Evolution. Der ordnende Mechanismus „ist jener kausale Zusammenhang, der die Gesetzmäßigkeiten der Makroevolution sowie der geordneten, vorhersehbaren Mannigfaltigkeit der organischen Gestalten zur notwendigen Folge hat; ja der die Ursache dafür ist, daß das Lebendige nicht ein unbeschreibliches Wirrwar, sondern eine beschreibbare Ordnung bildet, welcher selbst von den Mustern unseres Denkens ... entsprochen wird". Schon aus diesem Zitat wird deutlich, daß die Darstellung tief in erkenntnistheoretische Probleme eindringt. Sie geht daher folgerichtig im 1. Kapitel vom Erkennen von Gesetz und Ordnung aus, beschreibt im folgenden Abschnitt die Ausmaße und Formen der Ordnung des Lebendigen (u. a. Ordnung als Energie, als Unwahrscheinlichkeit der Zustände und als Umfang möglicher Voraussagen; Homologie, Ähnlichkeitsgrad, Ordnungsmuster) und untersucht im 3. Kapitel die molekularen Ursachen der Ordnungsmuster. Hier wird nachgewiesen, daß es vier Schaltmuster in den Gen-Wechselwirkungen gibt, denen auf der biologischen Ebene die vier strukturellen Grundmuster Norm (Kapitel IV), Hierarchie (V), Interdependenz (VI) und Tradierung (VII) entsprechen. Sie ermöglichen und kanalisieren, fußend auf einem Rückkoppelungs-Mechanismus, der Gen-Entscheidungen und, Phän-Ereignisse über die Erfolgschancen ihrer Änderungen in wechselseitige Abhängigkeit bringt den Aufbau der Ordnung des Lebendigen. Trend, Orthogenese und Typostasie, Biogenesegesetz und Homöostase, Homologie und Typus sowie die Realität der Systemgruppen und des Natürlichen Systems erscheinen als Fälle eines gemeinsamen Kausalzusammenhanges. Morphologie und Systematik, heute oft als, weil „nicht experimentell", Wissenschaft abgeschrieben, finden ihre kausale Grundlage. Das Denken projiziert nicht die Ordnung in die lebendige Welt; diese Ordnung ist ihr vielmehr immanent, und das Denken von Ordnung erscheint als Ausfluß ihrer Realität.

Die Evolution der Ordnung beruht auf einer Strategie der notwendigen Gesetzmäßigkeit gegen die Wahllosigkeit des Zufalls; sie stabilisiert sich selbst nach Selektionsgesetzen der „Selbst-Zielsetzung", des „self-design". Die Evolution ist folglich in hohem Maße dem Zufall entzogen, dies als notwendige Folge einer Selektion, die nicht nur von den Umweltbedingungen, sondern auch von den funktionellen Systembedingungen in der Organisation der Organismen selbst diktiert wird.

Mutation und Selektion lenken in ein System von Wechselabhängigkeit ein. Die stete Folge der ungleichen Chancen von Zufall und Notwendigkeit führte zur Determination. Die Evolutionstheorie Riedls ist folglich Systemtheorie und Determinationstheorie (im Sinne von Vorhersehbarkeit) zugleich.

Es ist gar nicht möglich, im einzelnen auf die Fülle der behandelten Probleme und Beweisführungen einzugehen. Festzuhalten bleibt, daß diese erweiterte, streng naturwissenschaftliche Konzeption der Evolutionstheorie geeignet erscheint, der Forschung neue Impulse zu verleihen und die „feindlichen Brüder" (Neo-Darwinisten und Neo-Lamarckisten) wieder zum gemeinsamen Gespräch zu bringen. Auch wenn das Buch sein Material vorwiegend der Zoologie entnimmt — schließlich ist der Verfasser Vorstand des 1. Zoologischen Instituts der Universität Wien —, so wird es auch der Botaniker nicht nur mit Gewinn lesen, sondern — hoffentlich auch — seine Gedankengänge nachzuvollziehen suchen. Ein außerordentliches, weil neue Wege erschließendes Buch, für das Autor und Verlag unser Dank gilt. S. J. Casper

Eclogae Geologicae Helvetiae Vol. 63 (1): Geochronology of Phanerozoic Orogenic Belts. (Hrsg.: Schweizerische Geologische Gesellschaft). Basel: Birkhäuser Verlag 1970. 388 S., 156 Abb., 71 Tab. Geh. sFrs. 80,—

Der vorliegende Band 63/1 von Eclogae Geologicae Helvetiae veröffentlich die Vorträge „Kolloquium on the Geochronology of Phanerozoic Belt" vom 23. 8.—4. 9. 1969 in Bern und Zürich.

Neben zusammenfassenden Beiträgen wie z. B. Afanass'yev „Certain Key Data for the Phanerozoic Time Scala" und Afanass'yev et. al. „Summarized Analyses of Standards" befaßt sich der Hauptteil der Arbeiten mit K-Ar und Rb/Sr-Altersbestimmung von Graniten und Granitmineralien. In methodischer Sicht sind besonders die Arbeiten über Argonverluste durch Diffusion und Erwärmung in Metamorphiten und die Anwendung der Fission-track-Methode auch auf Mineralien der Sedimente erwähnenswert.

Durch die internationale Zusammensetzung des Kolloquiums wird nicht nur eine Datensammlung über Altersbestimmungen an Gesteinen rund um unseren Globus geboten, sondern auch ein guter Überblick über den derzeitigen Stand der absoluten Altersbestimmung. G. PROFT

Dr. H. G. BRONNS Klassen und Ordnungen des Tierreichs. V. Band: Arthropoda, I. Abteilung: Crustacea. Hrsg. von H.-E. GRUNER. 2. Buch, IV. Teil, Ostracoda, 4. Lieferung. Bearb. von GERD HARTMANN. Jena: VEB Gustav Fischer Verlag 1975. Seiten 569–786, 101 Abb., 7 Tab. Brosch. M 62,–

Die vorliegende 4. Lieferung des Ostrakodenbuches in BRONNS traditionsreichem Werk, „Klassen und Ordnungen des Tierreiches" behandelt die Ökologie und Biologie sowie einen Teil der Systematik dieser arten- und formenreichen, auch paläontologisch bedeutsamen Tiergruppe.

Im Kapitel Ökologie bespricht der als hervorragender Kenner der Tiergruppe ausgewiesene Autor die verschiedenen, von Ostrakoden besiedelten Lebensräume und die in ihnen auftretenden morphologischen Anpassungen. Das Zusammenspiel von Form und Funktion bei der Einpassung in den Lebensraum wird dadurch eindrucksvoll belegt. Bei den limnischen und Brackwasser-Ostrakoden wird auch die ökologische Bedeutung abiotischer Faktoren kurz beleuchtet.

Der Abschnitt Biologie behandelt erschöpfend die Bewegungsformen, Sinnesleistungen, Fortpflanzung, Periodizität und Lebensdauer, Schutzmittel, Nahrung und Nahrungsaufnahme sowie die wechselseitigen ökologischen Beziehungen der Ostrakoden zu anderen Organismen. Der Stoff wird in beiden Kapiteln im allgemeinen nur bis zu den Gattungen hin dargeboten, was aber bei dem gegenwärtigen Kenntnisstand kein Mangel ist. Dem Leser werden dadurch vielmehr die Kenntnislücken bewußt, auf die der Autor an verschiedenen Stellen überdies ausdrücklich hinweist. Es schließen sich knappe, aber sehr aufschlußreiche Ausführungen über die große ökonomische Bedeutung an, die fossile Ostrakoden bei der geologischen Erkundung von Erdöllagerstätten erlangt haben.

Der umfangreichste Teil der Lieferung befaßt sich mit dem System der Ostrakoden, das nach dem neuesten Stand der Kenntnisse unter Berücksichtigung der bislang bekannten natürlichen Verwandtschaftsverhältnisse abgehandelt wird. Veränderungen ergeben sich dabei besonders in den Gruppierungen von Familien, Unterfamilien und Tribus. Die ausgezeichnete Darstellung umfaßt die Ordnungen Beyrichiida, Myodocopida und von den Podocopida im wesentlichen die umfangreiche Familie der Cytheridae. Rein fossile Ordnungen sind in dem hier vorgelegten System nicht näher berücksichtigt. Klare und übersichtliche Bestimmungsschlüssel, die sowohl Merkmale der Schale als auch des Weichkörpers berücksichtigen, führen bis zu den Unterfamilien. Die höheren systematischen Einheiten (bis zur Tribus) werden überdies ausführlich diagnostiziert. Listenmäßig angeführt sind sodann sämtliche (auch fossile) Gattungen mit kurzen Angaben von Verbreitung, Vorkommen und Zahl der bekannten Arten. Auf unvollkommen beschriebene Gattungen wird hingewiesen. Wo angebracht, finden sich kritische Ausführungen zu den gegenwärtigen Tendenzen im Ausbau des Systems, der vor allem durch die Tätigkeit der Paläontologen nahezu ins Uferlose getrieben worden ist. Rezent- und Fossilsystematik drohen so mehr und mehr auseinander zu fallen, wodurch die Verwirrung groß ist. HARTMANNS vorsichtig abwägendes Urteil in diesen schwierigen Fragen will zur Einsicht mahnen und sollte auch von den Paläontologen gründlich überdacht werden. Zahlreiche klare Strichzeichnungen führen neben morphologischen Einzelmerkmalen die besondere Formenvielfalt der Ostrakodenschale vor Augen. Alles in allem liegt mit dieser Bearbeitung ein Standardwerk vor, das sich durch straffe Gliederung, klaren, flüssigen Stil und umfassende Information auszeichnet. Für alle an der Thematik Interessierten bildet es ein verläßliches, unentbehrliches Arbeitsmittel. D. FLÖSSNER

CLIFFORD GROBSTEIN: The strategy of life. 2nd ed. (A series of books in biology, editors: DONALD KENNEDY & RODERIC B. PARK). San Francisco: W. H. Freeman and Company 1974. 174 S., 90 Abb., Geb. $ 4.95

Der Autor versucht sozusagen als „extraterrestrischer Beobachter" mit dem Wissen unserer modernen Biologie das Leben auf der Erde zu schildern, wie ein Entdecker eine neue Insel. Er sieht das Leben, zusammengesetzt aus Materie und als Teil des Universums. Er findet natürlich keine „imaginäre Lebenskraft", jedoch bestimmte Erscheinungen wie Stoffwechsel, Vermehrung, Vererbung usw., die lebende Materie von toter unterscheiden. Solche spezifischen Erscheinungsformen der lebenden Materie wertet er als „Strategie des Lebens“ und verfolgt sie in ihrer zeitlichen Entwicklung und Optimierung. Die den Biologen immer wieder begeisternde Formenmannigfaltigkeit der Organismen faßt er in der „Taktik des Lebens“ zusammen. Die Organismen verwirklichen die Strategie des Lebens. Flagellat und Baum unterscheiden sich nur quantitativ, als Stationen der Entwicklung des Lebens auf der Erde. So gesehen erhält der Mensch in seiner gesellschaftlichen Bindung eine Sonderstellung, indem er als einziges Lebewesen begann, selbst planend in die Strategie des Lebens einzugreifen und das Leben im Universum zu verbreiten. Diese Eingriffsmöglichkeit in die Strategie des Lebens regt zum Nachdenken an, ist sie doch nicht so einfach überschaubar wie die Gewalt der Atombombe.

Das Faszinierende an dem Buch sind nicht die Details, sondern die rationale Darstellung der Entwicklung dieser Mechanismen. Letztlich ist es ein kurzer Überblick über die moderne Biologie, die jeden im Detail befangenen Spezialisten beglücken, jeden Lehrer begeistern wird. Aber auch der nur am biologischen Geschehen interessierte vermag dem einfach und spannend geschriebenen Text leicht zu folgen. Angaben der wichtigsten Literatur jeweils am Ende der 11 Kapitel und ein Fachwortindex im Schluß des Buches erleichtern ein weiteres Studium. G. Mothes

Manfred Haider: Leitfaden der Umwelthygiene. Bern, Stuttgart, Wien: Verlag Hans Huber 1974. 209 S., 19 Abb., 19 Tab. Brosch. sfr 32,— (DM 29,—)

Bei der Lektüre dieses Buches wird es einem „gelernten“ Hygieniker warm ums Herz. Der Autor hat es verstanden, die Anliegen der Umwelthygiene — eines der brennenden Probleme unserer Zeit — einem breiten Publikum verständlich zu machen. Dabei handelt es sich nicht etwa um eine populärwissenschaftliche Schrift, darauf angelegt, Laien in dieses Gebiet einzuführen, ihr Interesse zu wecken und sie womöglich zu aktiver Mitarbeit anzuregen. Das Buch wendet sich in erster Linie an Studenten, Ärzte und Techniker, und es kann wohl einem Buch dieser Art kein größeres Lob gezollt werden als dieses, daß es sowohl den Fachmann als auch den mit diesen Fragen nicht unmittelbar Befaßten anspricht, Fakten vermittelt und belehrend zugleich ist.

Das Inhaltsverzeichnis zeigt, daß sämtliche Problemkreise angesprochen sind, die in der Fachvorlesung „Allgemeine Hygiene“ vermittelt werden: Luft, Lärm, Licht, Strahlenhygiene, Klima, Wasser, Boden, Abwasser, Lebensmittel und Umweltgestaltung.

Ein gründliches Studium dieses vom Umfang her kleinen Buches gibt eine sachlich Basis für den Studenten der Medizin und der Biologie, aber auch für technische Berufe, die mit umwelthygienischen Fragen befaßt sind. Der Arzt wird andererseits mit Gewinn zu diesem Werk greifen, um Vergessenes aufzufrischen oder sich Antwort auf eine aktuelle Frage zu holen.

Der Verf. hat es verstanden, den multidisziplinären Charakter der Umwelthygiene deutlich zu machen. Das ist nicht mehr die Domäne eines Fachgebietes; Erfolg in der Lösung der vielfältigen Probleme verspricht nur die Gemeinschaftsarbeit von Ärzten, Chemikern, Biologen, Physikern, Technikern und vielen anderen.

Zu begrüßen, besonders für den breiteren Leserkreis, sind die Ausführungen über Küchenhygiene (S. 177) und Körperpflege (S. 193), da sie in die persönliche Sphäre jedes Menschen eingreifen und bei strikter Einhaltung der Hinweise viele Krankheiten vermeidbar sind. W. Köhler

Frank B. Golley & Ernesto Medina (Hrsg.): Tropical Ecological Systems. Trends in Terrestrial and Aquatic Research (Ecological Studies **11**). Berlin, Heidelberg, New York: Springer-Verlag 1975. XV u. 398 S., 131 Abb., 72 Tab. Ln. DM 60,80 (US $ 26.20)

Mit Unterstützung der Internationalen Gesellschaft für Tropenökologie (ISTE) und der Internationalen Vereinigung für Ökologie (INTECOL) fand 1973 in Caracas, Venezuela, ein Symposium unter der Schirmherrschaft des Venezolanischen Instituts für Wissenschaftliche Forschungen (IVIC), Abteilung für Ökologie, über tropische Ökosysteme statt, bereits das dritte dieser Art in rascher Folge (1971 tagten die Tropenökologen in New Delhi). Diese Tatsache, verbunden mit der immer mehr anwachsenden Zahl von Beiträgen aus den tropischen Ländern selbst, belegt die Bedeutung dieser Forschungsrichtung für die Entwicklungsländer. Die Tropenforschung ist längst nicht mehr eine Domäne der ehemaligen Kolonialländer. Unter den 72 Autoren, die an den 25 Artikeln beteiligt sind, stammen 21 aus Lateinamerika, Afrika oder Indien. 6 Beiträge — die über Physioökologie der Tropen (F. Pannier), über Populationsdynamik (J. R. Rabinovick), über Struktur und Funktion des Ökosystems der tropischen Savanne (M. Lamotte), über Tropenflüsse

(H. SIOLI), über tropische Flußmündungen (R. RODRIGUEZ) und Inselökosysteme (D. MÜLLER-DOMBOIS) — behandeln grundsätzliche Probleme der Tropenökologie in angewandter und theoretischer Sicht. Von den übrigen Untersuchungen seien die über Tierreservate von J. TERBORGH und über land- bzw. forstwirtschaftlich genutzte Böden in Costa Rica besonders genannt. Der Bogen spannt sich von populationsdynamischen Fragestellungen über Analysen der Wälder und Savannen hin bis zu den Binnengewässern, erfaßt also tatsächlich, wenn auch nicht vollständig, terrestrische und aquatische Ökosysteme gleichermaßen.

Den Limnologen interessieren außer den bereits erwähnten Beiträgen von H. SIOLI und G. RODRIGUEZ die Untersuchungen über Produktivität, Biomasse und Populationsdynamik im Amazonas von E. J. FITTKAU et al. sowie über die täglichen Photosynthese-, Respirations- und Transpirationsraten in den Mangroven Südfloridas von E. A. LUGO et al.

Wie alle Bände der Ökologie-Reihe des Verlages zeichnet sich diese „Tropenökologie" durch gediegene Ausstattung und vorzüglichen Druck aus. Zweifellos wieder ein begrüßenswerter Beitrag zur Entwicklung der ökologischen Wissenschaft. S. J. CASPER

BOGDAN STUGREN: Grundlagen der allgemeinen Ökologie. Jena: VEB Gustav Fischer Verlag 1974. 2. unveränd. Aufl., 223 S., 104 Abb., 3 Tab. Ln. M 20,90

Der „STUGREN" ist seit seinem Erscheinen schnell zu einem Begriff geworden, ist er doch das erste zusammenfassende deutschsprachige Ökologiebuch, das nicht einseitig die botanische oder zoologische Seite betrachtet, sondern in einer zusammenfassenden, kurzen, aber prägnanten Form die wichtigen Beziehungen zwischen Statik und Dynamik des Stoffwechsels und Energiehaushaltes der einzelnen Ökosysteme darlegt. Der Schwerpunkt liegt bei terrestrischen Ökosystemen, die limnischen kommen leider etwas zu kurz. Sowohl dem Studenten wie dem Ökologen wird mit diesem Buch eine wichtige Grundlage in die Hand gegeben, die besonders auf die oft vergessenen großen Zusammenhänge innerhalb unserer Biosphäre hinweist. D. RONNEBERGER

ULRICH FÖRSTNER & GERMAN MÜLLER: Schwermetalle in Flüssen und Seen als Ausdruck der Umweltverschmutzung. Berlin, Heidelberg, New York: Springer-Verlag 1974. XIV u. 225 S., 83 Abb., 59 Tab. Geb. DM 36,00

Die Vergiftungen durch Quecksilber und Cadmium in Japan lieferten den Beweis für die Gefährlichkeit von Schwermetallen im Trinkwasser. Leider wissen wir noch viel zu wenig über das Verhalten der Spurenelemente im Wasser, über die Mechanismen ihrer Anreicherung und über ihre Wirkung auf Ökosysteme. Einen Baustein zu diesem Problemkreis stellt das vorliegende Werk dar, das vorwiegend die Untersuchungsergebnisse an Flüssen der BRD aus den letzten Jahren zusammenfaßt. Eine Fülle von Daten wird in übersichtlichen Tabellen und gelungenen graphischen Darstellungen geboten.

Prozesse der Schwermetallanreicherung in Sedimenten, Sediment—Wasser-Wechselwirkung, Bindungsart, Anreicherung in Fischen, anthropogene Herkunft, „background" und weitere Probleme werden in 24 Kapiteln abgehandelt. Besonderes Interesse für die Praxis verdienen die Abschnitte über Eliminierung von Schwermetallen bei der Wasserreinigung und Trinkwasseraufbereitung sowie bei der künstlichen Grundwasseranreicherung durch Uferfiltration. Auch der in vielen Wasseranalysen vernachlässigte, in den Schwebstoffen enthaltene Metallanteil, der durchschnittlich 50% der Gesamtfracht beträgt, wird analysiert. Gegenüber den Flußwasseruntersuchungen tritt der Bereich der Seen und Talsperren zurück. Obwohl die Literaturangaben am Ende des Buches 16 Seiten umfassen, wird keine Vollständigkeit erreicht. Das Buch ist ein wertvoller Beitrag zur Bewältigung der Umweltprobleme unserer Zeit. G. PROFT

JOSEF HOFMANN: Der Teichwirt. Anleitung zur Zucht und Haltung des Karpfens im Haupt- und Nebenbetrieb, einschließlich der Nebenfische. 4., neubearb. u. erw. Aufl. Hamburg, Berlin: Verlag Paul Parey 1975. 312 S., 223 Abb. Lam. DM 36,—

Der bereits 1958 auf Grund von Erfahrungen aus 40 Dienstjahren als Fischereirat in Oberfranken herausgegebene Leitfaden liegt nun in einer um mehr als 200 Seiten erweiterten 4. Auflage vor. Der Verfasser kennt die Entwicklung der Teichbetriebe in allen Einzelheiten und weiß durch seine Beratungstätigkeit, welche Probleme und Schwierigkeiten bei der Einrichtung und

Bewirtschaftung von Teichen auftreten können. In 19 Kapiteln werden Fragen der Karpfenteichwirtschaft, wie Lebensweise, Fütterung und Zucht des Karpfens sowie Düngung, Besetzung und Abfischung der Teiche behandelt, wobei besonders den Ausführungen über Teichbau und Teichpflege ausführliche Darstellungen gewidmet werden, weil das Buch auch dem Kleinteichwirt und dem Angler ein verständlicher Ratgeber sein soll. In knapper Form werden die wichtigsten Regeln der Überwinterung, des Fischtransportes, der Verhütung und Bekämpfung von Fischkrankheiten und der Fischsterben durch Abwässer behandelt. Das Buch vermittelt in ausführlicher und leicht verständlicher Weise Grundwissen für die Anlage und Bewirtschaftung von Karpfenteichen, wie sie im ehemaligen Aufgabengebiet des Verfassers in großer Zahl verbreitet sind. Auf die produktionsbiologischen und ökonomischen Probleme einer modernen, intensiv betriebenen Großteichwirtschaft nach Art der Anlagen in der DDR geht es nicht ein. H. Müller

Lorenz Magaard & Gerhard Rheinheimer: Meereskunde der Ostsee (Hochschultext). Berlin, Heidelberg, New York: Springer-Verlag 1974. VII + 269 S., 130 Abb., 30 Tab. Geh. DM 39,80

Das Buch, aus dem Manuskript einer Ringvorlesung, die an der Universität Kiel zum obigen Thema gehalten wurde, hervorgegangen, zeigt in 21 Kapiteln an vielen Beispielen und unter verschiedenen Gesichtspunkten, wie sich das „Ökosystem" Ostsee bis zum augenblicklichen Zustand entwickelt hat, erklärt die allgemeinen Gesetzmäßigkeiten in einem brackigen Randmeer ebenso wie ihre individuellen Besonderheiten. Gleichzeitig erhält der Leser einen Einblick in die Durchführung moderner Meeresforschung.

40% des Umfangs behandeln physiographische Aspekte wie Entstehungsgeschichte, Relief, Meteorologie, Wasserhaushalt, Strömungen, Temperatur, Salzgehalt, Eisverhältnisse, Wasserstände, Seegang, Seiches, chemische Parameter. 50% des Umfangs beschreiben Vorkommen und Leistungen der Organismen, zusammengefaßt in Plankton, Phytobenthos, Bakterien und Pilze, Zoobenthos, Fische, und geben einen Einblick in die komplizierten Wechselwirkungen innerhalb des Ökosystems. Abgerundet wird das Werk durch die Analyse zivilisatorischer Beeinflussungen (z. B. Fischerei und Verschmutzung), durch die Geschichte der internationalen Meeresforschung im Ostseeraum und durch ein Sachverzeichnis.

Die Autoren sind: F. Defant, G. Dietrich, F. Gessner, K. Grasshoff, G. Hatje, G. Hempel, R. Köster, W. Krauss, J. Krey, L. Magaard, W. Nellen, G. Rheinheimer, G. Siedler, C. Schlieper, F. Schott, W. Schramm, H. Schwenke, H. Theede und F. Thurow.

Kleine Mängel wie die vielleicht etwas unglückliche Kürzung des Artikels Meteorologie, die etwas kurze Darstellung der wichtigsten Pflanzennährstoffe und die wenig konkrete Darstellung der Wasserverschmutzung, Schreibfehler wie *Limnea* (Gattung: *Lymnaea*, jedoch Leitform *Radix peregra ovata*, nicht *Limnea ovata*), *Litorina* (*Littorina*) oder die (scheinbaren) Definitionsschwierigkeiten der Benthosvegetation (Makrophytobenthos) schmälern den Wert des Buches ebenso wenig wie die überwiegende Verwendung von Anschauungsmaterial aus der westlichen Hälfte der Ostsee. Das Anliegen des Buches richtet sich ja weniger auf die Darstellung des Verteilungsmusters physikalischer und chemischer Parameter, sondern will das Verständnis für die Entstehung dieser Muster, von den verschiedenen Umweltsituationen induziert, wecken. Manche Frage bleibt offen, wo die Forschung noch nach Antwort sucht: „Sichere" Korrelationen wie z. B. die Abhängigkeit der Sauerstoffzehrung und Schwefelwasserstofffreisetzung in den Senken von der zunehmenden Verschmutzung der Ostsee werden infrage gestellt (Salzgehaltszunahme in den letzten Dezennien ?).

Nach der Lektüre des Buches sieht man vor allem die Ostsee, aber auch weitere Brackwassergebiete mit anderen Augen, mit gewachsenem Verständnis an. Der Stoff ist streng sachbezogen, aber leicht leserlich geschrieben. G. Mothes

Gunther Franke: Nutzpflanzen der Tropen und Subtropen Band I. (2. erw. u. überarb. Aufl.). Leipzig S. Hirzel Verlag 1975. 468 S., 133 Abb., 138 Tab. Ln. M 55,–

Die zunehmende Anzahl von Fachleuten aus der DDR, die in den jungen Nationalstaaten der tropischen und subtropischen Klimazonen mithelfen, die landwirtschaftliche Produktion zu entwickeln, bedürfen genau so wie Studenten und andere Interessenten des notwendigen wissenschaftlichen Rüstzeuges. Das nunmehr in neuer, wesentlich verbesserter Auflage vorliegende bewährte Lehrbuch vermittelt in seinem 1. Band Kenntnisse über die wirtschaftliche Bedeutung, betriebsökonomische Probleme, Anzucht, Vermehrung, Anbau, Düngung, Pflege, Bekämpfung von Krankheitserregern und Schädlingen, Ernte und Erträge, Aufbereitung und Verwertung, Heimat, Verbreitung und Botanik wichtiger Nutzpflanzen. Der Herausgeber und seine Mitarbeiter H. Bruchholz, G. Fröhlich, W. Hain, H.-J. Heynoldt, W. Husz, K. Kaletta, S. Michaelis und A. Pfeiffer bearbeiten, ihrer jeweiligen Kompetenz entsprechend, Teilabschnitte der weitgesteckten Thematik, die genußmittel-, kautschuk-, gummi-, öl-, fett- und zuckerliefernde Pflanzen sowie Knollen- und Wurzelfrüchte umfaßt. Im einzelnen werden Kakao, Kaffee, Tee, Tabak, Parakautschuk, Guttaperchabäume, Gummiakazien, Ölpalme, Kokospalme, Ölbaum, Soja, Erdnuß, Sesam, Maniok, Yam, Batate, Kartoffel, Taro, Tania, Pfeilwurz, Erdmantel, Blumenrohr, Tahiti Arrowroot, Zuckerrohr, Zuckerrübe, Palmyrapalme, Zuckerpalme, Zuckerahorn und Zuckerhirse vorgestellt. Jedem Kapitel ist eine Literaturübersicht angefügt. Ein Sachverzeichnis erleichtert das Eindringen in die durch anschauliche Abbildungen und aufschlußreiche Tabellen wirkungsvoll unterstützte, klare Darstellung, die ihrem Anliegen vollauf gerecht wird. Vielleicht hätten einige der die taxonomischen Fragen der Botanik betreffenden Abschnitte etwas prägnanter gefaßt werden können. Insgesamt ein ausgezeichnetes, wertvolles, gut ausgestattetes und gedrucktes Buch.

S. J. Casper

Rudolf Heitefuss: Pflanzenschutz — Grundlagen der praktischen Phytomedizin. Stuttgart: Georg Thieme Verlag 1975. 270 S. 74 Abb., 23 Tab. Flex. Taschenbuch. DM 18,80

Wohl noch niemals stand der Pflanzenschutz so im Blickpunkt des öffentlichen Interesses wie heute. Dies ist ohne Zweifel eine Folge der stürmischen Entwicklung dieses Fachgebietes in den letzten Jahrzehnten, insbesondere der starken Zunahme der Anwendung chemischer Präparate in der Landwirtschaft und im Seuchenschutz sowie der daraus resultierenden Besorgnisse, die nicht selten den völligen Verzicht auf chemische Pflanzenschutzmittel fordern. Moderne Darstellungen von kompetenter Seite, die diesen Umständen Rechnung tragen, sind daher derzeit dringend erforderlich. Der Verfasser, Direktor des Instituts für Pflanzenpathologie und Pflanzenschutz der Universität Göttingen, hat sich dieser Aufgabe erfolgreich unterzogen.

In einem kurzen einführenden Kapitel (14 S.) werden zunächst Notwendigkeit und Bedeutung sowie die betriebswirtschaftliche Stellung des Pflanzenschutzes abgehandelt. Für den Leser aus der DDR besitzt dieser Teil wegen seiner fast ausschließlichen Orientierung auf kapitalistische Produktionsverhältnisse nur eine begrenzte Aussagefähigkeit. Daran anschließend erfolgt eine Einführung in die Gesetzmäßigkeiten der Schaderregervermehrung (Populationsdynamik, Epidemiologie, Massenwechsel, Gradationswahrscheinlichkeit u. a.) und darauf aufbauend die Darstellung der Befalls- und Schadensprognosemethoden unter Einbeziehung mathematischer Modellansätze (20 S.). Sodann bespricht der Autor die heute wieder stärker in der Diskussion befindlichen Möglichkeiten des vorbeugenden Pflanzenschutzes (über 30 S.). Dabei geht er zunächst auf den Einfluß verschiedener Kulturverfahren, Standortwahl, Fruchtfolge, Düngung, Sortenmehl, Ernteverfahren auf die Verbreitung von Schädlingen, Unkräutern und Krankheiten ein und diskutiert die sich daraus ergebenden Vorbeugemaßnahmen, wobei aber auch klar deren Grenzen unter den Bedingungen der modernen landwirtschaftlichen Produktion herausgearbeitet werden (z. B. verringerte Bedeutung von Fruchtfolge, mechanischen Pflegemaßnahmen, Stallmist usw.).

Der Hauptteil des Buches (über 130 S.) ist den direkten Bekämpfungsmaßnahmen gewidmet. Dabei werden die physikalischen Methoden sowie Bodenentseuchungs- bzw. Saatgutbehandlungsverfahren verhältnismäßig kurz abgehandelt, während der chemische Pflanzenschutz entsprechend seiner derzeitigen Bedeutung völlig zu Recht im Mittelpunkt der Betrachtungen steht. Einleitend werden hier die an ein Pflanzenschutzmittel — auch hinsichtlich Toxikologie und Umweltbelastung — zu stellenden Anforderungen, Wirkungswege, Selektivität, Zubereitungsformen, Verteilungsverhältnisse und Ausbringungsverfahren diskutiert und daran anschließend die wichtigsten Wirkstoffgruppen bzw. Vertreter der Fungizide, Insektizide, Akarizide. Nematizide und Herbizide jeweils in der Reihenfolge Wirkungsmechanismus, Anwendungsmöglichkeiten u. -grenzen sowie toxikologische Daten abgehandelt, wobei Formelbilder die Anschaulichkeit erhöhen. Bei den Herbiziden erfolgt noch eine gesonderte Besprechung der Selektivitätsursachen, der Abhängigkeit ihrer Wirksamkeit von Umweltfaktoren sowie ihres Verhaltens im Boden. Hier wäre es nach

Meinung des Rezensenten didaktisch richtiger, den Abschnitt „Inaktivierung und Abbau der Herbizide im Boden“ (S. 156) mit im Abschnitt „Abhängigkeit der Herbizidwirkung von den Umweltfaktoren“ (S. 153) abzuhandeln, in dem unter Punkt 1 der Boden als wirkungsbeeinflussender Faktor ohnehin bereits aufgeführt ist.

Ausführlich bespricht der Autor auch die Konsequenzen, Nebenwirkungen und Nachteile chemischer Pflanzenschutzmaßnahmen, wie sie vor allem bei regelmäßiger und großflächiger Ausbringung breit wirksamer, persistenter Mittel (wie der chlorierten Kohlenwasserstoffe) auftreten können, nämlich Einschränkung der noch verbliebenen Regulationsmöglichkeiten der Biozönose der Kulturlandschaft und die damit verbundene Erhöhung der Gradationswahrscheinlichkeit, Schädlingsresistenz, Rückstandsproblematik, Nahrungskettenanreicherung u. allgemeine Umweltbelastung (18 S.). Für eine spätere Auflage wären diesem Kapitel ausführlichere Angaben über die Auswirkungen der Pestizide (auch der wenig persistenten, aber akut toxischen) auf die freilebende Tierwelt (vor allem auch die Fische u. Gewässerlebewelt) sowie Hinweise über die akute Toxizität und den Umgang mit solchen Mitteln zu wünschen. — Vf. gelangt zu dem Schluß, daß diese Risiken trotz des größeren Stellenwertes anderer Zivilisationsauswirkungen für die Zukunft gewissenhafte Beachtung verdienen. Er untersucht deshalb in 2 weiteren Abschnitten die derzeitigen Möglichkeiten der biotechnischen (physikalische Reize, Lockstoffe, Repellents, Pheromone, Hormone) und biologischen Bekämpfungsmaßnahmen (Erhaltung, Förderung, Neueinbürgerung u. Massenaussetzung von Nützlingen, Anwendung von Bakterien u. Viren, Autozidverfahren) und gelangt dabei zu dem berechtigten Schluß, daß es trotz vielversprechender Versuche und Einzelerfolge nicht möglich ist, chemische Verfahren durch biologisch orientierte Methoden zu ersetzen, sondern daß es darauf ankommt — und das wird im letzten Kapitel aufgezeigt — anstelle von Routinemethoden im Rahmen des integrierten Pflanzenschutzes unter Berücksichtigung von Umweltfaktoren und Populationsdynamik der Schaderreger alle geeigneten (kulturtechnischen, chemischen, biologischen, biotechnischen) Bekämpfungsverfahren in guter Abstimmung im Sinne einer langfristigen Strategie zu kombinieren, um die Schädlingspopulationen unter der Schadensschwelle zu halten. Die damit zusammenhängenden Probleme und die (vielen) noch zu lösenden Fragen werden anhand von Beispielen diskutiert.

Das vorliegende Buch informiert in anschaulicher Form über die Grundlagen und Zusammenhänge des allgemeinen Pflanzenschutzes, wobei auch die Beziehungen zwischen den einzelnen Teilbereichen und zu einer Reihe von angrenzenden Disziplinen, wie Pflanzenernährung, Pflanzenzüchtung, Acker- und Pflanzenbau usw. Berücksichtigung finden. Eine vollständige Behandlung der verschiedenen Verfahren und Präparate sowie des speziellen Pflanzenschutzes war nicht beabsichtigt, entsprechende Einzelheiten müssen weitergehenden Darstellungen entnommen werden. Eine Reihe übersichtlicher Abbildungen, Graphiken und Tabellen erhöhen Anschaulichkeit und Informationsgehalt. Das gilt auch für die den meisten Kapiteln zugeordneten weiterführenden Quellenangaben, die sich allerdings nahezu ausschließlich auf westdeutsche (und z. T. englischsprachige) Veröffentlichungen stützen und die Fachliteratur des sozialistischen Lagers unberücksichtigt lassen (z. B fehlt auch das DDR-Standardwerk von Kurth „Chemische Unkrautbekämpfung“). Der Gesetzesanhang dürfte vorwiegen den Leser aus der BRD interessieren. Einige kleine Fehler in der Orthographie und Zeichensetzung (z. B. S. 42) sollten in einer kommenden Auflage beseitigt werden.

Trotz dieser Einschränkungen dürfte das Buch wegen seiner Klarheit, Übersichtlichkeit und Anschaulichkeit nicht nur als Einführung für Fachleute und Studierende landwirtschaftlicher Disziplinen geeignet sein, sondern darüberhinaus auch Vertretern anderer Wissensgebiete, wie z. B. Biologen und Naturschutzmitarbeitern interessante Informationen über den gegenwärtigen Stand des Pflanzenschutzes bieten. W. Merbach

Der Kaiserstuhl. Gesteine und Pflanzenwelt. (Die Natur- und Landschaftsschutzgebiete Baden-Württembergs 8). Ludwigsburg: Landesstelle für Naturschutz und Landschaftspflege Baden-Württemberg 1974. 241 S., 253 Schwarz-weiß- und 28 Farbabbildungen sowie eine Übersichtskarte auf hinterem Vorsatzblatt und Innenseite. Ln. DM 39,—

Wir hatten schon mehrfach Gelegenheit, in unserer Schriftenreihe auf die hervorragenden Gebietsmonographien der baden-württembergischen Landesstelle für Naturschutz und Landschaftspflege aufmerksam zu machen (vgl. *Limnologica* 1, 552—554, 1963; 5, 277—279, 1968, und 7, 495—496, 1970). Der jetzige Band über den Kaiserstuhl, jenes 100 km² große, lößbedeckte

Vulkangebirge in der Oberrheinischen Tiefebene mit seinen mannigfachen geologischen Bildungen und seiner Fülle an wärmeliebenden Pflanzen- und Tierarten, setzt diese bewährte Tradition unter etwas anderen Aspekten und einer die bisherigen Bände noch übertreffenden Ausstattung fort.

Im Gegensatz zu früheren Bänden wendet sich das großformatige Buch mehr an einen breiteren Leserkreis. Drei ausgezeichnete Kenner dieser einzigartigen Landschaft geben in wissenschaftlich prägnanter, aber dabei doch allgemeinverständlicher Form einen Überblick über Landschaft und Vegetation, Gesteine und Minerale sowie über Naturschutz und Landschaftspflege im Kaiserstuhl (die Darstellung auch der Tierwelt scheiterte leider an dem Mangel umfassender moderner Untersuchungen). Einführend schildert OTTILIE WILMANNS zunächst in Form einer gedrängten Übersicht Landschaftsbild und Klima (S. 9—32). WOLFHARD WIMMENAUER behandelt sodann die verwickelte Geologie, die mannigfachen Gesteine und Minerale dieser Vulkanlandschaft und gibt anschließend Hinweise für geologische Exkursionen (S. 33—71). Breiter Raum ist dann der Darstellung der reichhaltigen Vegetation, wiederum von O. WILMANNS, eingeräumt (S.72—206). Eingehend werden die Pflanzengesellschaften der Weinberge, der charakteristischen Löß-Hohlwege und -böschungen, der artenreichen Trockenrasen, der Wälder, der Talböden, der Dörfer, Straßen und Steinbrüche geschildert und abschließend ebenfalls Exkursionsvorschläge unterbreitet. GERHARD FUCHS umreißt dann (S. 207—216) kurz die Fragen des Natur- und Landschaftsschutzes im Kaiserstuhl, der hier durch die Erhaltung der vielen auf bestimmte Standorte spezialisierten Pflanzenarten und überhaupt der typischen Kaiserstuhllandschaft mit ihren ausgedehnten Trockenrasen, den Hohlwegen und Rebterrassen vor besondere Probleme gestellt wird.

Die verbalen Beschreibungen der genannten 3 Autoren begleitet eine Fülle von einzigartigen und auf dem guten Kunstdruckpapier hervorragend zur Wirkung kommenden Fotos von HELGA & KURT RASBACH. Erst durch sie wird dem Leser und vor allem demjenigen, der das Gebiet nicht aus eigener Anschauung kennt, die Schönheit des Kaiserstuhls und die Mannigfaltigkeit seiner Steine und Pflanzen so recht vor Augen geführt.

H.-D. KRAUSCH

BÖRJE KLASVIK: Computerized analysis of stream algae. (Växtekologiska studier 5). Uppsala: Svenska Växtgeografiska Sällskapet 1974. 100 S., 19 Abb., 8 Taf., 31 Tab. Brosch. (Available to scientific libraries and institutes in exchange for their own publications. It may also be purchased from publishers.)

Die Vernachlässigung der benthischen Algen der Fließgewässer in der limnologischen Forschung trifft auch für Schweden zu. B. KLASVIK hat in den Växtekologiska studier (die seit 1972 von der Schwedischen Pflanzengeographischen Gesellschaft herausgegeben werden) nach längerer Zeit wieder eine größere Arbeit über die schwedischen Flußalgen vorgelegt. Untersucht werden der Fluß Verkeån im südlichsten Schweden und zum Vergleich der Kaltisjokk-Bach am nördlichen Polarkreis in Schwedisch-Lappland.

Die mathematische Auswertung der Primärdaten beschränkt sich auf die Diskriminanzanalyse (Zuordnung der Algenpopulationen zu den Flußabschnitten) und die Varianzanalyse (Einfluß von Flußstrecke, Jahreszeit, Folien, Natursteinen, auf die Häufigkeit der Arten). Es wird Stellung genommen zur Methodik (exponierte Folien, in-situ-Untersuchungen), zur Einordnung der Flußabschnitte in das Saprobiensystem und zur Gliederung der Flüsse an Hand der Algenflora (*Vaucheria*- und *Zygnema*-Flüsse nach ISRAELSON 1949).

Sehr ausführlich und mit exakten Angaben wird der Einfluß der Strömungsgeschwindigkeit auf das Wachstum der Algenpopulationen behandelt. Im Gegensatz zur makroskopischen Algenflora sind Schwankungen der Fließgeschwindigkeit im Bereich von 0,3 ··· 0,8 m/s auf die Mikroalgen ohne deutlichen Einfluß. Der Grenzwert, bei dem strömungsliebende Arten eine Wachstumszunahme zeigen, liegt bei 1 m/s.

Die Arbeit von B. KLASVIK ist ein wertvoller Beitrag zur Ökologie der Fließgewässer-Algen.

W. SCHÖNBORN

KLAUS ODENING: Perspektiven der Cercarienforschung. Perspectives of Research on Cercariae (Parasitologische Schriftenreihe 21). Jena: VEB Gustav Fischer Verl. 1971. 205 S., 70 Abb., 11 Tab. Brosch. M 44,30

Unter der Redaktion von K. ODENING werden 26 Einzelbeiträge über Cercarien geboten, die im Anschluß an ein Cercariensymposium in Berlin entstanden (siehe *Limnologica* 7 (1970), 485 bis

488). Der Band enthält 4 Abschnitte: Lebenszyklus, Verwandtschaftsbeziehungen und Phylogenie der Trematoden; Determination, Klassifikation und Nomenklatur; Ökologie, Lebenszyklen und Faunistik; Empfehlungen der Teilnehmer des Symposiums. Ein Namens- und Adressenverzeichnis der Spezialisten schließt den Band ab.

Die Trematoden dürften sich aus Meeresturbellarien entwickelt haben, die sich in Mollusken ansiedelten (Entökie). Weitere Schritte sind die Ausbildung mehrerer Generationen, der Übergang zur Parthogenese bei der parasitischen Generation, Fähigkeit zur Anaerobiose, Einbeziehung des Wirbeltieres als zweiten Wirt. Die Spezialisierung zum 3-Wirte-Zyklus (mit 2. Zwischenwirt) haben nur die höchstentwickelten Trematoden durchgeführt, während die primitivsten die Mollusken als einzigen Wirt nutzen.

Systematische Divergenzen zwischen Cercarien- und Adultensystemen sind noch Provisorien. Sie werden durch neue Ergebnisse schrittweise überwunden.

Eine gute Entwicklung der Cercarien, ein hoher Molluskenbefall ergibt sich bei geringer oder fehlender Wasserbewegung und üppiger Makrophytenvegetation vor allem in den Verlandungsregionen der Gewässer, limnischer wie ozeanischer. Wirtsdichte und Kontinuität der Wirtsfolge spielen ebenso wie die Lebenszyklen der Wirte eine große Rolle. Solche Kenntnisse erlangen wirtschaftliche Bedeutung für die Tierzucht. Die unterschiedliche Temperaturpräferenz der Cercarien ist in Zusammenhang mit dem Wirtswechsel zu sehen. Eine Reihe faunistischer-ökologischer Beiträge schließt diesen Teil ab.

Die Empfehlungen zeigen, wie die Trematodenforschung koordiniert und intensiviert werden muß und welche volkswirtschaftliche Bedeutung ihr zukommt.

Jeder auf diesem Spezialgebiet wirkende Kollege wird an den hier nicht referierbaren Fakten kaum vorübergehen können. Uns Limnologen interessieren vor allem die Arbeit von Černogorenko über die Trematodenfauna in Stauseen der Dneprkaskade mit Forderungen für den Bau von Talsperren hinsichtlich der Minimierung einer Trematodenentwicklung und der Beitrag von Gottschalk mit Hinweisen zur Bewirtschaftung von Fischteichen, die zur Unterbindung parasitärer Entwicklungszyklen an der richtigen Stelle führen. G. Mothes

R. Gorenflot: **Précis de Botanique** (Enseignement supérieur). 1. **Protocaryotes et Thallophytes eucaryotes.** Paris: Doin éditeurs 1975. 184 S., 108 Abb. (bestehend aus 450 Schemata u. 71 Fotos), 39 Tab., 16 z. T. farb. Taf. Brosch. fr 98,—

Im deutschsprachigen Raume hat die Literatur der französischen Botanik seit den Zeiten der Klassiker Jussieu, Lamarck, De Candolle, Brongniart oder Adanson nicht mehr die ihr zweifellos gebührende Rolle gespielt, offensichtlich bedingt durch die vielfach unzureichenden Kenntnisse der französischen Sprache. Der Zugang zur angelsächsischen Literatur war und ist für die meisten leichter zu gewinnen. Dabei haben gerade in der Zeit nach dem Zweiten Weltkrieg die französischen Botaniker wichtige Beiträge auf vielen Teildisziplinen der Pflanzenkunde geleistet. Es sei in diesem Zusammenhang nur an die enzyklopädischen Darstellungen des Pflanzenreiches von Chadefaud und Emberger oder die Analyse der Feinststrukturen im Thallophytenbereich durch die Pariser Schule erinnert. Sind diesen Werken der eine oder andere Wissenschaftler hier und da begegnet, so hat eine Auseinandersetzung mit der französischen Lehrbuch-Literatur bei uns kaum stattgefunden. Der hier vorliegende 1. Band eines Kompendiums der Botanik ist als Unterrichtsmittel gedacht, und zwar für Studenten, aber auch für die Schüler der oberen Klassen der Normalschule. Es geht dem Autor nicht um eine Vollständigkeit anstrebende Vermittlung des vorhandenen Stoffes, sondern um die Herausarbeitung der für das Verständnis der großen allgemeinen Ideen, etwa der Evolution, notwendigen Grundlagen. So finden Themen wie Morphologie und Struktur, Fortpflanzung, Stellung in der Biosphäre, Phylogenie und Systematik eingehende Berücksichtigung.

Das Niveau der Darstellung ist außerordentlich hoch. Jeder Versuch, im Interesse einer angeblich leichteren Verständlichkeit unzulässig zu vereinfachen, ist glücklich vermieden worden. Der Text ist knapp, durchaus sachlich und drucktechnisch einprägsam gestaltet. Eine Fülle tabellarischer, schematischer, graphischer und photographischer Figuren ist so in den Text eingefügt, daß sie als ihm direkt zugehörig angesehen werden kann. 16 zweifarbige schematische Tafeln von Entwicklungszyklen eukaryotischer Thallophyten seien besonders wegen ihres didaktischen Wertes hervorgehoben. Alle Darlegungen zeichnen sich durch hohe Aktualität aus. Die Ergebnisse der elektronenmikroskopischen Forschung und der Biochemie finden gebührende Berücksichtigung und werden organisch dem traditionellen Material eingepaßt.

Der vorliegende Band umfaßt eine Übersicht über die großen Gruppen des Pflanzenreiches, eine ausführliche Analyse des Baues, der Vermehrung, der Ökologie, Phylogenie und Taxonomie der Protokaryoten, ein Sonderkapitel über die Morphologie und den Bau des Thallus bei Algen, Pilzen und Flechten, eine knappe, aber übersichtliche Einführung in die Fortpflanzungsverhältnisse im Thallophytenbereich und eine gründliche Erörterung und Charakterisierung der Rolle und Bedeutung der Thallophyten in der Biosphäre. Den Beschluß bilden Vorstellungen über die Phylogenie und die Systematik der behandelten großen systematischen Einheiten. Ein alphabetischer Index führt zu den erörterten Themen und Sippen.

Der Rezensent muß es sich versagen, auf Einzelheiten einzugehen, so interessant das in diesem oder jenem Falle auch wäre. Daß in vielen Fällen CHADEFAUD Pate gestanden hat, ist angesichts seiner Bedeutung für die französische Botanik nicht verwunderlich und erklärt auch manch möglicherweise subjektivistische Schau. Desungeachtet erscheint der hier eingeschlagene Versuch, Studenten in das immer komplizierter werdende Gebiet der traditionellen Botanik einzuführen, nachahmenswert. Die aufgezeigten Wege sollten bei der Gestaltung des programmierten Studienmaterials an unseren Hochschulen, wie es etwa in der KLOPFERschen „Systematik der Pflanzen" des VEB Gustav Fischer Verlages in Jena vorliegt, auf ihre Eignung hin geprüft werden.

S. J. CASPER

WERNER JACOBS: Taschenlexikon zur Biologie der Insekten, mit besonderer Berücksichtigung mitteleuropäischer Arten. Unter Mitarbeit von MAXIMILIMAN RENNER. Stuttgart: Gustav Fischer Verlag 1974. 635 S., 1145 Abb. Brosch. DM 38,— (Lizenzausgabe im VEB Gustav Fischer Verlag Jena)

Dies Buch ist einzigartig. W. JACOBS ist es gelungen, die Mannigfaltigkeit an Formen und Lebensweisen der Insekten in einem Taschenlexikon darzustellen. Der Text ist klar, knapp und doch umfassend. Eine Fülle von Wissen wurde hier übersichtlich verarbeitet. Die vielen und hervorragenden Abbildungen wurden nach den verschiedensten Vorlagen fast alle vom Verf. selbst gezeichnet. Umfangreiche Kenntnisse, Fleiß und Gewissenhaftigkeit kennzeichnen dies Werk von W. JACOBS.

Zunächst mag man überrascht sein, in einem Lexikon über die Biologie der Insekten hauptsächlich die systematischen Gruppen als Stichworte zu finden. Doch erkennt man bald, daß eine gut gewählte und ausreichende Anzahl auch allgemeinbiologischer Begriffe aufgenommen wurde, unter denen man die entsprechenden Erläuterungen selbst findet oder die zu den Gruppen verweisen, bei denen die betreffende Erscheinung näher besprochen wird.

Im Vordergrund der Ausführungen steht die Lebensweise der Insekten, aber auch die morphologische Charakteristik fehlt nirgends. Unter den Stichworten der großen systematischen Kategorien (z. B. Coleoptera, Käfer) findet man Überblicke über Morphologie, Physiologie gut untersuchter Merkmalskomplexe, Entwicklungsstadien, Anpassungen und Lebenweisen. Einzelheiten werden dann bei der Behandlung der Familien geboten. Beispiel Dytiscidae, Schwimmkäfer: Nach der morphologischen Kennzeichnung folgen Angaben über Anpassungen an das Wasserleben, wichtige Daten, die Artenzahl, Größen, über Lebensformtypen, Physik des Schwimmens, Larven und ihre Lebensweise. Man erfährt etwas über das Wehrsekret von *Dytiscus*, welches das Steroidhormon Cortexon enthält (das auch in der Nebenniere der Wirbeltiere gebildet wird) und auf Fische lähmend wirkt. Schließlich erfolgt eine kurze Vorstellung wichtiger Arten. Erwähnt sei noch, daß in dem Buch die modernen Erkenntnisse der Verhaltensforschung an Insekten ausführliche Berücksichtigung gefunden haben.

W. JACOBS war es nicht mehr vergönnt, den Abschluß seines Manuskripts zu erleben. Sein Schüler M. RENNER hat das Manuskript fertiggestellt und die Herausgabe übernommen.

Dem vom Verlag sehr gut ausgestatteten Taschenlexikon, das auch auf Exkursionen und bei Freilanduntersuchungen mitgeführt werden kann, dürfte ein breiter Leserkreis sicher sein. Zoologen der verschiedensten Disziplinen, Land- und Forstwirtschaftlern, Parasitologen, Lehrern, Studenten und Hobby-Entomologen wird das Buch ein zuverlässiger Ratgeber sein.

W. SCHÖNBORN

WOLFGANG LEITHE: Die Analyse der organischen Verunreinigungen in Trink-, Brauch- und Abwässern. 2. durchges. Aufl. Stuttgart: Wissenschaftliche Verlagsgesellschaft mbH 1975. XII + 175 S., 22 Abb., 26 Tab. Ln. DM 52,—

Nach verhältnismäßig kurzer Zeit liegt die 2. Auflage dieses Buches vor, was auf den Bedarf und die freundliche Aufnahme durch die Fachwelt schließen läßt. Gegenüber der ersten Auflage wurde das Literaturverzeichnis um 40 Titel erweitert und damit auf den neuesten Stand gebracht. Im Text haben sich dadurch nur geringfügige Änderungen ergeben (z. B. beim CSB-Wert, beim BSB und bei den Pestiziden), so daß Umfang und Inhalt beibehalten wurden. Es kann somit auf die Besprechung der ersten Auflage in *Limnologica* **10** (1), S. 211, 1976, verwiesen werden. G. PROFT

KLAUS KLOPFER: Systematik der Pflanzen. Programmiertes Studienmaterial zur Wiederholung und Übung. Jena: VEB Gustav Fischer Verlag 1974. 156 S., 17 Abb. Brosch. M 7,80

Das Büchlein ist zum Selbststudium gedacht und soll der Wiederholung und Festigung der in den entsprechenden Lehrveranstaltungen an der Hochschule erworbenen Kenntnisse dienen. Benutzungshinweise (im Vorwort) und ein Literaturverzeichnis (10 Titel) helfen dem Studierenden bei der Bewältigung der Aufgabe.

Insgesamt umfaßt das Programm 12 Kapitel: Allgemeine Systematik, Bakterien, Blaualgen, Algen, Pilze, Flechten, Moose, Farnpflanzen, Gymnospermen, Allgemeine Merkmale der Angiospermen, Dikotyle und Monokotyle. Die Kapitel sind in Abschnitte (insgesamt 120) gegliedert, die in der Regel in eine Einführung in die Problematik des Stoffes, die Fragen- bzw. Aufgabenstellung, Literaturhinweise und Antwortkontrolle bzw. Lösung der Aufgaben unterteilt sind. Bei der Aufgabenstellung werden Fragen, Lückentexte, Tabellen, Systematische Übersichten, Zuordnungsaufgaben und Zeichnungen berücksichtigt.

Das Material ist übersichtlich ausgebreitet, sachlich, an der ausgewählten Literatur gemessen, einwandfrei dargestellt und didaktisch geschickt aufbereitet. Problematisch erscheinen dem Rezensenten allerdings einige der Lückentexte, vor allem solche, die allgemeine Probleme behandeln (z. B. 1.1., 1.11.). Hier besteht die Gefahr einer begrifflichen Gängelung. S. J. CASPER

KURT KUNZE: So bewirtschaftet man Karpfenteiche. 2. neubearb. u. erw. Aufl. Hamburg, Berlin: Verlag Paul Parey 1975. 147 S., 14 Abb., Linson flex. DM 8,80

Ein Buch, dem man anmerkt, daß es mit Begeisterung geschrieben wurde; jede Zeile in ihm entspringt der eigenen Erfahrung des Autors. Es ist relativ schnell durchgelesen und so recht für den Anfänger geschrieben, dem es nicht nur eine Fülle von Wissen und praktischen Winken, sondern auch Enthusiasmus vermitteln kann.

Es geht dabei nicht um die Anlage großer produktiver Fischgewässer, sondern um Teiche, die beangelt oder die aus Freude an der Fischzucht bewirtschaftet werden sollen.

Der Inhalt des Bändchens sei nur in Stichworten angedeutet: Auswählen des Gewässers oder Geländes zum Anlegen eines Teiches, Bau der Abflußeinrichtung, Abdichten des Teiches, Wasserpflanzen, Bestimmung der Wasserbeschaffenheit, Nährtiere, Fischbesatz (Raubfische, Friedfische), Besatzmenge und biologisches Gleichgewicht, Bewirtschaftung nichtablaßbarer Gewässer, Verwendungszwecke, Ertragsverbesserungen, Pflege des Teiches, Zusatzfütterung, Abangeln, Abfischen, Transport, Größenklassen, Rentabilität, Verhütung von Fischkrankheiten usw.

Die vorliegende 2. Auflage berücksichtigt auch die neuen Erkenntnisse über „Futtermittel, Futterautomaten, chemische Wasserpflanzenbekämpfung, Fischtransport, Absatzmöglichkeiten und Fischgrößen“.

Das Bändchen ist nicht nur dem angehenden „Hobby-Kleinteichwirt“, sondern auch noch dem erfahrenen Praktiker zu empfehlen. W. SCHÖNBORN

WINIFRED PENNINGTON: The History of British Vegetation. Second Edition. London: Hodder & Stoughton Educational 1974. VIII u. 152 S., 32 Abb., 30 Fototaf. Ln. £ 3.75

Die englische Floristik und Vegetationskunde haben seit jeher in der Botanik eine hervorragende Rolle gespielt. Es ist daher leicht zu verstehen, daß Flora und Vegetation Großbritanniens, ihr gegenwärtiges Erscheinungsbild und ihre Geschichte, vergleichsweise sehr gut bekannt sind, nicht zuletzt dank solch klassischer Werke wie TANSLEYS „The British Islands and their Vegetation“ (1939), GODWINS „History of the British Flora“ (1956) oder BURNETTS „Vegetation of Scotland“ (1964). Sie machen die pflanzengeographische Sonderstellung der Britischen Inseln deutlich, die durch ihre — im Vergleich mit dem europäischen Kontinent — Armut an Arten und durch ihre spezifisch atlantische Ausprägung zum Ausdruck kommt. Allerdings kann heute kaum noch von einer natürlichen Vegetation gesprochen werden. Mit Ausnahme der Gipfelregion der Gebirge und einiger weniger küstennaher Bereiche hat der Mensch tief in das Gefüge der Pflanzendecke eingegriffen und sie stark verändert.

Das Buch WINIFRED PENNINGTONS, erschienen in der von J. G. WELLS herausgegebenen „Modern Biology Series“, will diesen Wandel deutlich machen. Ausgehend von der durch die Paläobotanik erschlossenen tertiären Flora Europas führt die Darstellung über die Veränderungen unter dem Einfluß der quartären Glazial- und Interglazial-Epochen zu einer kurzen Charakteristik des gegenwärtigen Erscheinungsbildes der Vegetation, wobei die Einflüsse der neolithischen Revolution und damit der gesellschaftlichen Tätigkeit des Menschen auf die Pflanzendecke eingehende Berücksichtigung finden.

Das handliche Bändchen ist als eine Einführung in die oder besser als eine Hinführung zu den obenerwähnten profunden Untersuchungen TANSLEYS und GODWINS gedacht, entstanden aus Lehrveranstaltungen an der Universität Leicester, die versuchten, die quartäre Paläobotanik und Stratigraphie mit der Taxonomie und Ökologie zu verknüpfen. Daraus erklärt sich auch die Betonung pollenmorphologischer Aspekte, die ihren Ausdruck in einer großen Zahl aussagekräftiger Pollendiagramme finden. Das zuerst 1969 erschienene, 1970 unverändert nachgedruckte Buch liegt hier in neuer, etwas veränderter Form vor, dabei die wesentlichsten jüngsten Forschungsergebnisse berücksichtigend, ohne seinen Charakter als Einführung zu verleugnen. Der übersichtlich gegliederte und leicht lesbare Text ist ausgezeichnet geeignet, den Studenten und den kontinentalen Botaniker mit den Problemen der Vegetations- und Florengeschichte der Britischen Inseln bekanntzumachen. Hinweise auf weiterführende Literatur, ein kurzes Glossarium sowie ein ausreichender Sach- und Sippenindex vervollständigen die vorzüglich gedruckte und durch charakteristische Schwarzweiß-Photographien bereicherte Darstellung, die ihren Zwecke in vorbildlicher Weise erfüllt.

S. J. CASPER

ROBERT H. PARKER: **The Study of Benthic Communities.** A Model and a Review (Elsevier Oceanography Series 9). Amsterdam, Oxford, New York: Elsevier Scientific Publishing Co. 1975. 279 S., 92 Abb., 13 Tab. Lw. Dfl. 78,— (US $ 32.50)

Innerhalb der Elsevier Oceanography Series, die bisher in 8 Bänden erschienen ist und Probleme der modernen Meeresforschung behandelt, hat der nun vorliegende 9. Band das Studium der benthischen Lebensgemeinschaften des Meeres zum Gegenstand. Der Autor, ein erfahrener Ökologe, berichtet über ökologische Untersuchungen, die er in ausgewählten Biotopen der Woods Hole Region über längere Zeit durchführen konnte. Ausgehend von diesen Untersuchungen und in Verbindung mit den dabei gewonnenen Ergebnissen, wird der Versuch gemacht, ein Modell für weitere moderne Forschungsarbeiten auf dem Gebiet der Benthosökologie zu entwickeln. Das gelingt dem Autor auch gut, indem er sich bei der Darstellung seiner Ergebnisse mit zahlreichen methodischen Fragen und wissenschaftlichen Problemen auseinandersetzt. So wird u. a. auf die bei der Verwendung moderner Meßsysteme anfallenden sehr zahlreichen Daten eingegangen. Dabei wird festgestellt, daß das umfangreiche Datenmaterial nur unter Nutzung von Computern aufgearbeitet und verdichtet werden kann, um ein den reellen Bedingungen entsprechendes Abbild der abiotischen Umweltparameter in ihrer gesamten Komplexität zu erhalten. Daneben werden produktionsbiologische und allgemeine ökologische Gesichtspunkte diskutiert. Auch Fragen umweltbedingter Veränderungen innerhalb von benthischen Lebensgemeinschaften und deren Abgrenzung sind Gegenstand einer Betrachtung.

Das Buch gliedert sich in 7 Kapitel. Im ersten Abschnitt, der als Einleitung gedacht ist, wird kurz auf frühere Untersuchungen im ausgewählten Gebiet eingegangen. Das zweite Kapitel behandelt vor allem methodische Fragen der Probenentnahme und der Datensammlung und -verarbeitung. Dem folgt die Darstellung und Diskussion der Ergebnisse der physikalischen und chemischen Messungen (Redox-Potential, Salzgehalt, Temperatur, Strömung, Sedimentqualität u. a.). Die biologischen Untersuchungsergebnisse sind in Kapitel 4 beschrieben. Sehr viel Raum wird dem Vergleich der eigenen Ergebnisse mit denen anderer Untersuchungen gelassen. Dadurch, daß der Autor in einem gesonderten Abschnitt seines Buches auf die Nutzung der „benthic community studies“ unter dem Gesichtspunkt der Meeresnutzung und der damit verbundenen Verschmutzung eingeht, wird die Arbeit aufgewertet. Wenn auch der vorliegende Band auf den ersten Blick den Eindruck einer herkömmlichen umfangreichen wissenschaftlichen Abhandlung vermittelt, kann er jedoch nach näherer Einsichtnahme als ein Standardwerk auf dem Gebiet der Benthosökologie angesehen werden.

G. Arlt

Reinhard Mocek: Wilhelm Roux — Hans Driesch. Zur Geschichte der Entwicklungsphysiologie der Tiere („Entwicklungsmechanik“). (Biographien bedeutender Biologen Bd. 1). Jena: VEB Gustav Fischer Verlag 1974. 229 S., 2 Portraits. Brosch. M 26,—

Die Entwicklungsphysiologie, ihre Vorgeschichte und ihre Begründung durch W. Roux und H. Driesch werden vom Standpunkt des dialektischen Materialismus einer kritischen Analyse unterworfen. Der philosophischen Untersuchung liegt ein eingehendes und sachgerechtes Studium der entsprechenden naturwissenschaftlichen Literatur zugrunde. Hervorzuheben sind das umfangreiche Literaturverzeichnis, die Zusammenstellung der Schriften von Roux und Driesch und ein Verzeichnis der nachgelassenen Briefe von H. Driesch.

Die übliche philosophische Etikettierung von Roux als „Mechanizist“ und Driesch als Vitalist wird als zu simpel abgelehnt. Beide Forscher haben bereits weitgehend „naturdialektische Sachverhalte“ erfaßt, sind jedoch infolge „Unkenntnis des dialektischen Materialismus“ zu einseitigen Auffassungen gekommen.

Dabei wird dargelegt, daß Drieschs Vitalismus nicht mit seinen biotheoretischen Erkenntnissen identisch, sondern diesen nur „aufgestockt“ ist. Auch A. Weismann wird wieder den „progressiven naturwissenschaftlich-materialistischen Traditionen“ zugeordnet.

Sehr ausführlich und für viele Biologen sicherleich neu sind die Ausführungen über die fortschrittliche politisch-liberale Haltung von H. Driesch. Sie bringt uns auch den späten Driesch wieder näher, dessen Parapsychologie dem heutigen Naturwissenschaftler so fern liegt.

W. Schönborn

Journal of Fish Biology. Vol. 1, no. 1, January 1969. London, New York: Academic Press: 1969. 84 S. $ 14.50 (für 4 Hefte)

Die im Oktober 1967 gegründete Fisheries Society of the British Isles bringt im ersten Heft ihrer neuen Zeitschrift. 9 wissenschaftliche Arbeiten, 2 Buchbesprechungen und eine Instruktion für Autoren. Folgende Arbeiten werden abgedruckt:

Brown, V. M., D. H. M. Jordan & B. A. Tiller: Die akute Giftigkeit für Regenbogenforellen durch schwankende Konzentrationen und Mischungen von Ammonium, Phenol und Zink; Kenndy, C. R.: Tubificiden (Oligochaeta) als Nahrung des Hasel (*Leuciscus leuciscus* L.); Maitland, P. S., & C. E. Price: *Urocleidus principalis* (Mizelle, 1936), ein nordamerikanischer monogener Trematode neu für die britischen Inseln, wahrscheinlich mit dem Forellenbarsch *Micropterus salmoides* (Lacépède, 1802) eingeschleppt; Mawdesley-Thomas, L. E.: Furunkulose beim Goldfisch *Carassius auratus* (L.); Banks, J. W., & W. Irvine: Eine Notiz über das Photographieren von Fischschuppen, Kiemendeckeln und Otolithen durch Gebrauch eines Vergrößerers; Cross, D. G.: Wasserunkrautbekämpfung durch Verwendung von Graskarpfen; Mann, R. H. K., & D. R. O. Orr: Eine vorläufige Studie der Nahrungsverwandtschaft von Fischen in Hartwasser- und Weichwasserflüssen im südlichen England; Maitland, P. S.: Ein vorläufiger Bericht über die Kartierung der Verbreitung der Süßwasserfische auf den britischen Inseln. Cragg-Hine, D., & J. W. Jones: Das Wachstum von Hasel *Leuciscus leuciscus* (L.), Plötze *Rutilus rutilus* (L.) und Döbel *Squalius cephalus* (L.) im Willow Bach, Northamptonshire. Außerdem werden die Bücher von F. R. Harden Jones „Fischwanderungen“ und von A. Netboy „Der Atlantische Lachs. Eine aussterbende Art?“ besprochen.

Die Ausstattung der neuen ichthyologischen Zeitschrift bezüglich der Wiedergabe von Text und Abbildungen läßt keine Wünsche offen.

H. Müller

Karl Probst † & Jürgen Lange: Das große Buch der Meeresaquaristik. Stuttgart: Verlag Eugen Ulmer 1975: 219 S., 60 Zeichnungen, 155 Farbfotos. Ln. DM 98,—

In gewissem Sinne ist das vorliegende Werk ein Pendant zu dem im gleichen Verlage erschienenen „Großen Aquarienbuch". Hier wie dort ging es Autoren und Verlag darum, ein schönes und sachlich einwandfreies Buch zu schaffen. Es erhält seinen besonderen Wert durch die Tatsache, daß ein erfahrener Praktiker auf dem Gebiete der Meeresaquaristik sein in 50 Jahren erworbenes Wissen weitergibt. Karl Probst hat aus tropischen und subtropischen Gewässern viele Tiere selbst lebend nach Hause gebracht, er hat ihre Gewohnheiten und Ansprüche, ihre Lebensräume an Ort und Stelle studiert. Seine souveräne Kenntnis der meeresbiologischen Grundlagen erlaubte ihm, ein Werk zu schaffen, daß aquarienfähige Tiere, d. h. also Fische und Wirbellose, aus den tropischen bis kalten Zonen und aus den seichten bis tieferen Regionen behandelt. Leider hat der Autor die Fertigstellung seines Buches nicht mehr erlebt. Jürgen Lange, Kurator am Wilhelma-Aquarium in Stuttgart, hat die Bearbeitung übernommen und dabei die großen Erfahrungen seines Institutes beigesteuert. Es ist ein Buch entstanden, das den Leser bei keiner wichtigen Frage im Stich läßt.

Der Textteil bietet Informationen über die Meerestiere und ihre Lebensräume, ihre Beschaffung und Haltung, außerdem im allgemeinen Teil Hinweise zur Einrichtung eines Meeresaquariums, Angaben über Lichtverhältnisse, Beleuchtung und Temperaturregelung, über Methoden der Wasserpflege und über natürliches und künstliches Meerwasser. Über Meeresalgen und über die Gestaltung spezieller Biotope finden sich bemerkenswerte Darlegungen.

Die Anordnung der erwähnten Tiere folgt streng wissenschaftlichen Prinzipien. Schwämme, Nesseltiere, Gliederwürmer und -füßler, Weichtiere, Kranzfühler, Stacheltiere, Manteltiere und Fische, insgesamt rund 900 Arten, werden beschrieben. Den technischen Apparat vervollständigen Listen der Schauaquarien in Europa (hier vermissen wir das Meeresaquarium von Stralsund), ein Literaturverzeichnis, ein Bildquellennachweis und ein allgemeines Register.

Auch wenn der Aufbau des Buches darauf abgestellt ist, dem Praktiker zu dienen, so beruhen doch alle Ratschläge und Hinweise auf dem Grundsatz, daß es dem Meeresaquarianer vor allem darum gehen muß, sich über die natürlichen Bedingungen an den typischen Lebensstätten der Tiere zu informieren, wenn er die Anforderungen seiner Pfleglinge unter künstlichen Bedingungen befriedigen will. Es ist daher außerordentlich zu begrüßen, daß die Autoren den ökologischen Grundlagen der Meeresaquaristik einen breiten Raum gewidmet haben. Autoren und Verlag haben ein hervorragend ausgestattetes, prachtvoll illustriertes Buch geschaffen, daß sich ohne weiteres mit dem „Großen Aquarienbuch" vergleichen kann und der Meeresaquaristik viele Freunde verschaffen wird.

S. J. Casper

R. Peterson, G. Mountfort & P. A. D. Hollom: Die Vögel Europas. Ein Taschenbuch für Ornithologen und Naturfreunde über alle in Europa lebenden Vögel. Übersetzt und bearb. v. Prof. Dr. Günther Niethammer. 10. völlig neu bearb. und erw. Aufl. — Hamburg, Berlin: Verlag Paul Parey 1973. 443 S., 1780 Abb., davon 830 farbig, einschl. 8 farb. Eiertaf. Ln. DM 30,—

In zahllosen Besprechungen wurde der „Peterson" schon gelobt und gewürdigt. Die Tatsache, daß 19 Jahre nach dem Erscheinen der ersten deutschen Ausgabe schon die 10. Auflage vorliegt, spricht für sich.

Im Hauptteil werden 481 Arten beschrieben und abgebildet, im Kapitel Irrgäste werden 139 Arten und eine sehr abweichende Rasse des Neuntöters beschrieben. Daran schließen sich einige kurze Angaben über Volierenflüchtlinge an.

Dem Buch vorangestellt sind eine Einführung von Julian Huxley, ein Vorwort der Verfasser und eines vom Übersetzer. Der Text beginnt mit Benutzungshinweisen, Angaben zur Topographie des Vogels und Erörterungen über das Rassenproblem. Danach folgen eine Liste der Vögel Europas sowie die Anschriften von 15 ornithologischen Gesellschaften in 14 Ländern Europas.

Den 74 Tafeln folgen ein nach Ländern geordnetes Literaturverzeichnis (107 Titel) und ein Index (18 S.), der die deutschen und die wissenschaftlichen Namen enthält; im Text sind außerdem bei jeder Art noch die landessprachlichen Bezeichnungen in Englisch, Französisch, Holländisch, Italienisch, Schwedisch und Spanien angegeben.

Gegenüber der 9. Auflage wurden der Text überarbeitet und teilweise erweitert, die 380 Verbreitungskarten verbessert sowie 10 Arten neu in den Hauptteil aufgenommen. Außerdem sind 13 erstmals in Europa nachgewiesene Irrgäste beschrieben.

Dieses großartige Gemeinschaftswerk, an dem zahlreiche genannte u. ungenannte Ornithologen vieler Länder mitarbeiteten, ist einfach geschrieben; es dient dem Laien wie dem Fachmann.

. Für die nächste Auflage seien folgende Hinweise gestattet: Auf Tafel 66 könnten auch Bestimmungspfeile angebracht werden. Im Literaturverzeichnis fehlen u. a. zwei Arbeiten von W. Makatsch: „Verzeichnis der Vögel Deutschlands", 2. Aufl. 1957 und „Wir bestimmen die Vögel Europas", 2. Aufl. 1969. D. Storch

Jun Tokida & Hiroyuki Hirose: Advance of Phycologgy in Japan. Jena: VEB Gustav Fischer Verlag 1975. 355 S., 161 Abb. 30 Tab. Ln. M 99,—

Die beiden japanischen Algologen Tokida und Hirose sind die Herausgeber des vorliegenden Buches. Es enthält 20 Beiträge von insgesamt 22 verschiedenen Autoren in englischer Sprache. Die Thematik der Einzelbeiträge ist so ausgewählt worden, daß möglichst viele Gebiete der Algenkunde zur Darstellung gelangt sind. Die Abhandlungen befassen sich zum großen Teil mit marinen Algen und basieren in dieser Hinsicht vorwiegend auf japanischen Originalarbeiten, die — sofern in der Landessprache publiziert — für uns ja praktisch unzugänglich sind. Es werden sowohl rein wissenschaftliche als auch technologische Probleme der Algenzucht behandelt. Bedingt durch verschiedene Schwierigkeiten datieren die einzelnen Beiträge aus den Jahren 1968 bis 1972, von denen die älteren im allgemeinen auf den neuesten Stand gebracht wurden. Jede Arbeit ist mit ausführlichen Literaturangaben versehen. Natürlich konnten in dem Buch nicht alle Gebiete der Algologie behandelt werden. Es wäre also wünschenswert, wenn später ein weiterer Band folgen würde, der noch Fehlendes ergänzen bzw. neue Fortschritte in der Algenforschung bringen könnte.

Das Buch enthält ein Vorwort der Herausgeber sowie 20 Beiträge, die in 7 Kapitel gegliedert sind: 1. Phylogenie der Algen (Feinstruktur der Chloroplasten, Zusammensetzung der Zellwand, photosynthetisch wirksame Pigmente; diese Themen sind alle unter phylogenetischen Gesichtspunkten dargestellt); 2. Morphologie der Algen (Feinstruktur der Schuppen und Geißeln der Chrysophyten, Feinstruktur der Diatomeenschalen, zytologische Studien an Phaeophyten, Rhodophyten und Chlorophyten), 3. Entwicklungszyklus der Algen (life-history), dargestellt an den Rhodophyten; 4. Physiologie einzelner Algenarten wie *Porphyra, Fucus, Chlorella* und farblose Algen; 5. Ökologie mariner Algen; 6. Geographische Verbreitung mariner Algen in Japan; 7. Praktische Bedeutung der Algen in Japan (Stickstoffbindung durch Blaualgen, Kultivierung von *Porphyra* und *Undaria* als Nahrungsmittel, medizinische Bedeutung der Algen).

Die Anregung zur Herausgabe des vorliegenden Werkes geht auf den Kieler Algologen und Hydrobotaniker F. Gessner zurück. Es ist zu begrüßen, daß damit Fortschritte in der wissenschaftlichen und angewandten algologischen Forschung eines für uns Mitteleuropäer so weit entfernten und sprachlich kaum zugänglichen Landes wie Japan allgemein erschlossen worden sind. Von besonderen Interesse erscheint mir das letzte Kapitel über die praktische Bedeutung der Algen und ihre Verwendung, z. B. als Nahrung für den Menschen. Es gibt in Japan einen ganzen Industriezweig, der sich speziell mit der Zucht und Kultivierung bestimmter Rotalgen für diese Zwecke befaßt. Für die Darstellung dieses Kapitels sind fast ausschließlich in japanischer Sprache gedruckte Originalarbeiten verwendet worden.

Das Buch wird mit 3 Registern beschlossen (Autornamen, Sachbezeichnungen und Namen der taxonomischen Kategorien), Jeder Artikel ist mit instruktiven Zeichnungen, Grafiken oder Fotos versehen. Druck und Ausstattung des gesamten Werkes entspricht der vom Verlag gewohnten Weise. Jeder algologisch Interessierte wird den Beiträgen neue bzw. zusammenfassende Informationen entnehmen können. H. Heynig

Paulette Gayral: Les Algues. Morphologie, cytologie, reproduction, écologie. Paris: Doin éditeurs 1975. 166 S., 84 z. T. mehrfarb. Abb. (242 Schemazeichnungen u. 51 Fotos), 3 Tab. Brosch. fr 120,—

Die französische Algologie besitzt Weltgeltung. Die Namen Thuret, Sauvageau, Lamouroux, Decaisne, Derbés, Bornet, Flahault, Gomont in der Vergangenheit und Bourrelly, Chadefaud, Magne, Dangeard und Feldmann in der Gegenwart sollen dafür stehen. Die vorliegende Algologie ist eine Einführung in ein wichtiges Teilgebiet der Morphologie, Zytologie, Fortpflanzung und Ökologie des Pflanzenreiches. Sie gliedert sich in drei Abschnitte von sehr unterschiedlicher Länge. Zunächst werden die allgemeinen morphologischen, zytologischen und fortpflanzungsbiologischen Merkmale bei Protokaryoten und Eukaryoten vorgestellt. Im Hauptteil erfahren die großen Gruppen der Algen, als die hier die Cyanoschizophyten, die Rhodo-, Pyrrho-, Chryso-, Eugleno-, Raphido- und Chlorophycophyten anerkannt werden, ihre biologische Charakteristik. Dabei liegt das Schwergewicht auf der Vermittlung der Zytologie, Morphologie und der Entwicklungszyklen wichtiger Vertreter. Relativ kurz, aber dennoch aussagekräftig kommen die Ökologie und Physiologie der Algen weg, die dabei als Phytoplankter, als Luft,- Boden-, Litoral-, Benthos- und Meeresalgen bekannt gemacht werden. Eine synoptische Tabelle der behandelten systematischen Gruppen, die bis zum Niveau der Ordnungen hinabreicht, eine kurze Bibliographie, ein Glossarium der Termini technici sowie ein Sippenindex bilden den technischen Apparat.

Die Darstellung ist übersichtlich, leicht verständlich geschrieben, ausreichend und gut illustriert und vermeidet Weitschweifigkeit. Jede Gruppe wird am Ende des ihr gewidmeten Textes noch einmal kurz zusammenfassend charakterisiert.

Unter Berücksichtigung des didaktischen Anspruches und der Ausrichtung auf einen begrenzten Leserkreis ist das Büchlein wohl gelungen. Freilich ist der in der Einleitung erhobene Anspruch, daß auch physiologische Tatsachen Berücksichtigung finden, nur begrenzt gültig. Die traditionellen Elemente der Morphologie und Zytologie, verbunden mit modernen Erkenntnissen der Feinstrukturforschung und der Biochemie dominieren eindeutig. Das ist indes nicht unbedingt ein Nachteil, wenn man berücksichtigt, daß der Verlag in der gleichen Reihe eine dreibändige Pflanzenphysiologie herausgegeben hat. Auch die hier relativ kurz behandelte Ökologie hat in der Biologie-Reihe des Verlages eine Sonderstellung gefunden, die algologische Aspekte von der Ökosystemtheorie her beleuchtet (vgl. unsere Rezension *Limnologica* **11**, 1, S. 179).

Über das dem Werke zugrundeliegende System der Algen wäre freilich allerhand zu sagen. Hier hätte die Verfasserin gut daran getan, in einem oder anderem Falle etwas unverbindlicher zu formulieren, um den Eindruck zu vermeiden, als ob das System bereits vollendet wäre. Sachlich nicht vertretbar sind die Angaben über die Verteilung der Pigmente bei den eukaryotischen Algen. Hier haben sich in der Gruppe der Pyrrhophycophyten, die in der Fassung der Autorin auch die Cryptomonaden einschließen, schwerwiegende Fehler gehalten. Auch in anderen Algengruppen entspricht die Charakterisierung der Pigmentierung nicht dem gegenwärtigen Stande. Gleichfalls ungenügend ist die Analyse der Bewegungsapparate bei Algen. Abgesehen von diesen Mängeln, die das Buch leider mit vielen vergleichbaren Unternehmen teilt, ist diese Einführung in die Algologie durchaus brauchbar.

S. J. Casper

D. Povoledo & H. L. Golterman (eds.): Humic Substances. Their structure and function in the biosphere. Proceedings of an international meeting held at Nieuwersluis, the Netherlands, May 29—31, 1972. Wageningen: Centre for Agricultural Publishing and Documentation 1975. 368 S., 177 Abb., 102 Tab. Brosch. Dfl. 75,—

Das Problem der Humussubstanzen unserer Binnengewässer im Rahmen einer internationalen Tagung zu diskutieren, griffen die beiden Herausgeber D. Povoledo & H. L. Golterman auf, wofür ihnen sehr zu danken ist. Mit Unterstützung des IBP fand im Limnologischen Institut Vijverhof 1972 eine Tagung statt, in deren Ergebnis 26 Beiträge nun in gedruckter Form vorliegen.

Die Schwierigkeiten bei der Analyse von Humussubstanzen sind noch keinesfalls überwunden, neben Extraktionsmethoden unter verschiedenen pH-Bedingungen werden folgende Analysenmethoden eingesetzt: Spektrophotometrie vom UV bis IR, Massenspektrometrie, paramagnetische Resonanzspektrometrie, Gaschromatographie und Gelchromatographie. Die kolorimetrischen Vergleichswerte in Pt-Einheiten sollten durch Tyrosin-Äquivalente ersetzt werden (H. de Haan). Der Einfluß von Humussubstanzen auf die Organismen, die biologische Umwandlung durch Bodenmikroben und Pilze sowie der Einfluß auf die Produktion der Gewässer werden in verschiedenen Beiträgen diskutiert.

Altersbestimmungen mit der ^{14}C-Technik führte H. W. Scharpenseel an verschiedenen Bodenhorizonten aus, um die Umwandlungsgeschwindigkeiten zu erfassen. In einem größeren Beitrag „Biogeochemistry of humic Compounds" gibt F. M. Swain einen Überblick über die geologische Entwicklung und Biochemie dieser Verbindungen im Verlauf der Erdgeschichte. Wenn Povoledo im Schlußwort ausführt: „The complexity of the problems involved is so great that it has eluded for a long time the efforts of aquatic and soil chemists", unterstreicht er die Bedeutung und die weiter nötige Forschung auf diesem Gebiet.

G. Proft

E. MICHAEL, B. HENNIG † & H. KREISEL: Handbuch für Pilzfreunde, Band VI: Die Gattungen der Großpilze Europas. Bestimmungsschlüssel und Gesamtregister der Bände I bis V. Herausgegeben von H. KREISEL. Jena: VEB Gustav Fischer Verlag 1975. 291 S., 5 Abb. 16 Taf. Ln. M 19,— (nichtsoz. Ausland. M 24,—).

Der MICHAEL—HENNIG hat seit dem Erscheinen des 1. Bandes im Jahre 1958 unter Mykologen und Anhängern der „Pilzjagd" viele Freunde gefunden. Bis 1970 auf 5 Bände angewachsen, hat sich das Unternehmen zu einem „Handbuch für Pilzfreunde" gemausert, das die Mehrzahl der in unserem Gebiet vorkommenden Pilzarten in Wort und Bild zu erfassen sucht. Trotz der vielfältigen Bemühungen HENNIGS und seiner vielen Freunde ist es bei einer Auswahl der auf 5300 Arten geschätzten Zahl der Großpilze in Europa geblieben. Leider konnte die Freude über die Vollendung des Werkes nicht ganz ungetrübt sein: Ein nur mühsam auszumachender „roter Faden", Inkonsequenzen und Fehler im Text- und Abbildungsteil erschwerten besonders dem Laien die Benutzung der Bücher.

H. KREISEL hat sich im 1. Teil des vorliegenden Bandes der Aufgabe unterzogen, alle europäischen Großpilze mindestens bis zur Gattung, vielfach sogar bis zur Art aufgeschlüsselt vorzuführen. Dabei versteht er unter Europa das Gebiet von Irland bis zum Ural, von der Arktis bis zum Mittelmeer und erfaßt als Großpilze alle diejenigen Pilze, deren Fruktifikationen mindestens 4 mm Durchmesser erreichen. Insgesamt werden 620 Gattungen aufgeschlüsselt. Die Schlüssel sind so gestaltet, daß sie auf möglichst schnellem, sicherem Wege zur Bestimmung der Gattung führen. Sie verwenden makroskopische und makrochemische Merkmale wie Form, Farbe, Konsistenz, Geruch oder Geschmack, außerdem leicht und schließlich auch schwieriger feststellbare mikroskopische Merkmale. Worterklärungen erleichtern den Gebrauch der Schlüssel. Ein alphabetisches Verzeichnis aller berücksichtigten Gattungsnamen ergänzt den Text. Dabei werden zu jeder Gattung die ihr übergeordnete Familie, die Typusart, die Anzahl der Arten in Europa und ihre Lebensweise, die der Gattung entsprechende Abbildung in den Bänden 1—5 sowie die wichtigsten Synonyme angegeben. Das Verzeichnis umfaßt insgesamt 930 Namen. Die sich daran anschließende Übersicht über das natürliche System der Pilze stellt eine Weiterentwicklung des KREISELschen Systems aus dem Jahre 1969 dar und ist lediglich als Information gedacht. Auf Tafeln werden 32 Pilzarten nach Photos von K. HERSCHEL und H. JAHN wiedergegeben, die in den Bänden 1—5 nicht abgebildet sind, darunter ein Photo der vieldiskutierten Lorchel (*Discina fastigiata*). Ein Verzeichnis der Abkürzungen der Autornamen vervollständigt die Darstellung.

Im 2. Teil haben ILSE & URSULA HENNIG ein Gesamtregister der wissenschaftlichen und der deutschen Namen der Bände 1—5 zusammengestellt.

H. KREISEL hat dem Band eine Würdigung BRUNO HENNIGS und seines „Handbuches für Pilzfreunde" vorangestellt und den schwierigen Werdegang des Werkes verdeutlicht. Dabei hat er sich auch nicht gescheut, die notwendige Kritik in angemessener Form vorzubringen. Der Rezensent ist überzeugt, daß es ihm gelungen ist, mit dem vorliegenden abschließenden 6. Band den MICHAEL—HENNIG in seinem Niveau dem gegenwärtigen Stande der Mykologie anzugleichen. Es ist daher durchaus berechtigt, wenn sich der Verlag entschlossen hat, aus dem „Zwei-Männer-Buch" ein „Drei-Männer-Buch" zu machen und so die Verdienste H. KREISELS um das Gesamtwerk zu verdeutlichen. Wenn wir das linke Vorsatzblatt richtig lesen, so dürfen wir hoffen, daß H. KREISEL in absehbarer Zeit das Handbuch weiterführen, d. h. die Bände überarbeitet vorlegen wird.

Wir wünschen ihm und dem Verlag dazu viel Erfolg. Denn die Pilzkunde ist, wie H. KREISEL in seiner Einleitung mit Recht hervorhebt, zu einem wissenschaftlichen Hobby breiter Bevölkerungskreise geworden und das Sammeln von Pilzen heute mehr und mehr zur anspruchsvollen Freizeitbeschäftigung vieler Menschen in unserer Gesellschaft. Mögen vom MICHAEL—HENNIG—KREISEL Impulse für eine Weiterentwicklung der Mykologie in unserer Republik ausgehen.

S. J. CASPER

LUDWIG RODEWALD-RUDESCU: Das Schilfrohr (Die Binnengewässer Band XXVII). Stuttgart: E. Schweizerbart'sche Verlagsbuchhandlung 1974. 302 S., 38 Fig. auf 11 Taf., 126 Abb. u. 94 Tab. im Text u. auf 2 Beilagen. Ln. DM 97,60

Das zunehmende Interesse, welches das Schilfrohr nicht zuletzt wegen seiner Verwendungsmöglichkeiten in der Zellulose- und Papier-Industrie in vielen Ländern findet, äußert sich auch in einer wachsenden Zahl von Publikationen über diese Art. Umfassende, z. T. monographische

Darstellungen verfaßten u. a. HÜRLIMANN (1951), BITTMANN (1953) und BJÖRK (1967; vgl. unsere Besprechung in *Limnologica* 7, 496, 1970). Auch der rumänische Limnologe LUDWIG RODEWALD-RUDESCU, seit langem als subtiler Kenner der Schilfbestände und deren Nutzung im Donaudelta ausgewiesen, veröffentlichte zusammen mit weiteren Mitarbeitern eine umfassende Monographie des Schilfes im Donaudelta. Leider war dieses Standardwerk (vgl. unsere Besprechung in *Limnologica* 5, 273, 1967) aber ausschließlich in rumänischer Sprache abgefaßt, wodurch seine Benutzung und Auswertung durch auswärtige Interessenten auf Schwierigkeiten stieß.

So muß man es besonders begrüßen, wenn der Autor das Thema jetzt in der Reihe „Die Binnengewässer" auch in deutscher Sprache abhandelt. RODEWALD-RUDESCU faßt seine vielschichtigen eigenen Untersuchungen im Donaudelta und anderen rumänischen Landesteilen und seine weltweiten Beobachtungen an Schilfbeständen mit den Ergebnissen zahlreicher anderer Forscher zu einer umfassenden Monographie des Schilfrohres zusammen. Das Werk gliedert sich praktisch in zwei Teile. Der erste behandelt die botanischen und ökologischen Grundlagen und bringt ausführliche Angaben und Daten zur Systematik, Morphologie, Anatomie, Physiologie, Aut- und Synökologie des Schilfrohrs sowie über dessen chemische Eigenschaften. Der zweite Teil ist der Schilfrohrkultur, der maschinellen Ernte und der wirtschaftlichen Bedeutung des Schilfrohrs gewidmet und geht abschließend auf Probleme des Schutzes von Schilfbeständen ein. Das Literaturverzeichnis enthält über 500 Titel.

Wenn der Verfasser auch in besonderem Maße auf eigene Befunde im Donaudelta zurückgreift, so bemüht er sich doch stets, überall die Angaben anderer Schilfforscher gebührend zu berücksichtigen, so daß weithin eine recht abgerundete und ausgewogene Darstellung erzielt werden konnte. Das Buch gibt somit einen eindrucksvollen Überblick über die gegenwärtigen Kenntnisse vom Schilfrohr, zeigt aber zugleich die Lücken auf, die hier noch bestehen, und fordert dadurch zu weiteren Untersuchungen auf.

Daß bei einem so umfassenden Thema vom Autor natürlich nicht alle einschlägigen Arbeiten zu erlangen waren und ausgewertet werden konnten (wie etwa die pflanzensoziologisch-ökologische Studie von SCHMALE in Verh. bot. Vereins Prov. Brandenburg 79, 1939), versteht sich von selbst. Trotz derartiger, bei einem solchen Werk nun einmal unvermeidlicher Lücken bietet das Buch jedoch gerade zur Autökologie des Schilfrohrs eine sehr eingehende und alle Bereiche berührende Schilderung. Weniger gelungen erscheint uns dagegen die Darstellung der Synökologie. Der Autor meint, die vielfältigen Untersuchungsbefunde zahlreicher Vegetationsforscher zur Vergesellschaftung des Schilfrohrs übergehen zu können, „da keine wesentlichen ökologischen Schlußfolgerungen daraus zu ziehen sind, somit sich nur eine lange Liste von Artnamen ergeben würde, die nichts weiter aussagen". Diese Auffassung dürfte sicher auf den Widerspruch aller derjenigen stoßen, die sich näher mit der Soziologie von *Phragmites communis* beschäftigt haben und wissen, daß gerade bei dem so anpassungsfähigen und eine weite ökologische Standortsamplitude einnehmenden Schilfrohr ohne die Hilfe von Begleitarten keine eindeutige ökologische Kennzeichnung und Abstufung der verschiedenartigen Schilfbestände zu erreichen ist. Auch der Abschnitt „Die Pflanzengesellschaften der *Phragmites*-Bestände" spiegelt den gegenwärtigen Stand der phytozönologischen Forschung nur ungenügend wider und zeigt, daß Interesse und Stärke des Autors eben doch auf anderen Gebieten liegen.

Wenn hinsichtlich Synökologie und Soziologie des Schilfrohrs letztlich auch noch einige Wünsche offenbleiben, so kann das gut ausgestattete und sauber gedruckte Buch jedoch jedem empfohlen werden, der sich mit dem Schilf und insbesondere mit dessen Nutzung befassen muß oder will.

H.-D. KRAUSCH

F. E. ROUND: **Biologie der Algen.** Eine Einführung. Deutsche Übersetzung und Bearbeitung von K. DRUMM. 2. Aufl. Stuttgart: Georg Thieme Verlag 1975. VIII + 342 S., 77 Abb., 12 Taf., 7 Tab. Flex. Taschenb. DM 22,—

Die neue Auflage (Original bei E. Arnold Publishers, London 1973) des gelungenen Taschenbuchs über die Biologie und Ökologie der Algen (vgl. Rezension in *Limnologica* 6 (1), 224, 1968) ist wesentlich erweitert worden. Die Zahl der Abbildungen ist um 6, die der Tafeln um 7 und die der Seiten um 27 vermehrt worden. Die bewährte Anordnung des Stoffes ist geblieben. Einzelne Kapitel haben, den Fortschritten der Algologie in den letzten Jahren entsprechend, Korrekturen und Verbesserung erfahren. Besonders zu begrüßen ist, daß der Übersetzer auch neue Ergebnisse deutscher Arbeitsgruppen eingearbeitet hat, da diese nur allzu oft im angelsächsischen Raum „übersehen“ werden. Außer rund 450 Literaturzitaten, auf die der Text direkt Bezug nimmt, ist noch eine Zusammenstellung von wichtigen Werken aus dem Zeitraum der letzten 10 Jahre zu vermerken, die als Orientierungshilfe gedacht ist. S. J. CASPER

W. I. ROMANENKO & S. I. KUSNEZOW: Die Ökologie der Mikroorganismen in Süßwässern. Eine Laboranleitung. Leningrad: Verlag „Nauka“ 1974. 194 S., 42 Abb., 13 Tab. Geb. Rbl. 1,02 (russ.)

Die Autoren sind in Fachkreisen bereits durch einige Standardwerke zur Mikrobiologie der Binnengewässer bekannt geworden. Erinnert sei an die umfassenden Werke von S. I. KUSNEZOW „Die Rolle der Mikroorganismen im Stoffkreislauf der Seen“, daß 1952 in Moskau und 1959 in einer überarbeiteten Fassung in deutscher Sprache (VEB Deutscher Verlag der Wissenschaften, Berlin) erschien, und „Die Mikroflora der Seen und ihre geochemische Tätigkeit“ (russ., Leningrad 1970). Das nunmehr vorliegende gewässermikrobiologische Methodenbuch ist ein weiterer Erfolg. Es baut auf der 1963 von S. I. KUSNEZOW & W. L. ROMANENKO im gleichen Verlag herausgegebenen Laboranleitung „Die mikrobiologische Erforschung der Binnengewässer“ auf. Neue Erkenntnisse (z. B. Einsatz radioaktiver Isotope, Bestimmung der heterotrophen und autotrophen Produktion, elektronenmikroskopische Analyse) sind mit bewährten älteren Arbeitsanleitungen sinnvoll gekoppelt. Die Verfasser sind dabei ihrem Grundsatz treu geblieben, durch eine sachlich klare und einfache Darstellung in die verschiedensten mikrobiologischen Methoden einzuführen und die Anwendung für einen breiten Interessentenkreis zu vereinfachen.

Das Buch gliedert sich in 7 Abschnitte: 1. Grundlegende Prinzipien der Nährbodenherstellung und Kultivierung von Mikroorganismen (9 S.), 2. Nährbodenrezepte zur Auslese und Kultivierung von Mikroorganismen (35 S.), 3. Lebendstudien der Mikroorganismen und Färbemethoden für Bakterien (13 S.), 4. Kultivierungsmethoden und Methoden zur Herstellung von Reinkulturen (44 S.), 5. Einige spezielle Methoden der bakteriologischen Wasser- und Sedimentanalyse in Seen und Staugewässern (25 S.), 6. Produktion und Destruktion des organischen Materials in Gewässern (39 S.) und 7. Einige spezielle Methoden zur chemischen Wasseranalyse (11 S.). Ein ausgewogenes Literaturverzeichnis (163 Nachweise) schließt sich an.

Das Büchlein erhebt keinen Anspruch auf Vollständigkeit gewässermikrobiologischer Untersuchungsmethoden. Es fehlen z. B. methodische Angaben zur Messung der mikrobiellen N_2-Fixierung, die Ascomyceten, Myxomyceten, Phycomyceten und Viren bleiben unberücksichtigt, auf die Analyse von Zellinhaltsstoffen wird ebenfalls nicht eingegangen. Der Gesamteindruck wird aber dadurch keinesfalls geschmälert.

Das Buch ist eine wertwolle Bereicherung der sehr schmalen methodischen Darstellungen in der Gewässermikrobiologie. R. KOSCHEL

EBERHARD SCHMIDT: Ökosystem See (Biologische Arbeitsbücher 12). Heidelberg: Quelle & Meyer 1974. 170 S., 37 Abb., 28 Tab. Flex. Ln. DM 19,50

Das vorliegende Heft bringt Arbeitsmaterial zur Durchführung eines limnologischen Kurses im Biologieunterricht und in Schüler-Arbeitsgemeinschaften an Oberschulen sowie im Rahmen der schulpraktischen Ausbildung von Studenten an Pädagogischen Hochschulen. Die Darstellung will in das komplexe Ordnungsgefüge des Ökosystems unserer Binnenseen einführen. Sie geht dabei von auffälligen Erscheinungen aus (Ufervegetation, größere Tiere), um den Schülern an leicht zugänglichen Objekten Einsichten in Struktur und Funktion von stehenden Binnengewässern zu vermitteln und damit zugleich das so dringend notwendige Verständnis für ökologische Zusammenhänge zu wecken. Das Konzept des Kurses muß dabei aus der Unterrichtssituation (Stadtschule!) heraus dem Schülerversuch den Vorrang vor Freilanduntersuchungen geben, die über eine obligate Arbeitsexkursion hinaus aber als Sonderaufgaben für speziell interessierte Schüler angestrebt werden können. Nach einer kurzen Vorstellung von Grundbegriffen und einem Hinweis auf die volkswirtschaftliche Bedeutung der Seen werden im einzelnen exemplarisch behandelt: die Ufervegetation, Tiere am Ufer und im Uferbereich, die Zone des freien Wassers, der Lebensbereich des Seegrundes, der Kreislauf der Stoffe, Seentypen und Altern der Seen sowie die Eutrophierung und Abwasserbelastung der Seen. Ein kurzer Vergleich von See und Fluß beschließt das inhaltsreiche, didaktisch ausgezeichnet aufgebaute Arbeitsbuch. Jedem Kapitel sind Hinweise auf die wichtigste, leicht erreichbare Literatur beigefügt. Das Buch vermittelt in gut faßlicher und übersichtlich gegliederter Form eine Vielfalt von grundlegenden, dem neuesten Kenntnisstand entsprechenden Informationen aus dem Bereich der Seenkunde.

Lobend hervorzuheben ist besonders die sich auf eine umfangreiches Faktenmaterial stützende Diskussion aktueller Fragen des Gewässerschutzes. D. Flössner

R. G. Kessel & C. Y. Shih: Scanning Electron Microscopy in Biology. A Students' Atlas on Biological Organization. Berlin, Heidelberg, New York: Springer-Verlag 1974. X u. 345 S., 22 Fig., 132 Taf. Ln. DM 48,50 (US $ 19.80)

Das Transmissionselektronenmikroskop hat uns das „Innere" der Zelle neu sehen gelehrt. Das Rasterelektronenmikroskop schickt sich an, uns das „Äußere" der Organismen, der Organe, Gewebe, Zellen und Organellen ebenfalls neu sehen zu lehren. Rasterelektronenmikroskopische Aufnahmen z. B. von mit festen Hüllen oder Schalen bedeckten Flagellaten oder Algen, von Samen, Pollen oder Sporen haben bereits zu bedeutenden Erkenntnissen für die Systematik und für das Verständnis bestimmter Zusammenhänge von Bau und Funktion geführt. Der vorliegende Atlas will dem Studenten die biologische Organisation rasterelektronenoptisch nahebringen. Es ist ganz erstaunlich, welche Möglichkeiten sich dieser neuen Sicht erschließen.

Die Darstellung beginnt mit einzelligen Organismen (Ciliaten, Flagellaten und Amöben), geht über zur Behandlung der Zellen in Kultur und wendet sich schließlich Bakterien und Blaualgen, Schleimpilzen, Pilzen, Algen und Kormophyten sowie deren Organsystemen (Wurzeln, Sproßspitze, Stengel, Blatt, Blüte und Samen) zu. Im Anschluß finden die vielzelligen Tiere, ihre Gewebe und Organsysteme sowie die Befruchtung bei Seeigeln und die Embryologie des Frosches eingehende „bildnerische" Berücksichtigung.

In der Regel ist jeder Tafelseite eine Textseite gegenübergestellt, die die Abbildungen erläutert und den abgebildeten Organismus systematisch einordnet. Nicht alle der vorzüglichen Mikrofotos sind für die Wissenschaft von Belang, die übergroße Mehrzahl aber doch so aussagekräftig, daß man sich wünschen muß, daß unsere Lehrbücher der Botanik und Zoologie mehr als bisher Gebrauch von dieser Art der „optisch" wirksamen Ausstattung machen. Am Rande sei erwähnt, daß den rasterelektronenmikroskopischen Photos in einigen Fällen transmissionselektronen- und lichtmikroskopische Aufnahmen zur Orientierung und vertiefenden Interpretation beigegeben sind. Ein kurzes einführendes Kapitel in Theorie und Praxis der Rasterelektronenmikroskopie und ein Vergleich mit anderen Methoden der Ultrastrukturforschung sowie ein Sachverzeichnis runden die einfache und leicht verständliche Darstellung ab. Die bibliographischen Angaben sind auf das Notwendigste beschränkt.

Der Atlas beweist, daß die Rasterelektronenmikroskopie eine vollgültige Methode der Strukturanalyse darstellt. Seine Nützlichkeit für die Studenten der Botanik oder Zoologie als Hilfsmittel für das Verständnis des in Vorlesungen und Praktika behandelten Stoffes auf morphologischem, anatomischem und zytologischem Gebiete ist evident. S. J. Casper

Kamil Rybníček: Die Vegetation der Moore im südlichen Teil der Böhmisch-Mährischen Höhe. (Vegetace ČSSR A 6). Praha: Academia 1974. 243 S., 79 Abb. (Fotos und Diagramme) im Text, 74 Tab., 18 Karten u. Profile. Ln. Kčs 150,—

Mit dem vorliegenden Band wird die bereits gut bekannte und eingeführte Schriftenreihe des Prager Akademie-Verlages über die Vegetation der Tschechoslowakei (vgl. unsere Besprechungen in *Limnologica* 5, 273, 1967, und 7, 242, 1969), in gelungener Weise fortgesetzt.

Die Untersuchung Kamil Rybníčeks ging aus dessen Dissertation hervor und wurde bereits Anfang 1967 zum Druck gegeben. Die lange Zeit der Drucklegung bringt es leider mit sich, daß neuere Arbeiten und Ansichten noch nicht berücksichtigt und manche Vorstellungen und Diskussionen bereits wieder überholt bzw. gegenstandslos geworden sind. Freilich: so bedauerlich diese Tatsache, für die der Autor natürlich nicht verantwortlich gemacht werden kann, in mancherlei Hinsicht auch sein mag, dem Kern der Arbeit wird dadurch kaum Einbuße getan. Hier liegt eine gründliche, auf umfangreichem Material fußende Untersuchung vor, deren Ergebnisse von bleibendem Wert sind und unsere Kenntnisse über die Moorvegetation Mitteleuropas wesentlich erweitern und vertiefen helfen.

Nach einleitenden Ausführungen über die Gegebenheiten des Untersuchungsgebiets sowie über methodische Grundsätze nimmt der Hauptteil des Buches die Darstellung der Moorgesellschaften, ihrer Klassifikation und synökologischen Charakteristik ein. Die meisten der behandelten Gesellschaften (19) gehören zu den Flach- und Zwischenmooren (Klasse Scheuchzerio-Caricetea fuscae mit den Ordnungen Tofieldietalia, Scheuchzerietalia und Caricetalia fuscae), 2 zu den Borstgrasrasen und Feuchtheiden, (Nardo-Callunetea) und nur eine zu den eigentlichen Hochmooren (Oxycocco-Sphagnetea).

Jede einzelne Gesellschaft wird eingehend beschrieben und belegt, wobei insbesondere auch den Standortsverhältnissen, der Verbreitung der Assoziation und dem Literaturvergleich breiter Raum eingeräumt werden. Da die Klassifikation der mitteleuropäischen Moore noch keineswegs, wie auch hier deutlich betont wird, einen befriedigenden und allseits akzeptierten Stand erreicht hat, unterbreitet der Autor an vielen Stellen Vorschläge zu einer Weiterentwicklung und Verbesserung des Systems der Vegetationseinheiten. Hier auf Einzelheiten einzugehen, muß der fachlichen Diskussion überlassen bleiben, die, wie gesagt, in diesen Fragen noch keineswegs zu einem Abschluß gelangt ist. Beachtung verdient auch die Gepflogenheit, bei jeder Vegetationseinheit den nomenklatorischen Typus anzugeben. So begrüßenswert ein solches Vorgehen im Hinblick auf eine eindeutige Definition der betreffenden Einheit auch sein mag, so darf andererseits doch nicht übersehen werden, daß gerade dieser Punkt noch der weiteren grundsätzlichen Diskussion bedarf. Das Buch behandelt dann die syndynamischen Beziehungen der Moorgesellschaften, ihre synökologischen Zusammenhänge und die standörtliche Differenzierung (mit einem beachtenswerten Bestimmungsschlüssel der vorhandenen Moorgesellschaften), gibt weiterhin eine Übersicht über die Nutzungsmöglichkeiten der Moore im Untersuchungsgebiet, wobei Fragen des Naturschutzes und der Landeskultur eine wichtige Rolle spielen, und schließt mit Zusammenfassungen in deutscher, tschechischer und englischer Sprache sowie einem ausführlichen Literaturverzeichnis. Die reichliche Ausstattung des Bandes mit Fotos, Diagrammen, Profilen, Karten und ausführlichen Tabellen verdient besonderes Lob. H.-D. Krausch

Agnes Ruttner-Kolisko: Plankton Rotifers. Biology and Taxonomy (Übersetzung von „Die Binnengewässer XXVI, Teil 1, Kap. „Die Rotatorien“ durch G. Kolisko). Stuttgart: Schweizerbart'sche Verlagsbuchhandlung (Nägele u. Obermiller) 1974. VII + 143 S., 53 Abb., 2 Tab., 4 Taf. Steifer Einband. DM 46,80

Die englische Übersetzung des Rotatorienbeitrages zum „Zooplankton der Binnengewässer“ erschien bereits 2 Jahre nach der deutschsprachigen Auflage. Dies zeigt, wie stark die Nachfrage im englischsprachigen Raum ist. Die Zusammenfassung der Planktonrotatorien mit zahlreichen ausgezeichneten Darstellungen vieler Einzelheiten wie Kauer, Dauereier oder ganzer Formenkreise füllt eine lange bestehende Lücke aus. Der englischen Ausgabe ist eine ebenso weite Verbreitung in ihrem Sprachbereich zu wünschen wie der deutschsprachigen. D. Ronneberger

Horst Kurt Schminke: Evolution, System und Verbreitungsgeschichte der Familie Parabathynellidae (Bathynellacea, Malacostraca). (Mikrofauna des Meeresbodens 24, Redaktion: Peter Ax, Göttingen). Wiesbaden: Verlag der Akademie der Wissenschaften und der Literatur, Mainz, in Kommission bei Franz Steiner Verlag GmbH 1973. 192 S., 48 Abb., Brosch. DM 46,—

Die Bathynellacea sind eine uralte Malakostrakengruppe, die im grundwasserführenden Lückensystem von Kiesen und Sanden, dem limnischen Mesopsammal, ein Refugium gefunden hat, an das sie durch Zwergwuchs, wurmförmige Körperstreckung, Rückbildung der Pleopoden und Fehlen von Augen hochgradig angepaßt ist. Durch ihr hohes phylogenetisches Alter beanspruchten sie seit ihrer Entdeckung vor ungefähr 90 Jahren das besondere Interesse der Zoologen und Evolutionsbiologen, das in der Zwischenzeit durch die Auffindung zahlreicher neuer Arten in den Grundwasserströmen fast aller Kontinente ständig wuchs. Lediglich aus dem australo-pazifischen Raum war die Gruppe bislang nicht bekannt. Diese Kenntnislücke zu schließen erscheint bei der bekannten biogeographischen Sonderstellung der notogäischen Region besonders dringlich, zumal sich im Falle des dortigen Vorkommens Möglichkeiten zu einer vertieften Analyse der Evolution und Ausbreitungsgeschichte dieser reliktären Grundwasserkrebse und zur Aufhellung anstehender biogeographischer (Rolle Antarktikas als Evolutions- und Ausbreitungszentrum) und erdgeschichtlicher (Kontinentzusammenhänge) Probleme eröffnen würden.

Die vorliegende breitangelegte Studie befaßt sich mit diesem Themenkomplex auf der Grundlage eines reichen Fundmaterials, das der Autor während einer Forschungsreise nach Neuseeland, Neukaledonien und Australien aus insgesamt 308 Grundwasserproben gewinnen konnte. Im Mittelpunkt der Untersuchung steht eine umfassende Revision der Familie Parabathynellidae, zu der die überwiegende Mehrzahl der Arten aus der Ordnung Bathynellacea gehört. Als Voraussetzung hierfür werden in einem ersten umfangreichen Abschnitt durch eine sorgfältige Analyse und Bewertung der wichtigsten morphologischen Merkmale die Grundlagen der Systematik der Bathynellacea erarbeitet, wobei das Hauptaugenmerk auf der Klärung der Homologieverhältnisse liegt. Es gelingt der Nachweis, daß einerseits Merkmale, die bislang für die systematische Gliederung der Gruppe verwendet wurden, neotenischen Prozessen unterliegen und damit systematisch belanglos sind und daß zum anderen entgegen früheren Auffassungen die Struktur des 8. Thorakopoden des ♂ ein systematisch bedeutsames Merkmal darstellt. Anschließend legt der Autor ein neues System der Parabathynellidae vor, das auf der Korrelation der Merkmalskomplexe Th. VIII ♂ — Mundwerkzeuge — Antenne II beruht. Die Familie umfaßt nach der neuen Gliederung 60 Arten, die sich auf 19 (davon 10 neuen) Gattungen verteilen. Alle Gattungen werden ausführlich beschrieben, diagnostiziert und die Verwandtschaftsbeziehungen der Arten innerhalb der einzelnen Gattungen eingehend diskutiert. Es folgen Ausführungen zu allgemeinen Aspekten der Evolution der Bathynellacea, in denen der Versuch zu einer Rekonstruktion der hypothetischen Stammart unternommen und auf die Rolle der Neotenie bei der Entstehung der Sandlückenfauna hingewiesen wird. Der Abschnitt über die Chorologie enthält eine kartographische Dokumentation der Verbreitungsareale der verschiedenen Gattungen der Parabathynellidae, Bemerkungen zur sogenannten Progressionsregel nach HENNIG sowie eine eingehende Darstellung der Verbreitungsgeschichte der Bathynellacea, die durch die Untersuchungen des Autors wesentlich an Aussagekraft gewonnen hat. In einem abschließenden taxonomischen Kapitel werden 11 Arten der eigenen Ausbeute und eine weitere Art aus Brasilien neu beschrieben sowie 8 schon bekannte Arten bis in alle Einzelheiten ergänzend charakterisiert. Zahlreiche klare Abbildungen illustrieren die Diagnosen und ausführlichen Beschreibungen.

Diese ausgezeichnete Abhandlung bildet einen wertvollen Beitrag zur weiteren Entwicklung des Systems der Bathynellacea, der über das enge Fachgebiet hinaus von allgemeiner evolutionstheoretischer und biogeographischer Bedeutung ist. D. FLÖSSNER

KAREL RATAJ: Revizion of the genus *Echinodorus* RICH. (Studie ČSAV č. 2, 1975). Praha: Academia 1975. 156 S., 54 Abb. Brosch. Kčs 37,—

Der Verfasser, bekannt durch zahlreiche Einzelveröffentlichungen zur Taxonomie der Alismataceen-Gattung *Echinodorus* und Leiter der Abteilung für tropische Wasserpflanzen des Botanischen Instituts der ČSAV in Šumperk, legt eine gründliche Revision der Gesamtgattung vor. In seiner Fassung gehören 47, und zwar nur neuweltliche Arten zum Geschlecht. Zwei südeuropäische und nordafrikanische Arten werden zur Gattung *Baldellia* PARLATORE, zwei asiatische und afrikanische Arten zur Gattung *Ranalisma* STAPF gerechnet. *Echinodorus* unterscheidet sich von *Baldellia* (6 Staubgefäße, terminale Griffelinsertion) durch den Besitz von mehr als 6 Stamina und laterale Griffelinsertion, von *Ranalisma* (flacher oder konvexer Blütenboden) durch einen verlängerten Blütenboden.

Zwei Untergattungen — *Helianthium* mit 7 und *Echinodorus* mit 40 Arten — sowie 10 Sektionen werden unterschieden und damit zum ersten Mal eine infragenerische Gliederung des Geschlechtes vorgenommen (eine Gattungsdiagnose fehlt übrigens in der Arbeit). Alle Sippen werden ansprechend zeichnerisch, unter Hervorhebung der diagnostisch wichtigen Merkmale, dargestellt. Eine Liste der species excludendae, Indizes der Taxa, der untersuchten Sammlungen und der Herbarien sowie ein (allerdings sehr knappes) Literaturverzeichnis vervollständigen die Untersuchung, die, sieht man von den leider zahlreichen Druckfehlern (bereits im Titel: Revizion!) ab, durchaus gelungen ist.

Neue Arten: *E. angustifolius*, *E. pellucidus*, *E. cylindricus*, *E. macrocarpus*, *E. glaucus*. Neue Varietäten: *E. quadricostatus* var. *xinguensis*. Neukombinationen: *E. latifolius* (SEUBERT) RATAJ (= *Alisma tenellum* f. *latifolia* SEUBERT in MARTIUS). *E. quadricostatus* FASSETT var. *magdalenensis* (FASSETT) RATAJ (= *E. magdalenensis* FASSETT). S. J. CASPER

HARALD SIOLI (Hrsg.): Ökologie und Lebensschutz in internationaler Sicht (Ecology and Bioprotection. International Conclusions). Freiburg: Verlag Rombach 1973. 548 S., 35 Abb., 23 Tab. Ln. DM 132,–

Die Begriffe Umweltschutz und Ökologie rücken heute immer näher in die Öffentlichkeit, zuweilen in beängstigender Weise strapaziert oder auch zu Geschäftszwecken mißbraucht. Dabei steht die Frage im Vordergrund, führen die Veränderungen in den Ökosystemen der Natur letzthin zum Schwund der Existenzmöglichkeiten der Menschen oder sind gezielte Aktivitäten in der Lage, die Biosphäre auf Dauer als Lebensgrundlage für ihn zu erhalten.

Das Anliegen dieses Buches ist es, die tieferen Gründe und nicht nur die immer wieder beklagten augenfälligen Symptome für den heutigen Zustand der Biosphäre darzustellen. Mit den Worten des Herausgebers versucht das Buch „möglichst objektiv und klar einige Hauptpunkte der heutigen Umweltsituation darzulegen, den Grundprinzipien nachzugehen, die zu dieser Situation geführt haben, sowie Möglichkeiten und Grenzen von Voraussagen abzustecken und einige unerläßliche Schutzmaßnahmen zu entwickeln". Die 15 Beiträge wurden von Autoren aus der BRD, UdSSR, Niederlande, aus Brasilien, Spanien und den USA sowie von Mitarbeitern der UNESCO, IUCN und des UNDP verfaßt. Sie nehmen zu nahezu allen aktuellen Umweltproblemen Stellung. Lediglich die Umweltkomplexe Lärm und Müll werden nicht in einem speziellen Artikel betrachtet. Den Limnologen interessieren insbesondere der Beitrag von A. D. HASLER über „Causes and Correctives of Man-Made Eutrophication" (23 S.), der Bericht von F. BÄR über „Die toxikologische Situation in der modernen Zivilisation" (31 S.) und die Darstellung „The Ocean as Final Recipient of the End Products of the Continent's Metabolism. Pollutions of the Oceans: Situation, Consequences and Outlooks to the Future" (49 S.) durch P. KORRINGA.

Jeder, der sich in Forschung oder Verwaltung mit der Erhaltung oder Wiederherstellung einer gesunden Umwelt zu beschäftigen hat, findet in dem Buch eine Vielzahl von Fakten, Daten und Zahlen über die Verschmutzung und Vergiftung der Atmosphäre, des Bodens und des Wassers, über den Tierschutz, über den Einfluß moderner Landwirtschaftsformen, über die Rolle der Nationalparks in der modernen Zivilisation oder über das internationale von der UNO begründete Programm „Man and the Biosphere".

In mehreren Beiträgen wird eindringlich auf die Fehler und Gefahren einer falschen, vielfach auf Profit ausgerichteten Ausnutzung der Naturressourcen hingewiesen, und es ist nicht zuletzt eine Frage der Gesellschaftsordnung, ob es zu einer wirklichen Ökokatastrophe kommen muß, von der in einem Beitrag die Rede ist, oder ob die menschliche Gesellschaft die keinesfalls zu verniedlichende Gefährdung durch Zusammenwirken ihrer Kräfte erkennt und überwindet. Um breite Kreise in die Probleme des Umweltschutzes einzubeziehen, sind die meisten der Beiträge sehr geeignet.

Die Artikel sind in Deutsch, Englisch, Spanisch und Französisch in einem Schlußkapitel nochmals zusammengefaßt. G. BREITIG

DIETMAR & RENATE AICHELE & HEINZ-WERNER & ANNELIESE SCHWEGLER: Lebensraum Alpen. Ein Kosmos-Biotopführer. 120 Tiere und Pflanzen in Farbe. Stuttgart: Kosmos-Verlag, Franckh'sche Verlagshandlung 1975. 72 S., 79 Abb., 120 Farbfotos. Kart. DM 8,80

Die „Bunten Kosmos-Taschenbücher" sind als Vorstufe und Ergänzung zu den Kosmos-Naturführern gedacht; als Vorstufe, weil sie auf einen umfangreichen Textteil verzichten, als Ergänzung, weil sie in ausgesuchten typischen Farbfotos Pflanzen, Tiere und ihre Umwelt zeigen. Ästhetische und wissenschaftliche Gesichtspunkte halten sich die Waage. „Biotopführer" — das heißt, daß im Vordergrund des Interesses des aufmerksamen Naturfreundes oder Wanderers nicht die Pflanze oder das Tier als isoliertes Einzelwesen stehen, sondern daß ihn das Beziehungsgefüge Lebewesen — Umwelt fesseln soll. Mit dem Führer steigt er durch die Laub- und Mischwaldstufe, die Nadelwaldstufe, die Krummholzstufe und die Hochstaudenfluren in die Regionen der Schneeböden und -tälchen, der Magerrasen und Matten, der Felsbänder und Grate auf. Er lernt vikariierende Arten, Endemiten kennen, begreift den Zusammenhang von Pflanzengestalt und Standort, erfährt Wissenswertes über Pflanzenwanderungen zwischen Alpen und Vorland und begegnet schließlich der Tierwelt in ihren mannigfachen Gestalten. Auch das Leben im Schnee oder in den Gebirgsbächen bleibt ihm nicht verborgen. Jeder Textseite mit abgeschlossener Thematik (z. B. „Geweihträger") ist eine sachlich darauf abgestimmte Bildseite mit jeweils 4 Farbfotos (in unserem Beispiel

Rehbock, Hirsch, Schälstelle, Hirschfamilie) zugeordnet. Strichzeichnungen auf der Randleiste und auf drei Vorsatztafeln runden die klare, leicht lesbare und gut überschaubare Darstellung ab. Ein Abbildungsverzeichnis bietet zusätzliche Information. Die Farbfotos sind durchweg hervorragend.

Ein nützliches Buch, das nicht nur biologisch-ökologisches Wissen vermittelt, sondern auch dem Natur- und Landschaftssschutz dient.

S. J. Casper

Ju. I. Sorokin: **The use of ^{14}C in the study of nutrition of aquatic animals.** (Mitt. internat. Verein. Limnol. No. 16). Stuttgart: E. Schweizerbart'sche Verlagsbuchhandlung (Nägele & Obermiller) 1968. 41 S., 10 Abb., 10 Tab. Brosch. DM 10,40

Im verstärkten Maße haben sich Isotopenmethoden auch bei meeresbiologischen und limnologischen Untersuchungen bewährt. Vielfältige Einsatzmöglichkeiten, ein hohes Auflösungsvermögen und kurze Versuchszeiten (Physiologie) heben sie über andere vergleichbare Analysenverfahren.

Der Einsatz von markierten Substanzen ergibt auch für die quantitative und qualitative Erfassung der Sekundärproduktion potentiell weitreichende Möglichkeiten. Der Verfasser, der selbst über 2 Jahrzehnte mit radioaktiven Isotopenmethoden Untersuchungen zur Primär-, Intermediär- und Sekundärproduktion durchgeführt hat, beschreibt in der vorliegenden Mitteilung die „Kohlenstoff(^{14}C)-Methode" in ihrer Anwendung für Ernährungsuntersuchungen von Wassertieren. Es werden verschiedene Möglichkeiten zur Markierung der Nahrung (Bakterien, Algen, Detritus, Nahrungstiere) mit ^{14}C-Glukose, ^{14}C-Hydrolysat von Algen, ^{14}C-Bikarbonat, die Grundlagen zur Berechnung der Produktionsraten, Umsatzraten und Effektivitäten der Sekundärproduktion sowie in einigen praktischen Beispielen Ergebnisse der Bestimmung des Nahrungsgleichgewichtes, des Nahrungsspektrums und seiner Abhängigkeit von der Nahrungskonzentration behandelt. Als Empfindlichkeitsgrenze der Methode wird 10^{-5} mg C/Ind. angegeben, die aber bei Vermeidung von Flüssigkeitsszintillatoren sicher noch zu verbessern ist.

Eine Methode, die neue Maßstäbe setzt und deren Verbreitung es weiter anzuregen gilt.

R. Koschel

Karel Rataj: Revision of the genus *Cryptocoryne* Fischer (Studie ČSAV čislo 3). – Praha: Academia 1975. 176 S., 63 Abb., Brosch. Kčs 40,–

Kurz nach seiner Revision der Gattung *Echinodorus* Rich. legt Rataj wieder eine neue monographische Bearbeitung einer besonders von Aquarianern geschätzten Gattung amphibischer Gewächse, der Araceen-Gattung *Cryptocoryne*, vor. Die konkurrenzschwache, an Flußufern und in Flüssen unter- oder aufgetaucht, vornehmlich im Schatten wachsende Sippe bewohnt die Tropen des asiatischen Kontinents von Südwestindien bis Südchina, Philippinen und Neuguinea. In kritischer Auseinandersetzung mit den taxonomischen Auffassungen der Aquarianer wird die Gattung in vier Untergattungen – *Suturina*, *Myrioblastus*, *Terrestrina* und *Submersina* – gegliedert, wobei die geographische Verbreitung und die zwischenartlichen Beziehungen als Ordnungsprinzipien im Vordergrund stehen. Die alte Englersche Einteilung in 3 Sektionen wird als unnatürlich verworfen. Als Ursprungzentrum der Gattung wird Indien betrachtet.

Die Monographie beschreibt 52 Arten ausführlich, von denen *C. burmensis*, *C. bertelihansenii*, *C. korthausae*, *C. stonei* und *C. bogneri* für die Wissenschaft neu sind. Außerdem werden noch 7 neue infraspezifische Sippen und einige Neukombinationen vorgestellt. Alle Arten sind abgebildet (klare Strichzeichnungen). Die Verbreitung der Untergattungen ist kartographisch festgehalten. Angaben über die Morphologie, Vermehrung, Ökologie, geographische Verbreitung und die Methoden der Kultivierung ergänzen den taxonomischen Hauptteil. Alle Sippen sind aufgeschlüsselt.

Der technische Apparat umfaßt Listen der auszusondernden Arten, der Herbarien und ihrer im Text verwendeten Abkürzungen, der untersuchten Sammlungen und der Taxa sowie ein Literaturverzeichnis. Insgesamt eine sehr wertvolle Arbeit, die eine empfindliche Lücke im botanisch-taxonomischen Schrifttum schließt.

S. J. Casper

M. B. Usher & M. H. Williamson (eds.): Ecological Stability. London: Chapman and Hall 1974. 196 S., 43 Abb., 12 Tab. Ln. £ 3.60

Der vorliegende Sammelband ist das Ergebnis einer im Juli 1973 an der University of York veranstalteten Vortragstagung und zeigt, über welch breite Erfahrungsbasis die englischen Fachkollegen heute auf dem Gebiet der mathematischen Ökologie verfügen. Die Veranstalter und Herausgeber waren bestrebt, eine kleine Gruppe von Fachleuten auf den Gebieten der mathematischen Ökologie, Fischereibiologie, terrestrischen Ökologie sowie Ökologie der Mikroorganismen zusammenzubringen, die einerseits imstande sind, ein breites Spektrum theoretischer Konzepte vorzulegen, zum anderen aber auch bereit sind, ihren Beitrag nur als einen Spezialfall zu betrachten, an dem Prinzipielles demonstriert werden kann. In diesem Buch wird versucht, Grundsätzliches und Verallgemeinerungsfähiges am Beispiel möglichst vieler Organismenarten und mathematischer Methoden zu demonstrieren. Die Autoren sind sich darin einig, daß kaum in einem Fall alle Parameter verfügbar sind, die man brauchte, um ein ökologisches System wirklich vollständig beschreiben zu können. Zum anderen bietet dieser Symposiumsbericht einen ausgezeichneten und umfassenden Überblick über den gegenwärtigen Wissensstand. Die Herausgeber und Autoren verstehen unter Stabilität ein Maß für die Intensität der Annäherung an ein Gleichgewichtsniveau oder an einen Grenzzyklus. Damit hängt die Fähigkeit der Ökosysteme zusammen, Störungen zu kompensieren. Instabilität liegt dann vor, wenn die Systeme weder zum Ausgangszustand zurückkehren noch sich auf einen neuen Gleichgewichtszustand einstellen.

In der allgemeinen Einführung weist R. M. May darauf hin, daß die gegenwärtige Zielstellung der mathematischen Ökologie nicht mit den Exaktheitsansprüchen der Physik verglichen werden kann. Dort, wo der theoretische Physiker ein Gleichheitszeichen setzt ($A = B$), gilt in der Ökologie meistens nur: $A \sim B$, d. h., A nnd B rangieren in der gleichen Größenordnung. Mehrere Autoren demonstrieren die große Bedeutung der räumlichen Heterogenität für die Aufrechterhaltung einer großen Artenmannigfaltigkeit und der damit verbundenen Stabilität.

Mit dynamischen Prozessen innerhalb des gleichen trophischen Niveaus befassen sich die Beiträge von M. Williamson (Situation der Entwicklungszyklen von Populationen unter Berücksichtigung der Altersstruktur) sowie T. Wyatt (Entstehung von Red-Tides infolge Verringerung der Durchmischungstiefe). Die vom letzteren Autor vorgestellten Gleichungen erklären die bei Dinoflagellaten beobachtete exponentielle Zunahme der eigenbeweglichen und eine generelle Abnahme der nichteigenbeweglichen Phytoplankter.

Die meisten Beiträge befassen sich mit Räuber—Beute- sowie Wirt—Parasit-Beziehungen. R. M. Anderson überprüft die innere und äußere (durch Umweltfaktoren wie Wassertemperatur und Freßintensität der Zwischenwirte bestimmte) Stabilität des Entwicklungsablaufes parasitischer Würmer (Fischbandwurm *Caryophyllaeus laticeps*). D. J. Bradley zeigt, daß die Stabilität von Wirt—Parasit-Verhältnissen am besten durch eine Kombination von mathematischem Modell, Gewässeruntersuchungen und Laborexperimenten erfaßt werden sollte. Die Bedeutung der für die Stabilität maßgebenden Mechanismen geht daraus hervor, daß eine Verringerung des *Anopheles*-Bestandes auf weniger als 2% des Ausgangswertes keinesfalls zu einem auch nur annähernd ähnlich starken Rückgang des Malariabefalles führen muß. Als Beurteilungsgrundlage eines Varianten-Vergleiches hinsichtlich Molluskizidanwendung und direkter Medikamentierung befallener Rinder bzw. Schafe legt G. Gettinby ein wahrscheinlichkeitstheoretisch fundiertes Modell des Kontaktes von *Miracidium*-Larven und der Schnecke *Lymnaea truncatula* vor.

Selbst einfache Modelle produzieren oft ein so großes Spektrum von Fallbeispielen, daß deren Gültigkeit nur auf Grund eines umfangreichen Primärdatenmaterials beurteilt werden kann.

Während ein Mehrfach-Parameter-Modell für die Wechselwirkung von Wirt und Parasit eine fast unendlich große Vielfalt von Kombinationen produziert, kommen I. Rogers & S. Hubbard zu dem Ergebnis, daß die natürlichen Verhältnisse weniger kompliziert sind und eine wesentlich elegantere Lösung repräsentieren, besonders im Hinblick auf adaptive Komponenten. Die Beiträge von J. R. W. Harris sowie J. H. Lawten et coll. befassen sich mit der Kinetik der Polyphagie. Auf Grund von Chemostatkulturen mit *Escherichia coli* und des Amöbenstadiums des Schleimpilzes *Dictyostelium discoideum* gelangen M. J. Bazin et coll. zu der Schlußfolgerung, daß Räuber—Beute-Beziehungen, wenn sie realistisch dargestellt werden wollen, entweder ein sehr komplexes mathematisches Modell erfordern, oder die Wahl eines genügend einfachen ökologischen Experimentalsystems. Das gewählte Beispiel wies einen hohen Grad von Stabilität auf.

M. J. WEBBER zeigt, daß die Stabilität mit der Höhe des trophischen Niveaus abnimmt. Im Beitrag von J. STEELE wird die Bedeutung der lateralen Disposition als Faktor der „Wolkenbildung" des marinen Phytoplanktons und der dadurch bedingten großräumigen Stabilität untersucht.

Dieses Buch vermittelt allen, die sich für theoretische Probleme der Ökologie interessieren, eine Fülle von Anregungen. Die Benutzung des Werkes wird durch ein Stichwortverzeichnis erleichtert.

D. UHLMANN

WINWOOD READE & ERIC HOSKING: Vögel in der Brutzeit. Deutsche Übersetzung und Bearbeitung von Dr. KLAUS RUGE. Stuttgart: Verlag Eugen Ulmer 1974. 308 S., 48 Zeichnungen, 168 Farbfotos, 41 Schwarzweißfotos u. 19 Taf. der Vogeleier. Brosch. DM 18,—

An guten Vogelbestimmungsbüchern herrscht wahrlich kein Mangel. Wer sich jedoch über die Brutbiologie unserer Vögel unterrichten wollte, war bislang auf die Benutzung teurer Handbücher oder die oft nicht leicht zugängliche Originalliteratur angewiesen. Diesen unbefriedigenden Zustand zu überwinden, ist Ziel und Absicht des vorliegenden, aus dem Englischen übertragenen Taschenbuches.

In sehr gedrängter Form werden einleitend ein allgemeiner Überblick der Fortpflanzungsbiologie und begrüßenswerte Hinweise zum verantwortungsvollen Beobachten brütender Vögel gegeben. Der Hauptabschnitt des Textes bringt unter den Stichwörtern Brutperiode, Brutgebiet, Nest, Eier, Altvögel, Jungvögel und Hinweise eine übersichtlich gegliederte, informationsreiche Darstellung der wesentlichen brutbiologischen Daten und feldornithologischen Kennzeichen fast aller mitteleuropäischen Brutvögel. Leider sind eine Reihe von zwar seltenen und bei uns nur lokal verbreiteten Arten, wie Rothalstaucher, Moorente, Wespenbussard, Felsenschwalbe, Schlagschwirl, Seggenrohrsänger, Sperbergrasmücke, Sprosser und Grüner Laubsänger, ohne ersichtlichen Grund nicht behandelt worden. Sehr wertvoll für den Feldbeobachter sind die knappen, aber lebendigen Beschreibungen des Balz- und Brutpflegeverhaltens, die ihm das erfolgreiche Aufspüren der Brutstätten sicher erleichtern. Zugleich wecken sie mit ihrer streng wissenschaftlichen Betrachtungsweise echtes Verständnis für die Eigenheiten tierischen Verhaltens.

Besonders hervorzuheben ist die für ein Taschenbuch geradezu verschwenderische Ausstattung des Bildteils mit etwa 200 überwiegend farbigen Fotos des weithin bekannten Fotografen ERIC HOSKING, deren Güte in Aufnahmetechnik und biologischer Aussagekraft, von wenigen schlechten Druckwiedergaben abgesehen, über alle Kritik erhaben ist. Von 176 Arten werden außerdem die Eier farbig dargestellt. Hier wäre eine stärkere Ausrichtung auf mitteleuropäische Verhältnisse wünschenswert. Die von der Deutschen Sektion des Internationalen Rates für Vogelschutz herausgegebene „Rote Liste" bedrohter Vogelarten, kurze Literaturhinweise, eine Zusammenstellung der vogelkundlichen Vereinigungen und ein Index der deutschen und lateinischen Vogelnamen beschließen das in Text und Bild vorzüglich gestaltete Taschenbuch. Es besitzt einen hohen Gebrauchswert bei erstaunlich niedrigem Preis.

D. FLÖSSNER

M. SCHRETZENMAYR (Hrsg.): Der Wald. 2. Auflage. Unter Mitarbeit von D. BLOSSFELD, R. BÖSENER, H. JAEGER, H. LANDBECK, M. LÖSCHAU, W. NEBE, W. WAGNER & K. WEIDERMANN Leipzig, Jena, Berlin: Urania-Verlag 1975. 260 S., 30 Farb- und 90 Schwarzweiß-Fotos, 40 Aquarelle, 80 Zeichnungen u. 15 Tab. Ln. M 19,50

Etwa 27% der Oberfläche unserer Republik sind von Wäldern bedeckt. Dennoch haben in unserem hochindustrialisierten Lande nur noch wenige Menschen eine engere Verbindung zu ihnen; denn etwa die Hälfte unserer Bevölkerung lebt bereits in der Stadt. Ihre Begegnung mit dem Walde beschränkt sich in der Regel auf die Zeiten und Gelegenheiten, da sie zur Wiederherstellung ihrer Arbeitskraft in ihnen Erholung suchen und neue Kräfte sammeln. Die große Bedeutung des Waldes für das Klima und den Boden, für den Wasserhaushalt und für die Hygiene des Luftraumes, für Landeskultur und Umweltschutz ist seit langem bekannt und hat in unserer Republik schon frühzeitig gesetzliche Bestimmungen entstehen lassen, die seine Erhaltung durch eine geeignete Forstwirtschaft zum Gegenstande haben. Schutz und Pflege des Waldes sind eine vordringliche und immer von neuem zu bewältigende Aufgabe unserer sozialistischen Gesellschaft. Es ist daher sehr zu begrüßen, wenn der Urania-Verlag in Zusammenarbeit mit einem Autorenkollektiv unter Leitung von M. SCHRETZENMAYR ein weite Kreise unserer Bevölkerung ansprechendes, im besten Sinne des Wortes populärwissenschaftliches Werk über das „Ökosystem" Wald erarbeitet hat und bereits in 2. Auflage vorlegt.

Die Darstellung nimmt ihren Ausgang von einer begrifflichen Bestimmung dessen, was unter Wald zu verstehen ist, und beschreibt die Bedingungen, unter denen Wald gedeiht (Autor: W. NEBE). Die folgenden Kapitel stellen die Wälder der tropischen, temperierten und kühltemperierten Klimagebiete einander gegenüber (Autor: M. LÖSCHAU) und skizzieren die nacheiszeitliche Waldentwicklung in Mitteleuropa auf der Grundlage der Ergebnisse pollenanalytischer Forschungen (Autor: SCHRETZENMAYR). Ist dieser Prozeß der nacheiszeitlichen Wandlung des Waldbildes im großen ganzen ein naturbedingter Prozeß, so unterliegt der Wandel des Waldbildes in unserer heutigen Kulturlandschaft (Autor: W. WAGNER) dem mehr oder weniger planmäßigen Einfluß des Menschen, er ist wirtschaftsbedingt. Trotzdem hören die Gesetze der Waldentwicklung nicht auf zu wirken, vielmehr laufen heute naturbedingte und anthropogene Waldentwicklung und -geschichte nebeneinander her. Wenn das umfangreiche Kapitel über die wichtigsten Waldgesellschaften Mitteleuropas (Autor: SCHRETZENMAYR) die Wälder der Gebirge, des Hügellandes und der Ebenen, die Wälder entlang des fließenden Wassers sowie an und in Brüchen und Mooren vorstellt, so ist dabei stets diese enge Verknüpfung von natürlichen Gegebenheiten und gesellschaftlicher Einwirkung zu bedenken. Die „Naturkunde" des Waldes schließt das Kapitel über das Tierleben im Walde ab (Autor: R. BÖSENER). Die Darlegungen wenden sich nunmehr der wirtschaftlichen Bedeutung des Waldes zu. Organisation und Leitung der Forstwirtschaft in unserer Republik (Autor K. WEIDERMANN), die Forsteinrichtungen und der Waldbau (Autor: H. LANDBECK) stehen im Mittelpunkt des 2. Teiles des Bandes. Hauptaufgabe der Forstwirtschaft ist die Erzeugung von Rohholz und dessen ausreichende und qualitätsgerechte Bereitstellung zur Befriedigung des volkswirtschaftlichen Bedarfs. Die interessanten und aufschlußreichen Ausführungen über die Bewirtschaftungen des Waldes (Autor: O. BLOSSFELD) vermitteln tiefe Einblicke in wichtige, vielen Menschen kaum vertraute, volkswirtschaftlich bedeutsame Prozesse. Auch Nebenprodukte des Waldes wie Harz, Waldstreu, Pilze und Waldfrüchte werden nicht vergessen. Der umfang- und inhaltsreiche Band schließt mit der Würdigung der landeskulturellen Bedeutung des Waldes (Autor: H. JAEGER) ab. Er verdeutlicht, daß unser Staat dem Naturschutz im Interesse einer menschenwürdigen Gestaltung unserer Umwelt große Aufmerksamkeit widmet.

Der Text liest sich trotz der Vielzahl der Autoren relativ leicht und ist bei allem Bemühen, sich einem breiten Leserkreis verständlich zu machen, wissenschaftlich fundiert und sachgerecht. Die großzügige Ausstattung mit Zeichnungen, Aquarellen, Schwarzweiß- und Farbfotos sowie mit Schemata, Karten und Tabellen ist geeignet, Interesse zu wecken und Sachverhalte zu veranschaulichen. Den Autoren und dem Verlag kann bescheinigt werden, daß sie ihr Vorhaben mit Unterstützung befähigter Zeichner erfolgreich bewältigt haben. Der Rezensent hätte sich gewünscht, daß im Kapitel über die wichtigsten Waldgesellschaften Mitteleuropas den Pilzen mehr Aufmerksamkeit geschenkt worden wäre. Gehört doch die „Pilzjagd" heute zu einer der beliebtesten Freizeitbeschäftigungen vieler Menschen. Auf sie findet sich nur im Abschnitt über die Nebenprodukte des Waldes ein bescheidener Hinweis. Der Anhang weist ein Literaturverzeichnis, einen Bildquellennachweis und ein Namen- und Sachverzeichnis auf.

Druck und Ausstattung sind gut. Autoren und Verlag haben ein wertvolles Buch geschaffen, das geeignet ist, bei den Lesern Verständnis und Interesse für den Schutz und die Pflege ihrer Umwelt zu wecken.

S. J. CASPER

RICHARD A. VOLLENWEIDER (Ed.): A manual on Methods for Measuring Primary Production in Aquatic Environments (IBP Handbook Nr. 12). Oxford & Edinburgh: Blackwell Scientific Publications 1974. 2. Aufl. XVIII + 225 S., 16 Abb., 3 Tab. Steif brosch. £ 3.75

Im zunehmenden Maße kristallisiert sich die primärproduktionsbiologische Forschung als eine Schwerpunktaufgabe innerhalb ozeanologischer und limnologischer Grundlagenuntersuchungen heraus. Im weltweiten Maßstab kommt es dabei besonders darauf an, durch international standardisierte Meßmethoden direkt vergleichbare Ergebnisse vorzulegen, ein Anliegen, das besonders vom Internationalen Biologischen Programm (IBP) durch die Herausgabe von Handbüchern konsequent verfolgt wird. Bisher erschienen in dieser Serie 24 Handbücher, 4 wurden dabei von der Sektion „Productivity of Freshwater" des IBP bearbeitet. Die vorliegende 2. Auflage des von RICHARD A. VOLLENWEIDER unter Mitarbeit von J. F. TALLING und D. F. WESTLAKE herausgegebenen Handbuches zur Messung der Primärproduktion in Gewässern baut im wesentlichen auf einer nur geringfügig veränderten Wiedergabe der 1969 erschienenen 1. Auflage auf. Allerdings haben neue methodische Erkenntnisse auf dem Gebiet der Flüssig-Szintillations-Techniken, der ATP-Bestimmung und der fluoreszenzoptischen Messung von Chlorophyll (in vivo) sowie einige Kalkulationsbeispiele für Primärproduktionsmessungen zur Erweiterung einzelner Abschnitte geführt. Zum anderen konnte durch die Herausgabe der IBP-Handbuchs Nr. 23 („Techniques for the assessment of microbial production and decomposition in fresh waters") der Originaltext der 1. Auflage im Abschnitt „Mikrobielle Produktion" stark gekürzt werden. In dieser Auflage vermißt man einen kurzen Überblick über die immer mehr Verbreitung findenden autoradiographischen Methoden zur Bestimmung artspezifischer Primärproduktionsleistungen und einen Abschnitt über die Besonderheiten mariner primärproduktionsbiologischer Methoden und Probleme.

Der Inhalt des Buches ist straff gegliedert. Nach einer kurzen Einleitung werden die Technik der Probenentnahme und die Methoden zur Messung der Qualität und Quantität der Biomasse des Phytoplanktons, der Makrophyten und des Periphytons erläutert (27 S.). Es folgt ein Abschnitt über Methoden zur Erfassung von Produktionsraten (102 S.). Hier werden chemische und physikalisch-chemische Methoden, ^{14}C-Tracertechniken, die in-situ-Messungen an natürlichen isolierten und nicht isolierten Phytozönosen, die Berechnung der Primärproduktionsraten aus Veränderungen der Biomasse, die Messung der Photosynthese unter Laborbedingungen und die Interpretation der Ergebnisse beschrieben. Bei der bakteriellen Produktion werden nur einige ausgewählte Probleme behandelt (4 S.). Der sich anschließende Abschnitt geht auf für die Primärproduktion wichtige chemische und physikalische Umweltfaktoren ein (31 S.). Es werden dabei besonders die Methoden zur Strahlungsmessung in Gewässern hervorgehoben. Der Anhang enthält Bemerkungen über Algenkulturen und statistische Verfahren zur Auswertung und Beurteilung gewonnener Daten (19 S.). Ein ausgewogenes Literaturverzeichnis ermöglicht ein tieferes Eindringen in einzelne Probleme.

Das vorzüglich ausgestattete Handbuch wird seiner Aufgabe, Leitfaden auf dem Gebiet der Primärproduktionsforschung in Gewässern zu sein, voll gerecht. Allein die Mitarbeit von 22 namhaften Autoren aus 11 Ländern spricht für die Qualität. Das Anliegen dieses Handbuchs besteht dabei nicht nur in der Wiedergabe des letzten Standes der Methodologie auf diesem Fachgebiet, sondern auch in der Vermittlung kritischer Momente. Es ist uneingeschränkt zu empfehlen.

R. Koschel

Ivo Horvat †, Vjekoslav Glavač & Heinz Ellenberg: Vegetation Südosteuropas (Geobotanica selecta, Hrsg.: R. Tüxen, Bd. IV). Stuttgart: Gustav Fischer Verlag 1974. XXXII u. 768 S., 412 Abb., 153 Tab., 2 farb. Kt. Ln. DM 360,— (Lizenzausgabe im VEB Gustav Fischer Verlag Jena)

Dieses Werk, von dem jugoslawischen Geobotaniker Ivo Horvat begonnen und nach dessem frühen Tode von seinem Schüler Vejkoslav Glavač und dem Göttinger Ökologen Heinz Ellenberg zu Ende geführt, ist eine bewunderungswürdige Leistung. Es gibt einen modernen Überblick über die Vegetation dieses so außerordentlich vielgestaltigen und floristisch reichhaltigen Teils von Europa. Die Darstellung beschränkt sich dabei nicht auf verbale Beschreibungen, sondern legt in rund 150 zusammengefaßten Vegetationstabellen eine Fülle von Einzelmaterial vor, das eine weitgehende Einsichtnahme in Zusammensetzung, Gliederung und Struktur aller wesentlichen Vegetationseinheiten des südosteuropäischen Raumes gestattet.

Als Südosteuropa wird hier die eigentliche Balkanhalbinsel bis zur Donau, Save und Kupa im Norden verstanden. Der pannonisch – transsilvanisch – karpatische Raum bleibt somit außerhalb der Betrachtung (für ihn wünschte man sich späterhin eine ähnliche Darstellung). Voranges telltwird ein einführender Überblick mit ausführlichen Informationen u. a. zur Forschungsgeschichte, über Klima, Geologie, Böden, Vegetationsgeschichte und florengeographische Gliederung. Der Hauptteil des Buches behandelt dann, gegliedert nach Vegetationszonen, die einzelnen Formationen und Pflanzengesellschaften. Die Darstellung beginnt mit den immergrünen mediterranen Küstenzonen im S und äußersten W, behandelt die sich daran anschließende submediterrane winterkahle Laubmischwald-Zone, die kontinentalen Laubmischwald- und Steppenwaldzonen in den zentralen und nordöstlichen Ebenen und Hügelländern sowie die zentraleuropäische Laubmischwald-Zone im N, wendet sich dann den zentraleuropäischen montanen Buchen- und Tannenwald-Zonen sowie den montanen und submontanen Nadelwald-Zonen der Gebirge zu und endet mit den Zonen der Hochgebirgsvegetation. Der von zahlreichen Artenlisten und Vegetationstabellen durchsetzte und vertiefte Text wird von einer Fülle instruktiver Fotos, Diagrammen, Vegetationsprofilen und Verbreitungskärtchen begleitet, die dem Leser die Tatsachen und Probleme in nachahmenswert anschaulicher Weise nahebringen. Zwei farbige Vegetationskarten, und zwar eine Karte der zonalen Vegetationseinheiten 1:3 Mill. und eine Karte der Vegetation des nordwestlichen Karstgebietes von Jugoslawien 1:500000, tragen zur weiteren Verdeutlichung der Ausführungen bei. Das umfangreiche Literaturverzeichnis belegt die benutzte Literatur und gibt zugleich Hinweise auf die Einzelarbeiten; Register der gültigen Pflanzennamen und (bei der Mannigfaltigkeit der Literatur sehr wichtig!) deren Synonyma, der Pflanzengesellschaften, der geographischen Namen und der übrigen Sachen und Begriffe helfen, das Werk nach allen Richtungen hin zu erschließen.

Auf S. 319 wird der Name Dobrudscha als „Eichenwaldgebiet" erklärt. Hierzu sei auf das Russische etymologische Wörterbuch von Max Vasmer, Bd. 1, Heidelberg 1953, S. 356, verwiesen, das diesen Landschaftsnamen als osmanische Umgestaltung eines bulgarischen Adjektivs Dobrotiča (zum Namen eines Fürsten Dobortica) deutet.

Das Buch besticht nicht nur durch das immense Sachwissen, das in ihm niedergelegt ist, sondern auch durch eine überall spürbar werdende sorgfältige Bearbeitung, durch die ausgewogene, auch die vielen noch offen bleibenden Fragen und Probleme deutlich machende Darstellung und nicht zuletzt durch seine gute Ausstattung und den sauberen Druck. Mitteleuropäische wie dort heimische Vegetationsforscher werden es mit großem Gewinn benutzen, nicht nur als Zusammenschau des bisher Bekannten, sondern auch als verläßlichen Führer bei geobotanischen Exkursionen. Was die „Vegetation Südosteuropas" darüber hinaus wertvoll macht, sind die vielen Ansätze und Anregungen für weitere Untersuchungen, die es vermittelt, und welche dazu beitragen werden, die zwangsläufig noch verbliebenen Lücken zu schließen. H.-D. Krausch

Edward O. Wilson (ed.): Ecology, Evolution, and Population Biology. (Readings from Scientific American). San Francisco: W. H. Freeman & Co. 1974. VIII u. 319 S., ca. 315 Abb., 2 Tab. Brosch. £ 2.80 ($ 5.95)

Die neue Serie der „Readings" umfaßt „Essentials of Modern Biology" in vier Bänden, unter denen das vorliegende Sammelwerk angesichts des hohen Stellenwertes von Ökologie und Evolutionstheorie in der Biologie (und Naturwissenschaft) der Gegenwart großes Interesse beanspruchen kann. Zeigt sich doch immer mehr, wie E. O. Wilson in seiner kurzen Einleitung hervorhebt, daß Evolutionstheorie ohne Ökologie leer und daß umgekehrt Ökologie ohne evolutionistische „Geschichte" oder „Dynamik" von Populationen wissenschaftlich irrelevant sind. Insofern leisten die 34 ausgewählten, originalen Einzelbeiträge, die eine Zeitspanne von 23 Jahren (1950 bis 1972) umfassen, der Klärung der Beziehungen und wechselseitigen Anhängigkeiten der beiden Disziplinen einen guten Dienst. Die Artikel sind in vier Sachgebiete — 1. The Evolutionary Process; 2. The Multiplication and Dispersal of Species; 3. The Growth and Interaction of Populations; 4. Ecosystems — eingeordnet, denen jeweils eine ausführliche Einführung samt Hinweisen auf weiterführende Literatur (so bei den „älteren" Beiträgen den Anschluß an aktuelle Forschungen vermittelnd) vorangeht. Eine ausreichende Bibliographie und ein Index bilden den technischen Apparat des gut gedruckten und ausgezeichnet illustrierten Buches. Einige der Einzelbeiträge seien, natürlich wertfrei, genannt: Wills „Genetic Load"; Watts & Stokes „The Social Order of Turkeys; Lack „Darwin's Finches"; Kurtén „Continental Drift and Evolution"; Ehrlich & Raven „Butterflies and Plants"; Margulis „Symbiosis and Evolution"; Hutchinson „The Biosphere"; Woodwell „The Energy Cycle of the Biosphere"; Marples „Life on the Human Skin. S. J. Casper

J. Salánki & J. E. Ponyi (eds.): Limnology of Shallow waters. (Symposia Biologica Hungarica vol. 15). Budapest: Akademie Kiadó 1975. 304 S., 150 Abb., 59 Tab. Ln. Ft. 190,—

Die ungarische Akademie der Wissenschaften führte vom 3.—8. 9. 1973 im Biologischen Forschungsinstitut in Tihany am Plattensee das Symposium „Limnology of Shallow waters" durch.

Der vorliegende Band beinhaltet alle 29 dort gehaltenen Vorträge. Im Mittelpunkt stehen die Beiträge über die Verhältnisse im Plattensee und Neusiedler See (8 Beiträge). Die anderen Arbeiten verteilen sich auf Untersuchungen in Fischteichen, geschichteten Seen, Stau- und Hochgebirgsgewässern sowie anderen flachen Seen. Das Buch wird mit einem kurzen Vorwort der Herausgeber, der Teilnehmerliste und den Begrüßungsworten von J. SALÁNKI und B. ZÓLYOMI eröffnet. Ein erster Abschnitt beschäftigt sich mit der Primärproduktion (11 Originalarbeiten). Darin werden u. a. Mechanismen der Nährstofffreisetzung durch *Phragmites*-Gesellschaften (R. SCHRÖDER), die Phytoplanktonproduktion im Plattensee (S. HERODEK & G. TAMÁS), die Primärproduktion in der Litoralzone des Tahoesees (CH. R. GOLDMAN & E. DE AMEZAGA), Vergleiche zwischen pelagischer und litoraler Primärproduktion (J. KOMÁRKOVÁ & J. KOMÁREK) sowie die Auswirkungen veränderter Fischbestände auf Phytoplanktonstruktur, Biomasse und Produktion (I. SPODNIEWSKA) dargelegt.

Der zweite Abschnitt ist der bakteriellen Produktion und Dekomposition gewidmet (7 Originalarbeiten). Er beinhaltet u. a. Arbeiten zur Bakterienproduktion im Freiwasser und Schlamm des Neusiedler Sees (M. DOKULIL), zur Funktion des Metalimnions (J. OLÁH), zur Bedeutung von Schichtungsformen für die Ernährung der Organismengesellschaften des Bodens (B. T. HARGRAVE), und zur Soda- und H_2S-Produktion in alkalischen Gewässern (R. VÁMOS & R. TASNÁDI).

Im dritten Abschnitt werden Ergebnisse zur Sekundärproduktion vorgestellt (11 Originalarbeiten). Es sind u. a. Beiträge zum Einfluß von Temperaturschwankungen auf planktische Rotatorien (A. RUTTNER-KOLISKO), zur Zooplanktonbiomasse im Plattensee (J. E. PONYI), zur Nahrungausnutzung von *Eudiaptomus gracilis* (N. P. ZÁNKAI), zur Fischproduktion im Plattensee (P. BIRÓ) und über Entwicklungsrichtungen von Biozönosen in neuentstandenen Gewässern (K. HEUSS) zu finden.

23 der Originalarbeiten sind in englischer, 6 in deutscher Sprache abgefaßt. Sie enthalten z. T. ausführliche Literaturverzeichnisse. Eine Diskussion der Einzelvorträge fehlt.

Der Band liefert wertvolles Material für die weltweiten Bemühungen (z. B. die UNESCO-Programme IBP- und Man and Biosphere), quantifizierbare ökologische Aussagen zu erarbeiten, um die langfristigen Folgen „unbeabsichtiger" Eingriffe in die Gewässer abschätzen und der steuernden Kontrolle durch den Menschen unterwerfen zu können. Besonders für flachere Gewässer sind gegenwärtig derartige Grundlagenarbeiten dringlich. Flache Gewässer werden sehr schnell durch zivilisatorische Verunreinigungen in ihrem Stoff- und Energiehaushalt beeinträchtigt. In kürzester Zeit können für Wasserhaushalt, Wirtschaft und Natur folgenschwere irreparable Schäden erwachsen. Um so mehr ist es zu begrüßen, eine Vortragssammlung zur Hand zu haben, die Ergebnisse zur Primär-, Intermediär- und Sekundärproduktion in verschiedenen Typen von Flachgewässern darstellt. Ausstattung, Druck und Qualität der Abbildungen in diesem Symposiumsband sind hervorragend. R. KOSCHEL

W. J. VAN WAGTENDONK (ed.): *Paramecium* — A current Survey. Amsterdam, New York: Elsevier Scientific Publishing Co. 1974. XVIII u. 499 S., 304 Abb., 53 Tab. Ln. US $ 53.90 (Dfl. 140,—)

10 Autoren bringen in 8 Kapiteln eine gezielte Auswahl bisher nicht genügend gewürdigter Aspekte der *Paramecium*-Forschung: 1) E. VIVIER: Morphology, Taxonomy, Gener. Biol. 89 S., 44 Abb., 2) H. M. BUTZEL, jr.: Mating Type Determination and Development 40 S., 5 Abb., 3) I. FINGER: Surface Antigens 34 S., 7 Abb., 4) S. DRYL: Behavior and Motor Response, 54 S., 18 Abb., 5) M. KANEDA & E. G. HANSON: Growth Patterns and Morphogenetic Events in the Cell Cycle, 44 S., 22 Abb., 6) C. F. EHRET & E. W. MCARDLE: Structure [Molecular-, Organelle-, Cell-Level], 76 S., 166 Abb., 7) W. J. v. WAGTENDONK: Nutrition, 34 S., 7 Abb., 8) A. T. SOLDO; Intracellular Particles, 66 S., 8 Abb.

Das Bedürfnis, das stark angewachsene Wissen zu sichten und die Verlagerung der Forschungsschwerpunkte seit der letzten monographischen Bearbeitung von WICHTERMAN (1953) zu kennzeichnen, waren die Triebkräfte zu diesem unter dem Patronat des Special Publications Committee of the Society of Protozoologists stehenden Unternehmen.

Angelpunkt für die gestiegene Bedeutung von *Paramecium* als Objekt für die in den Vordergrund gerückten biochemischen und genetischen Untersuchungen ist die erfolgte Schaffung definierter axenischer Kulturmedien für einige *Paramecium*-Stämme, denen WAGTENDONK über 20 Jahre Arbeit widmete. Der Appell von SOLDO (S. 426) an künftige Experimentatoren, solche Medien zu benutzen, sollte nicht ungehört verhallen. Nur so wird z. B. das immer noch von Unsicherheiten getrübte Bild der Herkunft und Natur der inzwischen lt. SOLDO auf 10 (durch HECKMANN, 1975 mit dem Partikel Omikron auf 11) angewachsenen genetisch autonomen intrazellulären Partikel zu klären sein. Auch die weitere Erforschung der Mating Types, verschiedener Entwicklungsfragen der Antigenverhältnisse, neuromotorischer Verhaltensweisen und der strukturellen wie substrukturellen Eigentümlichkeiten wird dadurch gefördert werden.

Die prägnante und glänzend illustrierte morphologische Übersicht von Ehret & Ardle demonstriert deutlich sowohl die Fortschritte durch Einsatz aller neuen Forschungsmittel als auch die Möglichkeiten einer ökonomischen Buchgestaltung. Hierfür gibt man gern 1 Dollar pro 10 Seiten aus, wohingegen zahlreiche textliche Überschneidungen und mehrfach gedruckte Abbildungen den Leser etwas verstimmen. Mangelnde Koordination wird auch deutlich an den erfolglosen Bemühungen von Vivier und Kaneda um nomenklatorische Änderungen, die unberücksichtigt bleiben.

Während im allgemeinen der Résumé-Stil vorherrscht, trägt das Kapitel 5 von Kaneda & Hanson ganz den Charakter einer experimentellen Arbeit, der man ein früheres Erscheinen gewünscht hätte. Hier wird einer der Nachteile wissenschaftlicher Buchpublikationen besonders deutlich. Häufig sind wichtige Neuerungen nur in den Addenda zu finden, die jedem Kapitel nachgestellt sind und Literatur bis 1973 berücksichtigen, während die Kapitel selbst zwischen 1969 und 1972 abgeschlossen wurden.

Der bedeutendste Vorteil eines Buches, das Register, wird hier mit 7 doppelspaltigen Seiten der Faktenfülle nicht ganz gerecht. Ein systematischer Index fehlt trotz der zahlreichen Seitenblicke der Autoren auf andere Protozoen, und lediglich die *Paramecium*-Spezies werden mit je einer Seitenangabe aufgeführt.

Obgleich die Domäne der Protozoenforschung wohl jenseits des Atlantiks liegt, ist doch keine Isolierung von anderen Forschungsgruppen zu spüren. Zwei europäische Koautoren (Vivier, Frankreich, Dryl, Polen) und die offenbar regen Konsultationen mit sowjetischen Protozoologen dienen der Sache sehr. Die Statistik der deutschsprachigen Beteiligung ist recht mager: 148 von 1528 Referenzen der Kapitelautoren (~10%), davon nur 11 nach 1969. Nur ca. 3% sind es in der um Vollständigkeit bemühten *Paramecium*-Bibliographie des Herausgebers seit 1953 mit 1354 Zitaten. Diese heroische Fleißarbeit muß man loben. Satztechnisch günstigere Gliederung wäre dem Leser sehr nützlich gewesen. Hätten die Kapitelautoren die *P.*-Bibliographie mit genutzt, wären von den insgesamt ca. 115 Literaturseiten ein beträchtlicher Teil gespart worden, vorteilhaft für die Handlichkeit und den Preis, die ja nicht zuletzt die sehr zu wünschende Verbreitung des Buches beeinflussen. C. Schönfeld

W. D. Williams (ed.): Biogeography and Ecology in Tasmania. (Monographiae Biologicae vol. 25). The Hague: Dr. W. Junk b. v. Publishers 1974. X u. 498 S., 122 Abb., 1 Farbtaf., 48 Tab., 1 farb. Kt. Ln. Dfl. 140,—

In den zahlreichen tiergeographischen und evolutorischen Erörterungen über die australische Fauna wurde die durch die Bass-Straße vom Festland getrennte Insel Tasmanien zumeist vernachlässigt. Auf den ersten Blick scheint die Fauna des Festlandes auch mit der Tasmaniens weitgehend identisch zu sein. Durch Meeressenkungen kam es im Pleistozän mehrmals zu Landverbindungen, so daß ein Faunenaustausch stattfinden konnte. Doch besitzt Tasmanien eine Anzahl klimatischer und geographischer Eigenheiten, die aus seiner Inselnatur und seiner Entwicklung (es war z. B. im Gegensatz zum Festland im Pleistozän teilweise vom Eis bedeckt) resultieren und die einen deutlichen Niederschlag in seiner Fauna gefunden haben. Es gibt viele Rassenbildungen, besonders unter den Säugern, und Arten, die bereits auf dem Festland ausgerottet waren, konnten auf Tasmanien überdauern.

Die 16 Autoren des Bandes fassen das Wissen über die tasmanische Fauna und die relativ noch spärlich vorhandenen ökologischen Studien zusammen, diskutieren die spezielle Problematik dieser Tierwelt, die vieles enthält, das von allgemeinem Interesse ist und vermitteln auch viele neue Kenntnisse.

Für den Limnologen interessant sind die großen Seen Tasmaniens (und auch einige Klein- und Kleinstgewässer), in denen die evolutorisch wichtige paläozoische Krustazeen-Gruppe der Syncarida (Familie Anaspididae) ein Refugium gefunden hat. W. D. Williams gibt einen interessanten Überblick über diese Süßwasser-Krustazeen. P. H. Tyler berichtet über die Limnologie der großen Seen im zentralen Tasmanien, die noch in den Anfängen steckt. Nach ausführlicher Beschreibung einiger Seen werden Angaben über das Plankton gemacht. Es liegen auch schon erste Daten über die Primärproduktion im Sorell- und Crescentsee vor. In Tasmanien gibt es eine Anzahl von Seen, in denen Fische fehlen. Hier bieten sich vergleichende Produktionsuntersuchungen an, die zu wichtigen Erkenntnissen über die Rolle der letzten Konsumenten im pelagischen Ökosystem beitragen könnten.

Weitere Themen des Buches sind: Geomorphologie, Süßwassrfische, Biogeographie der Fauna des marinen Litorals, Oligochaeten, terrestrische Isopoden, Amphibien, Reptilien, Vögel, Säuger, Ureinwohner und ein Kapitel über Notwendigkeit und Möglichkeit der Erhaltung dieser Welt.

Ein wichtiges Werk der ökologischen und historischen Tiergeographie. W. SCHÖNBORN

Beiträge zur Meereskunde, Heft 34 (Hrsg.: ERICH BRUNS unter Mitwirkung von O. KOLP, G. SAGER, K. VOIGT Berlin: Akademie-Verlag 1974. 64 S., 23 Abb., 9 Tab. Br. M 20,—

Hydrophysikalische und hydrochemische Arbeiten großer Aktualität bilden den Inhalt dieses Heftes. H.-J. BROSIN berichtet von „Untersuchungen zur mittelmaßstäblichen horizontalen Diffusion mit Driftbojen in den Gewässern um Rügen". Mittels Schiffsradar wurden Driftkörper während geringer Windbewegung und geringem Wellengang beobachtet. L. BRÜGMANN stellt „Die Bestimmung von Zink, Kadmium und Blei in der Ostsee durch inverse Voltammetrie" vor. Eine stationäre Quecksilberelektrode von KEMULA-Typ ermöglicht durch inverse Voltammetrie, die labilen Konzentrationen der genannten Elemente zu bestimmen. Die tabellarisch und graphisch vorgeführten Ergebnisse stammen von Wasserproben aus der Ostsee. L. GOHS berichtet „Über einige hydrographische und hydrooptische Beobachtungen an der Westmole von Warnemünde 1972 und 1972". G. SAGER behandelt „Bestimmung der Gezeitenellipse aus zwei Stromvektoren".

H. FRANCK & W. MATTHÄUS bieten den „Vergleich zweier Methoden zur Berechnung von Jahresgängen aus nichtäquidistanten Meßwerten". Dieses bei langfristigen Messungen äußerst akute Problem wird durch Vergleich der Harmonischen Analyse und des Ausgleichsverfahrens an ozeanographischen Messungen (Bornholmbecken und Gotlandtief) diskutiert. Es werden 7 Empfehlungen für die Anwendung der Verfahren gegeben. Das Ausgleichsverfahren bietet offensichtlich größere Einsatzvorteile. P. HUPFER, CH. DRUET & O. KUSNEZOW stellen das „Internationale Experiment EKAM 73 in der ufernahen Zone der Ostsee bei Zingst" vor. Die vielfältigen „Einflüsse der Küste auf Atmosphäre und Meer" (EKAM) verlangten nach komplexen Untersuchungen, zu denen sich hier die UdSSR, die VR Polen und die DDR zusammenfanden. Die Meßprogramme werden ausführlich erörtert. Man darf auf die Ergebnisse sehr gespannt sein. L. KREY

WALTER CARLÉ: Die Mineral- und Thermalwässer von Mitteleuropa. Geologie, Chemismus, Genese. Bücher der Zeitschrift Naturwissenschaftliche Rundschau. Stuttgart: Wissenschaftliche Verlagsgesellschaft mbH. 1975. XXIV u. 643 S., 14 Abb., 1402 Analysentab., 15 Karten in besonderer Mappe. Geb. DM 265,— (Vorzugspreis für Bezieher der Zeitschrift DM 212,—)

Das inhaltsreiche Werk umfaßt 5 Teile. Die Einleitung (48 S.) enthält nach kurzen Bemerkungen über die Werksentstehung, die Abgrenzung der Aufgabe und des untersuchten Gebiets eine geologische Übersicht Mitteleuropas sowie einen kurzen Abriß der Mineralwasserkunde. Der nach geologisch-hydrologischen Einheiten in 6 Hauptabschnitte untergliederte Hauptteil (500 S.) enthält — ausgehend von den geologischen Verhältnissen — die Beschreibung der vorkommenden Wässer und ihrer Zusammensetzung sowie Ausführungen zur Genese einzelner Wässer oder Wassergruppen. Im 3. Teil (38 S.) finden sich Ausführungen zur regionalen Verbreitung und Genese der Mineral- und Thermalwässer. Der Band wird durch einen Verzeichnisteil (50 S.) abgeschlossen. Der Kartenteil wurde in einer besonderen Mappe beigegeben.

Das vorliegende Werk ist Ausdruck einer fast 30jährigen Beschäftigung mit der Mineralwassergeologie. Aus einem Gebiet von etwa 1 Million km², das teilweise oder ganz 15 Staaten einschließt, werden rund 2200 Vorkommen aufgeführt, d. h., der Verfasser hat nach Vollständigkeit gestrebt. Diese so umfangreiche Inventur aller dem Autor bekannt gewordenen Wässer, auch der nie oder nicht mehr genutzten, ist ein sehr wertvolles Nachschlagewerk für Hydrogeologen, Hydrologen,

Geologen, Limnologen, Balneologen und viele andere an der Thematik Interessierte. Die genetische Deutung war für den Autor eigentlich das wichtigste Ziel seines Werkes, dabei wurde auch die modernste Literatur berücksichtigt (das Literaturverzeichnis enthält noch im Jahre 1974 veröffentlichte Arbeiten).

Diesem hervorragenden, gut gedruckten Werk ist eine weite Verbreitung sicher. In einer zweiten Auflage müßten aber unbedingt Veränderungen hinsichtlich der Schreibweise von Ortsnamen erfolgen. Der Autor erklärt zwar ausdrücklich, daß die vorrangige Verwendung deutscher Namen „nicht aus chauvinistischen Gründen" erfolgt, „sondern um des Verständnisses für meine deutschen Landsleute willen, für die das Buch in erster Linie geschrieben worden ist". Trotzdem muß man nach Ansicht des Rezensenten alle heute gültigen Ortsnamen in den Karten und im Text angeben, wenn es sich um Orte handelt, die seit 30 oder mehr Jahren zu Polen, Ungarn, der Tschechoslowakei oder anderen Staaten gehören. Daß es mal Warschau und mal Warszawa heißt oder gar Preßburg (Karte I. 4) und Bratislava (Karte I. 9), ist sicher auf die verwandte alte Literatur zurückzuführen. Sehr bedenklich aber ist z. B., daß Bad Salzungen, Bad Liebenstein, Schmalkalden, Suhl und andere Orte unter der Überschrift „2.8. Hessisch-nordbayerisches Bergland" zu finden sind, Lobenstein dagegen unter „3.1.2. Nordost-Bayern". D. Storch

David P. Hoult (ed.): Oil on the sea. Proceedings of a symposium on the scientific and engineering aspects of oil pollution of the sea, sponsored by Massachusetts Institute of Technology and Woods Hole Oceanographic Institution and held at Cambridge, Massachusetts, May 16, 1969. New York—London: Plenum Press 1969. VII + 114 S., 33 Abb. Ln. $ 8.75

Die Verschmutzung unserer Gewässer durch Öl ist ein immer größer werdendes Umweltproblem. Welche katastrophalen Folgen Öleinbrüche hervorrufen können, wird hier am Beispiel der Santa Barbara Ölbohrung demonstriert.

Die Ölverunreinigung der Ozeane und deren Wirkung auf die marine Lebewelt, mögliche Gegenmaßnahmen bei Havarien, z. B. mit Dispersionsmitteln oder Abfackelung, die Ausbreitung und der Transport von Öl im Meerwasser bis hin zu Kontrollmaßnahmen, werden in 9 Einzelberichten behandelt. Da wenig zusammenfassende Darstellungen auf diesem Gebiet vorhanden sind, ist die Veröffentlichung der Symposiumsergebnisse sehr zu begrüßen. Die Form ist als gelungen zu bezeichnen. G. Proft

Eugen Seibold: Der Meeresboden. Ergebnisse und Probleme der Meeresgeologie. Berlin, Heidelberg, New York: Springer-Verlag 1974. XII + 183 S., 86 Abb., 10 Tab. Geheftet. DM 29,80 (US $ 13.00)

Der Keiler Meeresgeologe schrieb dieses in erster Linie als Hochschultext gedachte Buch nicht nur für Studenten der Geologie, sondern auch für Fachwissenschaftler der verschiedensten Disziplinen und für interessierte Laien. Die Fragen der Meeresgeologie haben in den letzten Jahren eine immer größere Bedeutung erlangt, nicht nur durch verbesserte Untersuchungsmethoden und neue geologische Hypothesen, sondern auch durch die weltweit einsetzende Rohstoffsuche auf und unter dem Meeresboden und die zunehmende Eutrophierung auch der Meere.

E. Seibold versuchte, „kein detailliertes Grundlagenwissen vorauszusetzen." Er bringt überwiegend Fakten und Erkenntnisse der jahrzehntelangen meeresgeologischen Forschungen, einschließlich neuester Untersuchungsergebnisse (das Literaturverzeichnis enthält u. a. 12 Titel aus den Jahren 1972—1974). Daran schließen sich die kritische Wertung des erreichten Kenntnisstandes und der Hinweis auf offene Probleme an. Auf eine ausführliche Beschreibung von Geräten und Methoden verzichtet der Autor bewußt, ebenso auf die breite Darstellung von Hypothesen und Theorien.

Nach der Einleitung (endogene und exogene Kräfte) werden in 7 Kapiteln folgende Themen behandelt: Formen des Meeresbodens (26 S.), Herkunft und Zusammensetzung der marinen Sedimente (30 S.), Meeresboden und Wasserbewegung (38 S.), Meeresboden und Organismen (21 S.), Meeresboden und Klimazonen (21 S.), Meeresboden und Rohstoffe (12 S.) und Zur Entstehung der Ozeane (19 S.). Dem Text folgen eine „Formationstabelle" (sollte durch einen moderneren Begriff ersetzt werden!), weiterführende Literatur (62 Titel), das Quellenverzeichnis der guten, instruktiven Abbildungen und ein Sachverzeichnis.

Das Buch bietet den an mariner Ökologie interessierten Biologen und Mikrobiologen einen hervorragenden Überblick über die behandelten Themen, d. h. die geologischen Grundlagen für ihre Untersuchungen. Der Mikrobiologe wird vielleicht Ausführungen über die Bedeutung der Bakterien für die Sedimentation und die Wechselwirkung zwischen Substrat und Wasserkörper vermissen. Die ökologischen Angaben sind für den Paläontologen bei palökologischen Fragestellungen wichtig, aber auch für paläolimnologische und limnologische Untersuchungen sind Anregungen enthalten.

Dieses wertvolle Buch ist mehr als nur eine „Einführung“, es zeigt Probleme und offene Fragen auf, die nur in interdisziplinärer Gemeinschaftsarbeit zu lösen sind. Es schließt mit einem Satz von G. KNETSCH (1972): „Die Erde beantwortet nur Fragen, die gezielt an sie gestellt werden, aber sie spricht in Sprachen von vielen Spezialfächern und nur die vielsprachige Gemeinschaft vermag sie ganz zu verstehen!“ D. STORCH

M. J. D. WHITE: Genetic Mechanisms of Speciation in Insects. Symposia Held at the XIVth International Congress of Entomology, Canberra, Australia August 22–30, 1972, sponsored by the Australian Academy of Science and the Australian Entomological Society. Dordrecht-Holland/Boston-USA: D. Reidel Publishing Company 1974. VII u. 170 S., 43 Abb., 13 Tab. Geb. Dfl. 45,– (US $ 18.00)

Der vorliegende Band enthält die Vorträge zweier Symposia, die anläßlich des 14. Internationalen Kongresses vom 22.–30. 8. 1972 in Canberra stattfanden. Wie bereits im Titel zum Ausdruck kommt, werden genetische Mechanismen, die bei einigen Insektengruppen zu einer großen Formenmannigfaltigkeit führten, besprochen und aus dem umfangreichen Untersuchungsmaterial teilweise recht orthodoxe Schlußfolgerungen gezogen. An mehreren Beispielen wird dargelegt, daß das als universal geltende allopatrische Entwicklungsmodell keine allgemeine Anwendung finden kann, sondern daß vielmehr unter speziellen Bedingungen auch eine sympatrische Artbildung zu beobachten ist.

Im ersten Beitrag (G. BUSH, Austin, Texas) wird auf Grund von Beobachtungsergebnissen an den parasitischen Tephritiden und ihren Wirten, der *Drosophila-virilis*-Gruppe, ein Modell der sympatrischen Artbildung entwickelt. E. CHADDOCK (Sydney) behandelt die chromosomale Evolution und Artbildung bei der in Eukalyptuswäldern Südostaustraliens endemischen *Didymuria violescens*, die rassenunterschiedlich eine große Variation der Chromosomenmorphologie und Anzahl als Spezifikum aufweist. Die hierdurch bedingten genetischen Barrieren können zu einer kompletten reproduktiven Isolation und evtl. zur Artbildung führen. Die Beiträge von K. H. KEY (Canberra) und M. J. WHITE (Melbourne) behandeln die Artbildung bei australischen Heuschrecken (Eumastacidae, endemisch in Australien) unter taxonomischen und ökologischen sowie unter cytogenetischen Gesichtspunkten.

Im zweiten Teil wird in 6 Beiträgen die Evolution der Drosophiliden von Hawaii und den benachbarten Inseln besprochen. Von den etwa 500 auf der Inselgruppe vorkommenden Spezies sind 279 endemisch. Die Vielfalt, mit der sich diese Gattung in einem zusagenden, abgeschlossenen Milieu entwickelt hat, kommt auch in der Besitznahme unterschiedlichster Brutsubstrate durch die Larven, wie verwesende Blätter, Stengel, Rinde, Blüten, Früchte, Pilze, Pflanzenhaare von *Cyrtandra* u. a., sowie der Größenausbildung, die bei der kleinsten endemischen Art von 1,5 mm bis 20 mm Flügelspannweite bei den größten Arten reichen, zum Ausdruck. Durch die eifrige Arbeit einer Gruppe von Wissenschaftlern wurde in den letzten 20 Jahren wertvolles Material zusammengetragen und die immer stärker auf Reliktstandorte zurückgedrängten ursprünglichen Vorkommen vor dem drohenden Aussterben einer taxonomischen und evolutionsgenetischen Bearbeitung zugänglich gemacht. Hierbei ergaben sich wesentliche Gesichtspunkte und allgemeingültige Argumente, die für eine Evolution der auf Hawaii endemischen *Drosophila*-Arten aus einer einzigen Art sprechen. W. BAUDISCH

JOSEF G. ZÖTL: Karsthydrogeologie. Wien, New York: Springer-Verlag 1974. X u. 291 S., 114 Abb., 2 Falttaf. Ln. öS 986,– (DM 138,–; US $ 56,30)

Die Errichtung von Wasserversorgungsanlagen, Talsperren, Kraftwerken und Tunnelbauten auch in Karstgebieten in den letzten Jahrzehnten lenkte die Aufmerksamkeit immer mehr auf die Karsthydrogeologie. Die Nichtbeachtung ihrer Erkenntnisse bzw. die Nichtanwendung ihrer Untersuchungsmethoden führten in der Vergangenheit mehrfach zu spektakulären Mißerfolgen selbst großer Bauvorhaben. Diese Situation wurde u. a. dadurch erleichtert, daß einerseits seit

40 Jahren keine zusammenfassende karsthydrogeologische Darstellung mehr vorliegt und andererseits die letzte von O. LEHMANN (1932) seit langem überholt ist. J. G. ZÖTL, seit 2 Jahrzehnten mit der Bearbeitung karsthydrogeologischer Probleme beschäftigt, hat dankenswerterweise seine Erfahrungen auf diesem Gebiet im vorliegenden Buch wiedergegegeben. Das ist umso wertvoller, weil er selbst viele Methoden ausgearbeitet bzw. entwickelt hat.

Das Buch gliedert sich in 4 Teile: 1. Die geologischen, chemischen und klimatischen Grundlagen der Karsthydrogeologie, 2. Die Wasserzirkulation in verkarsteten Gebieten, 3. Beispiele hydrogeologischer Erfahrungen im Zuge des Ausbaues technischer Objekte in Karstgebieten und 4. Zusammenfassung und Ausblick. Ein nach Abschnitten gegliedertes Literaturverzeichnis von 21 Seiten sowie ein Sach- und Ortsverzeichnis beschließen das inhaltsreiche, vorzüglich gestaltete Werk. Das Schwergewicht liegt bei der umfassenden Darstellung der Methoden der hydrogeologischen Forschung in Karstgebieten und der Musterlandschaften und Phänomene der Karstentwässerung. Dies entspricht der Ansicht des Autors, daß die Auseinandersetzung mit den Methoden „die Grundlage für eine erfolgreiche Weiterentwicklung karsthydrogeologischer Forschung" darstellt. Dem Autor darf bescheinigt werden, daß auch dieses Kapitel „voll auf der Höhe der Zeit" steht, reichen doch die verarbeiteten Versuchsergebnisse und die angeführten Literaturzitate bis in das Erscheinungsjahr. Erwähnenswert ist hier u. a. die Einfügung eines Abschnitts „Die elektronische Datenverarbeitung (EDV) als Hilfsmittel für den Hydrologen und Hydrogeologen" von D. BORMANN.

Das Wesen des Karstes, seine Genese und die mit ihm verbundenen technischen Probleme werden hervorragend dargestellt. Die Schilderung von Musterlandschaften sowie von hydrogeologischen Erfahrungen in Karstgebieten gibt allen, die mit gleichen oder ähnlichen Fragestellungen zu tun haben, wertvolle Anregungen. D. STORCH

PAUL MÜLLER: Aspects of Zoogeography. The Hague: Dr. W. Junk b. v., Publishers 1974. 208 S., 110 Abb. Kart. Dfl. 35,–

Der vorliegende schmale Band behandelt, wie der Titel sagt, nur bestimmte Aspekte des weitgespannten Forschungsgebiets der Zoogeographie. Es sind dies einmal die Großlebensräume und Tierregionen der Erde, zum anderen Probleme der historischen Tiergeographie, wie das Konzept der Ausbreitungszentren. Ein eigenes kurzes Kapitel ist der Großstadt als Ökosystem gewidmet, ein bisher in zusammenfassenden Darstellungen weitgehend vernachlässigtes Gebiet. Die marinen und limnischen Ökosysteme werden äußerst knapp nur mehr am Rande behandelt und das wichtige Teilgebiet der ökologischen Tiergeographie bleibt, abgesehen von sporadischen Hinweisen auf spezielle Adaptationen, z. B. bei Wüstenbewohnern, leider ganz unberücksichtigt. Nach einer Einleitung, in der die Stellung der Zoogeographie im Rahmen der Biogeographie dargelegt wird, folgen außerdem noch allgemeine Bemerkungen über das Verbreitungsareal und seine Dynamik, wobei zoogeographische Aspekte der Art und Unterart, der Substratrassen und Superspezies sowie die disjunkte und reliktäre Verbreitung, die Tierwanderungen und die Verbreitung der Tiere durch den Menschen besprochen werden. Bemerkenswert erscheint, daß die enge Verbindung zoogeographischer Fragen zu umfassenden Problemkreisen, wie der Evolutionstheorie, im Text immer wieder betont wird.

Bei der ziemlich ausführlichen Schilderung der Tierregionen und Biome der Erde werden auch charakteristische Beispiele aus dem Pflanzenreich angeführt. So entsteht ein sehr anschauliches Bild von den Großlandschaften der Erde und ihrer typischen Besiedlung mit Tieren, wozu wesentlich auch die reichliche und geschickt ausgewählte Bebilderung des Textes beiträgt. Es fällt allerdings auf, daß im Text in keinem Fall auf die Abbildungen verwiesen wird, so daß deren Zuordnung zu spziellen Fragen, wie etwa beim Industriemelanismus der Schmetterlinge, bisweilen erschwert wird. Ein umfangreiches Literaturverzeichnis ermöglicht dem Leser des Bandes, sich über spezielle Probleme zu informieren. Das Buch stellt eine knappe, aber gut gelungene Einführung in die Zoogeographie dar, das besonders Studenten mit Gewinn benutzen können. D. FLÖSSNER

Schriften des Naturwissenschaftlichen Vereins für Schleswig-Holstein, Band 43, 44 (Hrsg.: EKKE W. GUENTHER & HEINZ KLUG). – Kiel: Komm.-Verl. Lipsius & Tischer 1973, 1974. Bd. 43: 91 S., 28 Abb., 21 Tab. Brosch. DM 22,50; Bd. 44: 131 S., 66 Abb., 18 Tab. Brosch. DM 26,–

In den beiden Bänden wird eine breite Themenfülle behandelt, die hier nicht ausführlich gewürdigt werden kann. In Band 43 berichten D. J. WERNER über den Tornado im Raum Kiel am 18. 10. 1971, I. HENNING über die periglaziale Talentwicklung (hübsch ist der Druckfehler im Inhaltsverzeichnis „Talententwicklung") u. H. Klug über die Landschaft als Geosystem, wobei er über die Grundzüge eines quantitativ-geographischen Forschungsprojektes einer Küstenlandschaft der Kieler Außenförde berichtet, das „exakte Aussagen zu Problemen der Umweltbelastung ermöglichen soll". W. Wetzel stellt zwei altpaläozoische Geschiebe mit neuartigen problematischen Fossilien vor. M. Hoop und V. Haeseler legen in zwei Arbeiten Ergebnisse ihrer Sammeltätigkeit dar (Goldwespen und Stechimmen in Schleswig-Holstein) bzw. Aculeaten und Chrysiden Schleswig-Holsteins und angrenzender Gebiete. Nach pflanzensoziologischen Untersuchungen über die Unkrautvegetation der Äcker der dänischen Insel Fanö von H. Wedeck folgen zwei Beiträge über anthropologische und demographische Verhältnisse (B. Guenther-Peters bzw. I. Matzdorf) im Staate Puebla (Mexico) und 6 Rezensionen.

Genauso breit gestreut ist die Thematik in Band 44: A. Sigurdsson berichtet über die Benthosfauna bei Surtsey u. U. Treter über landschaftsökologische Einheiten in einem Stauchmoränengebiet Schleswig-Holsteins, wobei ihn mehr die Charakterisierung der ausgeschiedenen Einheiten interessiert als die Grenzziehung. K. Brehm & Th. Eggers skizzieren die Vegetationsentwicklung in nordfriesischen Speicherbecken von 1959—1974. M. Hoop bringt Angaben über 50 Neufunde holsteinischer Pflanzenwespen. Die Arbeit von Th. Ernst „Die Hohwachter Bucht. Morphologische Entwicklung einer Küstenlandschaft Ostholsteins" bringt neben der Erfassung der geomorphologischen Entstehung dieses Küstenabschnittes Angaben über den „Transgressionsablauf im Bereich der südlichen und südwestlichen Ostseeküste" während der letzten 6000 Jahre. H. Klug schildert den Einfluß eines Leitdammes und seiner geplanten Verlängerung auf die Morphologie der Watt- und Strandgebiete vor Cuxhaven, während C. Gehlhaar Untersuchungen über Alter und Wachstum von Meerforellen in zwei schleswig-holsteinischen Salmonidengewässern bringt. Am Ende des Bandes berichtet E. W. Guenther kurz über den am Strande der Flensburger Außenförde gefundenen Backenzahn eines tertiären (?) Rüsseltieres, woran sich 4 Buchbesprechungen anschließen.

Insgesamt gesehen, geben beide Bände einen guten Einblick in die vielseitige Tätigkeit der Mitglieder des Naturwissenschaftlichen Vereines für Schleswig-Holstein. D. Storch

Heinz-Hermann Reichenbach-Klinke: Bestimmungsschlüssel zur Diagnose von Fischkrankheiten. Stuttgart: Gustav Fischer Verlag 1969. 65 S., 6 Taf., Kart. DM 6,—

Der „Bestimmungsschlüssel" ist eine Sonderausgabe, die im wesentlichen auf dem Werk „Krankheiten und Schädigung der Fische" des gleichen Autors basiert. Er ist in erster Linie für den Praktiker gedacht, der sich schnell orientieren will. Das Büchlein beginnt mit einer allgemeinen Beschreibung der Krankheitssymptome. Es folgen in Tabellenform geordnet krankhafte Veränderungen der einzelnen Organe und ihre Ursachen. Von großer praktischer Bedeutung ist noch die Tabelle der wichtigsten Fischmedikamente, die Angaben zur Anwendung und Diagnose enthält. Häufige Fischparasiten sind auf Tafeln dargestellt. Die im Anhang befindliche Tabelle zum Bestimmen einheimischer Süßwasserfische wird vor allem für Lernende von Nutzen sein. L. Küchler

Jürgen Schwoerbel: Einführung in die Limnologie. Stuttgart: Gustav Fischer Verlag 1971. VIII u. 170 S., 46 Abb. Brosch. DM 12,80 (Lizenzausgabe im VEB Gustav Fischer Verlag Jena)

Das vorliegende Buch ist gekennzeichnet durch eine souveräne Wahl des Stoffes aus dem großen und heterogenen Wissensgebiet der Limnologie, was zugleich von der umfassenden Literaturkenntnis des Verf. Zeugnis ablegt.

Der Stoff ist übersichtlich gegliedert und didaktisch geschickt angeordnet. Er umfaßt den Wasserkreislauf, die Physik und Chemie des Wassers, die Lebensgemeinschaften und den Stoffhaushalt der Gewässer sowie die „Limnologie des Abwassers und der verunreinigten Gewässer". Es werden auch die Hydrobakteriologie und die Probleme des Abbaus der organischen Substanz gebührend berücksichtigt. Hervorzuheben ist die Ausgewogenheit zwischen klassischen und modernen Aspekten der Gewässerforschung, wie man es in neueren zusammenfassenden Darstellungen nur noch selten findet.

Der Verf. möchte mit seiner „Einführung" denen, die sich nur am Rande mit Limnologie beschäftigen und den Studierenden der Limnologie eine erste Grundlage bieten.

Das Buch ist so geschrieben, daß es den genannten Leserkreis anspricht, zum anderen aber durch seine konkreten Angaben und Beispiele auch von den „Fortgeschrittenen" mit Gewinn zu benutzen ist. Es ist ein Buch, das man uneingeschränkt empfehlen kann. W. Schönborn

Die LIMNOLOGICA erscheinen in zwangloser Folge in einzeln berechneten Heften wechselnden Umfangs. Etwa 30 Druckbogen (480 Seiten) bilden einen Band.
Die LIMNOLOGICA veröffentlichen Originalarbeiten aus allen Gebieten der wissenschaftlichen Erforschung der Binnengewässer und der benachbarten Lebensräume. Neben den Originalarbeiten erscheinen Übersichten der Fachliteratur osteuropäischer Staaten, Tagungsberichte und Rezensionen.

BEZUGSMÖGLICHKEITEN:

Sämtliche Veröffentlichungen unseres Verlages sind durch jede Buchhandlung im In- und Ausland zu beziehen. Falls keine Bezugsmöglichkeit vorhanden ist, wende man sich
in der Deutschen Demokratischen Republik an den Akademie-Verlag, 108 Berlin, Leipziger Straße 3–4,
in der BRD und Westberlin an KUNST UND WISSEN, Erich Bieber, 7 Stuttgart 1, Wilhelmstraße 4–6,
in Österreich an den GLOBUS-Buchvertrieb, Wien I, Salzgries 16,
bei Wohnsitz im übrigen nichtsozialistischen Ausland an den Buchexport, Volkseigener Außenhandelsbetrieb der Deutschen Demokratischen Republik, 701 Leipzig, Leninstraße 16.
Im sozialistischen Ausland können Bestellungen über die Buchhandlungen für fremdsprachige Literatur bzw. den zuständigen Postzeitungsvertrieb erfolgen.
Auf Wunsch sendet der Akademie-Verlag Interessenten bei Bekanntgabe der Anschrift und Fachgebiete unverbindlich Informationen über lieferbare und kommende Veröffentlichungen und gibt auch Bezugsquellen im In- und Ausland bekannt.

Erschienen im Akademie-Verlag 108 Berlin, Leipziger Str. 3–4

Lizenznummer: 202 · 100/507/76
P 71/76
Gesamtherstellung: VEB Druckerei „Thomas Müntzer“, 582 Bad Langensalza
Bestellnummer: 762 057 3 (2111/11/1) · LSV 1305
Printed in the GDR
DDR 42,– M

Bei Fragen zur Produktsicherheit wenden Sie sich bitte an:
If you have any questions regarding product safety,
please contact:

Walter de Gruyter GmbH
Genthiner Straße 13
10785 Berlin
productsafety@degruyterbrill.com